Stealth Construction

Stealth Construction: Integrating Practices for Resilience and Sustainability

BY

SEYI S. STEPHEN
University of Johannesburg, South Africa

AYODEJI E. OKE
University of Johannesburg, South Africa

CLINTON O. AIGBAVBOA
University of Johannesburg, South Africa

OPEOLUWA I. AKINRADEWO
University of the Free State, South Africa

PELUMI E. ADETORO
Malawi University of Business and Applied Sciences, Malawi

AND

MATTHEW IKUABE
University of the Witwatersrand, South Africa

emerald PUBLISHING

United Kingdom – North America – Japan – India – Malaysia – China

Emerald Publishing Limited
Emerald Publishing, Floor 5, Northspring, 21-23 Wellington Street, Leeds LS1 4DL.

First edition 2025

British Library Cataloguing in Publication Data
A catalogue record for this book is available from the British Library

ISBN: 978-1-83608-183-8 (Print)
ISBN: 978-1-83608-182-1 (Online)
ISBN: 978-1-83608-184-5 (Epub)

To God,
Who Made All Things Beautiful

Contents

About the Authors

Seyi S. Stephen is a Postgraduate Student at SARChI in Sustainable Construction Management and Leadership in the Built Environment, Faculty of Engineering and Built Environment, University of Johannesburg, South Africa. He is an accomplished author and a key figure in the Research Group on Sustainable Infrastructure Management Plus (RG-SIM+), where he serves as one of the team leaders. With over 80 publications, including research books and journals, he focuses on sustainable infrastructure management (SIM). His research encompasses various aspects such as sustainable construction, value and risk management, and smart construction within the context of the digital era.

Ayodeji E. Oke is an Associate Professor and a Research Fellow at cidb Centre of Excellence and Sustainable Human Settlement and Construction Research Centre, University of Johannesburg, South Africa and the Federal University of Technology, Akure, Nigeria. He is an author and a lecturer with several years of teaching and research involvement in higher learning institutions worldwide. He founded the Research Group on Sustainable Infrastructure Management Plus (RG-SIM+) and is currently the Team Leader. With more than 400 publications, including research books, his research interest is in sustainable infrastructure management (SIM), emphasising sustainable construction, value management, quantity surveying and construction in the digital era. He is an Associate Professor at the Department of Quantity Surveying, Federal University of Technology Akure, Nigeria.

Clinton O. Aigbavboa is a Professor in the Department of Construction Management and Quantity Surveying and Director of the DSI/NRF Research Chair in Sustainable Construction Management in the Built Environment and the cidb Centre of Excellence and Sustainable Human Settlement and Construction Research Centre, University of Johannesburg, South Africa. He completed his PhD in Engineering Management and has published several research papers on housing, construction, engineering management, and research methodology for construction studies. He has extensive knowledge in practice, research, training, and teaching.

Opeoluwa I. Akinradewo is a Researcher and Senior Lecturer at the Department of Quantity Surveying and Construction Management, Faculty of Natural and Agricultural Sciences, University of the Free State, Bloemfontein, South Africa. He received his PhD in Construction Management in 2023 from the University of Johannesburg where he also completed his MTech in Quantity Surveying with distinction in 2020. He has published over 100 research articles in reputable journals, book chapters, and conference proceedings. His research focuses on applying blockchain technology to streamline business processes in the construction industry.

Pelumi E. Adetoro is a Postgraduate Student at the Malawi University of Business and Applied Science through the prestigious African Sustainable Infrastructure Mobility (ASIM) scholarship. He specialises in academic consultancy, building information modelling (BIM), and general construction. He has worked as a construction cost engineer and has practical costing and project management knowledge. His commitment to advancing construction knowledge is reflected in his role as a co-author of the textbook *Risk Management Practice in Construction: A Global View* (Springer, 2023).

Matthew Ikuabe is a Senior Lecturer at the School of Construction Economics and Management, University of the Witwatersrand, South Africa. He obtained honours and a master's in quantity surveying from the Federal University of Technology, Akure, Nigeria. He was awarded his doctorate in Construction Management from the University of Johannesburg, South Africa. He had his postdoctoral research fellowship at the SARChI in Sustainable Construction Management and Leadership in the Built Environment, Faculty of Engineering and Built Environment, University of Johannesburg, South Africa. He has published research in accredited journals, book chapters, and peer-reviewed conferences. He is a corporate member of the Nigerian Institute of Quantity Surveyors (NIQS) and the Association of South African Quantity Surveyors (ASAQS). He has won many academic accolades, such as being the best-graduating student in his honours class and completing his master's degree with distinction.

Preface

This book is timely and necessary due to several concurrent global challenges: climate change, resource scarcity, and an urgent call for sustainable development and resilience. Historically known for its significant environmental footprint, the construction industry faces increasing pressure to evolve. This book proposes a transformative approach by integrating sustainability into construction methodologies, addressing these pressing issues head-on. Hence, the book emerges as a pivotal publication of construction innovation, responding to the urgent need for transformative practices in the face of evolving challenges. This work is a comprehensive exploration of a novel approach that seamlessly merges diverse construction practices (lean construction, value management, supply chain management, smart construction, etc.) and technologies in construction 4.0/5.0 into strategic construction stages (before, during, and after), aiming to redefine the industry's standards towards heightened resilience and sustainability, as termed stealth construction.

This book serves a crucial purpose in an era where the construction landscape grapples with climate change, resource scarcity, and the imperative for sustainable development. Developed from stealth, which refers to a strategy or approach characterised by secrecy, discretion, and avoidance of detection, it seeks to bridge the existing gap in the literature by providing an in-depth understanding of stealth construction that stems from stealth technology. This innovative methodology enhances the efficiency of construction processes and fortifies structures against future uncertainties. The purpose is two-fold: first, to equip industry professionals, researchers, and students with a detailed guide on implementing stealth construction, and second, to contribute to the ongoing discourse on reshaping construction practices for a more sustainable and resilient built environment.

The timing of this book is strategic, as the global construction industry is at a crossroads, navigating the intersection of technological advancements and the imperative for eco-friendly practices. The book addresses contemporary challenges, emphasising integrating practices for resilience and sustainability. The construction sector is undergoing a paradigm shift, and 'stealth construction' provides a roadmap for navigating this transformation. Furthermore, at its core, the book aims to present a holistic view of Stealth Construction, showcasing its theoretical underpinnings and practical applications for the construction industry. Through case studies and real-world examples, readers will gain insights into how it can be successfully implemented, demonstrating its efficacy in diverse contexts. Also, the book delves into the refinement of eco-friendly materials, energy-efficient designs, and adaptive strategies, ensuring that the publication is a valuable resource for professionals seeking tangible solutions to the sustainability challenges prevalent in the construction industry.

What sets this book apart is its unwavering focus on resilience and sustainability, addressing the twin imperatives that define the future of construction. Through its integration of diverse practices, the book empowers its readers to adapt to the changing landscape and actively contribute to shaping a more resilient and sustainable built environment. In essence, 'stealth construction' emerges as a timely and indispensable guide, fostering a paradigm shift that positions construction at the forefront of innovative, eco-conscious practices essential for the challenges and opportunities of the 21st century.

Seyi S. Stephen
Ayodeji E. Oke
Clinton O. Aigbavboa
Opeoluwa I. Akinradewo
Pelumi E. Adetoro
Matthew Ikuabe

Acknowledgements

This book is a product of the Sustainable Human Settlement and Construction Research Centre (SHSCRC), SARChI in Sustainable Construction Management and Leadership in the Built Environment, Faculty of Engineering and the Built Environment, University of Johannesburg, South Africa, under the leadership of Professor Clinton Ohis Aigbavboa. Also, special thanks to Professor Ayodeji E. Oke of the Research Group on Sustainable Infrastructure Management Plus (RG-SIM+), Department of Quantity Surveying, Federal University of Technology, Akure, Nigeria and Department of Construction Management and Quantity Surveying, University of Johannesburg, South Africa.

Chapter One

General Introduction to Stealth Construction

Abstract

The chapter introduced stealth construction as a transformative concept emerging in the construction industry, emphasising resilience and efficiency through innovative technologies, cutting-edge materials, and advanced methodologies. Drawing parallels to stealth aircraft, stealth construction aims to create buildings and infrastructure that blend seamlessly into their environments, minimising environmental impact and enhancing sustainability. This approach integrates smart materials, strategic planning, and covert practices throughout all construction stages, ensuring robust, adaptive, and discreet structures. Stealth construction represents a significant paradigm shift by prioritising energy efficiency, environmental responsibility, and resilience against natural disasters and cyber threats. It combines modern aerospace engineering principles with diverse construction practices to achieve construction invincibility, setting new standards for the industry regarding sustainability, security, and aesthetics.

Keywords: Stealth construction; resilient infrastructure; innovative technologies; sustainable building; covert construction practices

Introduction

In the ever-evolving construction world, an intriguing and transformative concept is quietly but steadily emerging – stealth construction. This concept, where construction meets resilience, represents a paradigm shift in how we approach and execute building projects. As we embark on this journey of exploration into stealth construction, we delve into the realm of innovative technologies, cutting-edge materials, and advanced methodologies that enable construction projects to

Stealth Construction: Integrating Practices for Resilience and Sustainability, 1–18

Published under exclusive licence by Emerald Publishing Limited
doi:10.1108/978-1-83608-182-120251001

not only be more efficient but also discreet and resilient, resembling the characteristics of stealth aircraft (Zohuri & Zohuri, 2020). This introduction sets the stage for a captivating journey through the fascinating world of stealth construction.

The Birth of Stealth Construction

Stealth construction is not a term that is commonly heard. However, its inception can be traced back to the growing demand for infrastructure that is resilient, sustainable, and, in some cases, concealed from prying eyes. Much like the development of stealth technology (ST) in aviation, stealth construction seeks to make buildings and infrastructure blend seamlessly into their environments. This approach goes beyond aesthetics; it creates minimally disruptive, environmentally sensitive, and highly functional structures while maintaining a low profile.

Stealth construction refers to an advanced approach in construction where deliberate efforts are made to conceal and discreetly execute building projects. This methodology involves the seamless integration of innovative technologies, strategic planning, and covert practices across all stages of construction, including pre-construction, construction, and post-construction, as shown in Fig. 1. The primary goal of stealth construction is to achieve a level of invisibility, blending the built environment seamlessly into its surroundings while employing cutting-edge technologies to enhance efficiency, security, and sustainability (Westwick, 2021). This concept encompasses the meticulous selection of materials, utilisation of smart construction methodologies, and the strategic deployment of technologies to ensure that the construction process is executed discreetly, minimising environmental impact and meeting stringent security and aesthetic requirements.

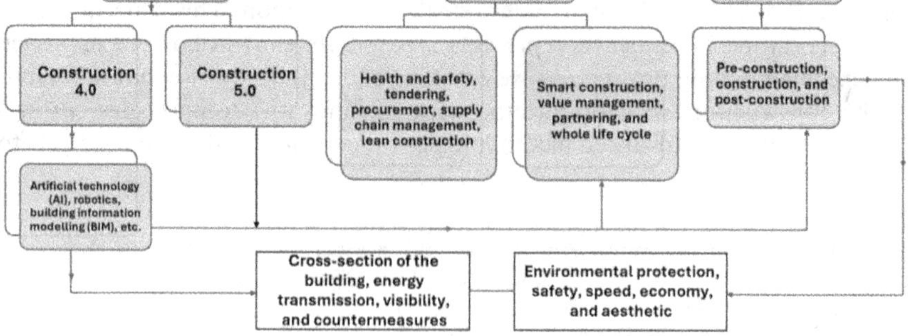

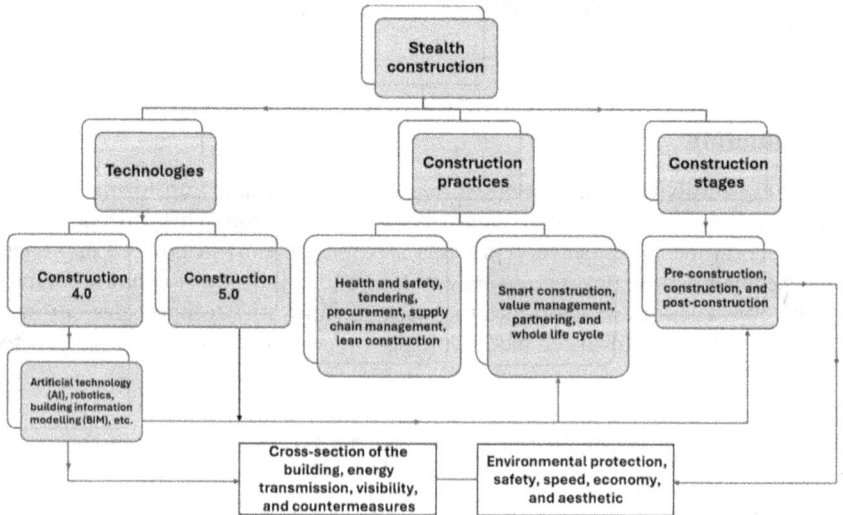

Fig. 1. Stealth Construction Framework.

As architects, engineers, and builders grapple with urbanisation, climate change, and resource scarcity challenges, stealth construction emerges as an elegant solution combining form and function. This concept results from converging ideas from various disciplines, including architecture, civil engineering, materials science, and sustainability. In essence, it represents the evolution of construction in the 21st century, reflecting our society's growing awareness of the need for responsible development.

Beyond the Surface

Stealth construction is not solely about appearances; it goes far beyond surface-level camouflage. It embraces the integration of smart materials, innovative engineering techniques, and cutting-edge design philosophies that enable structures to withstand natural disasters, cyber threats, and even physical attacks (Kim et al., 2023). In this era where resilience is paramount, stealth construction takes a holistic approach, ensuring that buildings and infrastructure are robust and adaptive to the ever-changing challenges of the modern world.

One of the fundamental principles underlying stealth construction is the notion of 'anticipatory design'. Buildings and infrastructure are designed to be proactive and capable of sensing and responding to their surroundings (Ahmad et al., 2019). This means incorporating sensors, artificial intelligence, and data analytics into the very fabric of construction. For example, a stealth construction building may adjust its energy consumption based on weather forecasts or enhance its security measures when it detects unusual activity nearby.

Embracing Sustainability

As Lima et al. (2021) explained, the influence of sustainability in construction is paramount, and it is at the core of the stealth construction movement; by designing structures that are energy-efficient, environmentally responsible, and capable of harmonising with their surroundings, stealth construction pioneers a new era of green building. It seeks to reduce the ecological footprint of construction while offering solutions that endure for generations. In this sense, stealth construction represents technological advancement and a profound commitment to the planet and future generations.

Green roofs, renewable energy integration, and sustainable building materials exemplify how stealth construction embraces sustainability. Beyond reducing environmental impact, this approach also considers the well-being of occupants. Stealth buildings often prioritise natural light, indoor air quality, and ergonomic design, fostering healthier and more productive environments.

The Unseen Benefits

While stealth construction may evoke images of secret military bases or covert installations, its applications extend beyond defence. From resilient infrastructure capable of withstanding natural disasters to urban developments that blend seamlessly into cityscapes, stealth construction can revolutionise how we conceive and execute projects in various sectors, including civil engineering, architecture, and urban planning (Zhu et al., 2020).

For example, in densely populated urban areas, stealth construction techniques can help reduce noise pollution, minimise disruption to communities during construction, and create visually pleasing structures that enhance the urban fabric. In disaster resilience, buildings engineered with stealth principles can better withstand earthquakes, hurricanes, and other catastrophes, safeguarding lives and critical infrastructure.

As we embark on this journey to explore the world of stealth construction, we will delve deeper into the fundamental principles, technologies, and real-world examples that illustrate how construction can meet invincibility. Together, we will discover the innovative approaches and ingenious solutions shaping the future of building and infrastructure, fostering a world where construction becomes more robust but also stealthier, sustainable, and innovative. In the following chapters, we will navigate the intricacies of stealth construction, unveiling the secrets of its success and its potential to redefine how we build and inhabit our world.

ST: The Art of Invisibility

ST represents the pinnacle of modern aerospace engineering. Evolving over several generations, from the 5th to the cutting-edge 6th generation, this practice has revolutionised how aircraft, drones (both manned and unmanned), and other military platforms interact with their environments (Rudyk et al., 2021). ST is a masterful amalgamation of advanced engineering principles and innovative technologies, all working in unison to confer a unique form of invincibility upon these machines. Remarkably, this invincibility exists not in the conventional sense of being impervious to attack but rather in the artful concealment of these platforms from enemy detection and targeting systems. It encompasses a comprehensive suite of technologies designed to improve multiple critical aspects of aerial operations, including speed, stealth, and dynamism, culminating in a paradigm shift in aerial warfare.

Speed: The Velocity of Stealth

One of the primary facets of ST is its relentless pursuit of speed (Westwick, 2021). The ability to reach astonishing velocities enables aircraft and drones to execute missions with unmatched agility and efficiency. These remarkable speeds can outrun adversaries and drastically reduce the time exposed to potential threats. Advanced propulsion systems, aerodynamic designs, and lightweight materials are seamlessly integrated into the aircraft's structure to achieve these incredible speeds. This need for velocity is not only about offensive capabilities but also about the ability to respond rapidly to emerging threats or execute precision strikes, making it a fundamental element of modern aerial warfare.

Stealth: The Art of Concealment

At the heart of ST lies its namesake principle – stealth. This aspect renders these aircraft and drones 'invisible' to enemy radar and detection systems. Stealth is achieved through ingenious design elements, including specialised shapes and contours, radar-absorbent materials, and advanced coatings that scatter and absorb radar waves (Kim et al., 2023). This technology allows these platforms to operate undetected deep within enemy territory, infiltrating hostile airspace with reduced risk of interception. Stealth also extends to minimising infrared and electromagnetic signatures, making these platforms challenging to detect by radar and other tracking mechanisms.

Dynamism: Adapting to the Battlefield

In the context of ST, dynamism represents the capacity for adaptability on the battlefield. These aircraft and drones have advanced avionics and sensor suites that allow them to collect and process vast amounts of data in real time. This wealth of information empowers operators to make split-second decisions, dynamically altering flight paths and mission objectives as situations evolve. Integrating artificial intelligence and machine learning further enhances this dynamic capability, enabling autonomous decision-making and course corrections. In essence, these platforms become living entities on the battlefield, capable of adapting to changing conditions and outsmarting adversaries.

ST is an intricate and awe-inspiring fusion of cutting-edge engineering and innovation. As it evolves into the realm of 6th-generation platforms, its impact on aerial warfare is poised to be more transformative than ever. By emphasising speed, stealth, and dynamism, ST has ushered in a new era of aerial invincibility, where aircraft and drones can operate undetected and execute missions with unparalleled precision and adaptability. This technology has redefined the art of war in the skies, making the invisible a formidable force to be reckoned with on the modern battlefield.

ST in Construction: Achieving Construction Invincibility

In an era where construction and infrastructure development are undergoing rapid transformation, the incorporation of ST marks a significant paradigm shift. This innovative approach, often called 'ST in construction,' harnesses 5th and 6th-generation ST's principles and practices to revolutionise how we conceive and execute construction projects (Kumar et al., 2019). Much like its pioneering applications in aviation and military sectors, ST in construction is a comprehensive system of technologies and methodologies working in synergy to elevate various essential aspects of construction endeavours. Speed, stealth, and dynamism collectively empower construction professionals to achieve unparalleled efficiency, security, and adaptability in their projects, thus reshaping the construction industry's landscape.

Speed: Accelerating Construction Efficiencies

Adopting ST in construction places a profound emphasis on speed as one of its fundamental pillars (Ahmad et al., 2019). This focus on rapidity is not solely about completing projects more quickly; it is about optimising every facet of construction processes to maximise efficiency. Advanced robotics, 3D printing, and prefabrication techniques expedite project timelines significantly. These innovations reduce construction durations, translating into cost savings, reduced resource consumption, and the capacity to tackle complex projects swiftly. Speed, therefore, emerges as a cornerstone of ST in construction, catalysing a transformation in the construction industry's project delivery capabilities.

Stealth: Concealing Construction Impact

Much like ST conceals military assets from radar detection, ST in construction revolves around concealment. However, in this context, concealment pertains to minimising the impact of construction projects on their immediate surroundings and neighbouring communities (Zohuri & Zohuri, 2020). Advanced construction methodologies, noise-reduction technologies, and the conscientious selection of eco-friendly materials contribute to this stealthy approach. By minimising disruptions and employing environmentally responsible practices, construction teams execute their work discreetly, ensuring that neighbouring areas experience minimal inconvenience. This harmonious coexistence with the environment and local communities is a hallmark of ST in construction.

Dynamism: Adapting to Project Variables

The dynamism inherent in stealth construction underscores its ability to rapidly adapt to evolving project variables. Construction projects are inherently susceptible to uncertainties, from unpredictable weather conditions to unforeseen obstacles encountered on-site. By harnessing real-time data analytics, machine learning, and artificial intelligence, construction teams armed with ST can make swift, informed decisions to navigate these challenges. This dynamic capability minimises delays and enhances safety and quality standards, setting an unprecedented standard for the industry's responsiveness to the ever-changing construction landscape.

In summation, ST in construction is an essential transformation within the construction industry. Much like its counterparts in aviation and defence, it capitalises on the principles of speed, stealth, and dynamism to usher in a new era of construction invincibility. Stealth construction empowers construction professionals to embark on projects with unparalleled precision and adaptability by optimising efficiency, minimising disruptions and expertly adapting to unpredictable variables. As the industry continues its evolutionary journey, this technological incorporation promises to redefine construction practices, ensuring that construction endeavours are faster and more efficient, considerate of their surroundings, and resilient in the face of unpredictability. This groundbreaking approach ushers in a new age where construction meets invincibility, revolutionising how we build the world around us.

Stealth Construction in Functional Construction Principles

In the contemporary construction and infrastructure development landscape, 'functional construction practices' have emerged as a guiding philosophy intricately entwined with innovative ST in construction, as Laurent and Leicht (2019) illustrated. The journey from traditional construction methodologies to cutting-edge, technology-driven approaches has necessitated a fundamental re-evaluation of how we conceive, design, and execute construction projects. Functional construction practices encapsulate a holistic approach that prioritises the result and how it is achieved. Inextricably linked to the broader theme of ST in construction, these practices underscore the critical importance of optimising every facet of construction to enhance functionality, efficiency, and sustainability. This introductory exploration delves into the essential aspects of functional construction practices, shedding light on their significance in achieving construction invincibility within the context of ST.

At the heart of functional construction practices lies a commitment to redefining the essence of construction excellence. For example, according to Wolman (2024), the Israeli Defence Force (IDF) deployed unmanned D9 bulldozers in Gaza, marking a significant use of ground robotics in combat. These 'Panda' bulldozers, weighing 50 tons each, perform excavation and obstacle removal independently without risking human lives. They can navigate rough terrain and adverse weather conditions, hence enhancing visibility. This transformative approach transcends traditional project parameters to encompass a broader perspective that integrates sustainability, safety, efficiency, and adaptability. Functionality becomes the expectancy upon which every decision and innovation hinges, from selecting construction materials to utilising advanced technologies. In his context, the construction practices stand as the bridge between conceptualisation and realisation, guiding construction professionals in pursuing projects that are structurally sound, environmentally responsible, unobtrusive to their surroundings, and capable of dynamic adaptation in the face of unforeseen challenges. In this era of construction transformation, functional construction practices become the lodestar that illuminates a path to achieving construction invincibility. These are defined in the principles:

- Environmental protection;
- Safety;
- Speed;
- Economy; and
- Aesthetic.

The concept of stealth construction represents a dynamic evolution in construction methodologies, transcending traditional boundaries to create a new paradigm for the construction industry. Within this innovative framework, stealth construction integrates several critical practices that redefine how we approach, design, and execute construction projects. These essential practices include environmental protection, safety, speed, economy, and aesthetics. By aligning these practices with the principles of stealth construction, we can envision a construction landscape that advances efficiency and elevates our commitment to sustainability, safety, cost-effectiveness, and visual appeal.

Environmental Protection

Stealth construction is a testament to the industry's commitment to environmental protection. This practice is deeply ingrained in the philosophy of stealth construction, emphasising responsible environmental stewardship at every phase of a project's life cycle. From selecting eco-friendly construction materials to implementing sustainable construction practices, such as minimising waste and adopting energy-efficient design principles, every effort is made to reduce the environmental footprint of construction projects. Furthermore, integrating advanced environmental monitoring and control systems ensures that construction activities have minimal adverse effects on the surrounding ecosystems. This practice, in this regard, is a vanguard for environmentally conscious construction practices, leaving behind a legacy of reduced environmental impact.

Safety

Safety remains paramount in stealth construction practices. The commitment to safety is a guiding principle and a fundamental ethos permeating every aspect of construction project planning and execution. The core components of stealth construction are advanced safety protocols, state-of-the-art safety technologies, and meticulous risk assessments. Construction teams work tirelessly to ensure that construction sites are efficient and secure environments for all workers and stakeholders. By maintaining a proactive approach to safety, stealth construction protects the well-being of construction workers and enhances overall project performance, reducing accidents and delays.

Speed

Speed is a central tenet of stealth construction, enabling swift and efficient completion of projects. This practice is achieved through meticulous project planning, precise scheduling, and the adoption of cutting-edge construction technologies. Lean construction principles are embraced to eliminate waste and optimise processes, accelerating construction timelines. The benefits of speed extend to reduced costs, minimised disruptions, and quicker occupancy of structures, aligning with the need for efficient project delivery in today's fast-paced world. Stealth construction ensures that projects are executed with unmatched efficiency without compromising quality.

Economy

The economy in stealth construction extends beyond mere cost efficiency; it encompasses a broader spectrum of value-driven practices. Innovative procurement strategies, precise project planning, and judicious resource allocation ensure that construction projects provide lasting value. By focusing on structures' long-term sustainability and operational efficiency, stealth construction achieves economic benefits extending beyond the construction phase. Projects completed under this practice demonstrate both cost-effectiveness and a commitment to delivering infrastructure that remains valuable and efficient throughout its lifecycle.

Aesthetics

While often overshadowed by functionality, aesthetics play a pivotal role in stealth construction. This is particularly evident in urban development and architectural projects where visual harmony with the environment is paramount. While efficiency and functionality remain critical, stealth construction endeavours to seamlessly blend structures into their surroundings, enhancing the overall visual appeal of the built environment. Innovative design concepts, integration with natural landscapes, and using aesthetically pleasing materials create highly functional and artistically appealing structures. These projects meet their functional objectives and contribute to the aesthetic enrichment of communities.

Stealth construction represents a harmonious fusion of these critical practices, demonstrating that construction can be conducted with an unwavering commitment to environmental protection, safety, speed, economy, and aesthetics. Adhering to these principles, stealth construction projects exemplify a new era of construction excellence that advances efficiency and elevates the industry's impact on the environment, the well-being of workers, and the overall quality of life for communities and end-users. Stealth construction is poised to shape the construction industry's future by providing a blueprint for sustainable, efficient, safe, and visually appealing projects that are a testament to human ingenuity and innovation.

Construction Practices for Stealth Construction: Crafting the Future of Building

Stealth construction is a revolutionary approach that has redefined the very essence of construction, calling for a fundamental shift in construction practices. To realise the full potential of this innovative concept, a new set of construction practices has emerged, dedicated to ushering in an era where construction projects are conducted with unprecedented efficiency, sustainability, and adaptability. These practices summarise a holistic construction approach that integrates stealth principles and strives to harmonise the built environment with its surroundings, as outlined in Table 1. From the meticulous selection of eco-friendly materials to the strategic use of advanced technologies, the construction practices for stealth construction encompass a spectrum of methodologies that collectively redefine the future of the building.

First and foremost, these construction practices prioritise the concept of 'stealth,' aiming to minimise the impact of construction on the environment and surrounding communities. The selection of construction materials focuses on sustainability, utilising eco-friendly and locally sourced resources wherever possible.

Table 1. Construction Stages and the Associated Construction Practices.

Construction Stages	Construction Practices
Pre-construction	Health and safety
	Tendering
	Procurement
Construction	Supply chain management
	Lean construction
	Smart construction
Post-construction	Value management
	Partnering
	Whole life cycle

Innovative building designs aim to reduce energy consumption and promote environmental harmony, resulting in structures seamlessly blending into their surroundings. Moreover, adopting construction methodologies that minimise noise and pollution ensures that stealth construction projects remain unobtrusive, minimising disruptions and preserving the tranquillity of nearby areas. As a result, the practices of stealth construction prioritise the environment's well-being and foster a harmonious coexistence between constructed structures and the natural world.

Additionally, these construction practices strongly emphasise safety, recognising it as a paramount concern. Robust safety protocols are integrated into every construction stage, ensuring workers' and stakeholders' protection. Cutting-edge safety technologies, such as wearable devices and real-time monitoring systems, enhance on-site safety. Risk assessments and proactive safety measures are meticulously implemented to prevent accidents and minimise delays. By prioritising safety, stealth construction projects are efficient and provide a secure and risk-mitigated environment, fostering a culture of well-being and ensuring that the construction process aligns with the highest standards of safety and excellence. The construction practices for stealth construction represent a transformative approach that promises to shape the future of construction. These practices align with stealth, sustainability, safety, and harmonious aesthetics, redefining how construction projects are conceived, planned, and executed. As stealth construction continues to gain momentum, these practices stand as a testament to the industry's commitment to environmental responsibility, safety, and efficiency, ushering in an era where construction is not merely an act of building but an art of harmonising with the environment and enhancing the quality of life for all.

Lean Construction: The Paradigm Shift Needed

Lean construction is a dynamic and transformative approach to project management that has revolutionised the construction industry. Rooted in lean thinking and continuous improvement principles, it aims to eliminate waste, optimise resources, and enhance efficiency throughout every stage of a construction project (Albalkhy & Sweis, 2021). Lean construction represents a departure from traditional project management paradigms, focusing on value delivery to clients, stakeholders, and end-users. This practice has gained momentum as the construction industry seeks innovative solutions to increase productivity, reduce costs, and improve project outcomes. At its core, lean construction seeks to identify and eliminate non-value-adding activities and processes, often called waste. These wasteful elements can manifest in various forms, including overproduction, excess inventory, unnecessary movements, and delays. Meticulously analysing each component of a construction project, lean construction practitioners can identify areas for improvement and implement strategies to streamline operations. The goal is to maximise the value delivered to the client while minimising the use of resources, time, and effort. Lean construction embraces a holistic view of project management, emphasising collaboration among all project stakeholders, from designers and contractors to suppliers and owners.

Lean construction leverages a range of tools and methodologies to achieve its objectives. These may include pull planning, the last planner system (LPS), visual management techniques, and just-in-time delivery of materials (Aslam et al., 2020). Pull planning, for instance, involves coordinating activities based on their dependency and priority, ensuring that resources are allocated efficiently and project timelines are met. The LPS empowers project teams to collaborate closely, set realistic goals, and adjust plans to adapt to changing circumstances. Visual management tools provide real-time insights into project progress, enabling teams to identify bottlenecks and areas for improvement promptly. Just-in-time delivery ensures that materials and resources are provided precisely when needed, minimising inventory costs and reducing waste. Lean construction represents a powerful paradigm shift in the construction industry, emphasising efficiency, collaboration, and continuous improvement. By identifying and eliminating waste, optimising processes, and fostering a culture of collaboration, lean construction enhances project outcomes and aligns construction practices with the evolving demands of the modern world. As the construction industry continues to embrace lean principles, it is poised to achieve higher productivity levels, cost-effectiveness, and client satisfaction, ultimately reshaping the future of construction project management.

Smart Construction: Building the Future with Digital Precision

Smart construction is at the forefront of the construction industry's evolution, leveraging digital technologies and data-driven strategies to transform how we design, build, and manage construction projects (Xu et al., 2022). This innovative approach embraces the principles of connectivity, automation, and real-time data analytics to enhance efficiency, accuracy, and sustainability. Smart construction represents a paradigm shift, bridging the gap between traditional construction practices and the demands of the digital age, ultimately redefining how we conceive, plan, and execute construction projects. Building Information Modelling (BIM) is at the heart of smart construction, a digital representation of the project's physical and functional characteristics. BIM serves as the digital backbone, offering a collaborative platform for architects, engineers, contractors, and stakeholders to visualise, design, and simulate every aspect of the construction process. This shared digital space fosters better coordination, reduces errors, and enhances decision-making, resulting in more precise construction projects. Moreover, the Internet of Things (IoTs) is pivotal in smart construction, with sensors and devices embedded in construction equipment, materials, and structures. These IoT sensors provide real-time data on project progress, equipment performance, and environmental conditions, allowing for proactive adjustments and timely decision-making.

One of the key benefits of intelligent construction is the ability to monitor and manage construction sites remotely (Ramesh et al., 2022). Real-time project monitoring, often facilitated by drones and advanced software, allows project managers to assess progress, identify bottlenecks, and ensure that projects stay on schedule and budget. Additionally, smart construction enhances safety by enabling predictive maintenance of equipment, early detection of hazards, and data-driven safety protocols. Sustainability is another core aspect, as smart construction leverages data to optimise energy usage, reduce waste, and enhance the environmental performance of construction projects. As the construction industry continues to embrace innovative construction practices, it is poised to create more efficient, safer, and environmentally responsible projects, ultimately shaping the future of construction.

Value Management in Construction: Maximising Returns through Strategic Planning

Value management is a pivotal practice within the construction industry, essential for optimising resources, controlling costs, and delivering projects that meet or exceed stakeholder expectations. At its core, value management seeks to align project objectives with the needs and priorities of stakeholders while ensuring that the project remains within budget constraints (Madushika et al., 2020). This practice involves a systematic approach that identifies value-adding components while eliminating unnecessary costs and maximising the return on investment.

A critical component of value management is the structured identification of project objectives and stakeholder requirements. By engaging stakeholders from the outset, value management ensures all parties understand the project's goals. This collaborative approach helps set clear priorities and establish a shared vision for the project. Subsequently, value management seeks to assess all aspects of a project to identify areas where value can be added or costs can be minimised. This process often involves value engineering, exploring design alternatives and cost-saving measures without compromising project quality.

The benefits of value management extend beyond cost control. It fosters a culture of efficient decision-making, where project teams continuously evaluate options and strive for improvements (Bennett & Mayouf, 2021). Value management encourages innovation and creativity in problem-solving, leading to optimised project outcomes. Moreover, by focusing resources on the most critical and value-adding aspects of a project, value management ensures that project objectives are met efficiently, delivering cost savings and projects that are highly functional, resilient, and capable of exceeding stakeholder expectations. In a construction industry characterised by complex challenges and evolving demands, value management is a guiding principle for achieving excellence in project delivery.

Whole Life Cycle in Construction: Beyond Building to Sustaining Excellence

Whole life cycle thinking is a fundamental practice in the construction industry that transcends the traditional project-centric approach. It extends from merely constructing a building to considering its lifespan, from conceptualisation and design through construction, operation, and eventual decommissioning or repurposing. This holistic approach acknowledges that buildings impact their users, environment, and communities in the long term. It emphasises the need to optimise the initial construction costs and ongoing operational and maintenance expenses, ensuring that buildings remain efficient, cost-effective, and environmentally responsible throughout their entire lifecycle (Hasik et al., 2019).

The whole life cycle thinking process begins during the design phase, where architects and engineers evaluate different building materials, construction techniques, and systems with an eye towards long-term sustainability and durability (Zhao & Yang, 2022). By selecting materials and systems that require minimal maintenance and are easy to repair or replace, construction teams can significantly reduce the operational costs and environmental impact over the building's lifetime. This practice aligns with the growing emphasis on sustainability in construction, as it helps minimise waste, conserve resources, and reduce the carbon footprint of buildings. During the operational phase, whole life cycle thinking remains a guiding principle. It involves continuously monitoring building systems, energy efficiency, and maintenance to ensure the structure remains efficient and functional. Proactive maintenance and upgrades can extend the building's life expectancy and optimise its performance, leading to lower operational costs and improved user comfort. As the building reaches the end of its useful life, whole life cycle thinking involves responsible decommissioning or repurposing to minimise waste and environmental impact. In an era of increasing environmental consciousness and resource scarcity, whole-life cycle thinking has become a cornerstone of responsible construction, guiding the industry towards buildings that stand as enduring testaments to efficiency, sustainability, and long-term value.

Supply Chain Management in Construction: Orchestrating Seamless Project Delivery

Supply chain management is a critical practice within the construction industry, essential for ensuring the timely and efficient flow of materials, equipment, and information from suppliers to construction sites. It encompasses a complex network of activities, from strategic sourcing and procurement to inventory management and logistics optimisation. Effective supply chain management is pivotal in construction projects, contributing to cost control, timely completion, and overall project success (Kim & Nguyen, 2022).

Strategic sourcing is at the core of supply chain management in construction. It involves identifying and selecting suppliers, negotiating contracts, and establishing relationships with vendors who can provide the necessary materials, equipment, and services at competitive prices. Procurement practices are instrumental in ensuring that materials and components meet quality and cost standards while minimising lead times. The procurement process also includes carefully considering lead times and ensuring that materials are available when needed to avoid costly project delays.

Inventory management and logistics optimisation are equally crucial components of supply chain management. These practices ensure that materials are efficiently stored, tracked, and transported to construction sites. Just-in-time delivery practices help minimise inventory costs while ensuring that materials are available precisely when required for construction activities. Effective supply chain management ensures construction projects can access the right resources at the right time, minimising costly delays and disruptions. In an industry where project timelines and budgets are paramount, supply chain management is a linchpin for successful project delivery, offering a systematic and coordinated approach to managing the complex flow of resources in construction projects (Kim et al., 2023).

Procurement in Construction: Strategic Sourcing for Project Success

Procurement is a critical practice in the construction industry, encompassing the systematic process of acquiring goods, services, and labour for construction projects. It is the foundation for ensuring that the necessary materials and resources are secured efficiently, cost-effectively, and with an unwavering commitment to quality (Kabirifar & Mojtahedi, 2019). Effective procurement practices are instrumental in achieving project objectives, controlling costs, and delivering construction projects that meet or exceed stakeholder expectations.

Strategic sourcing is a central aspect of procurement in construction. It entails identifying and selecting suppliers and subcontractors who can provide the required materials, equipment, and services at competitive prices and with the desired level of quality. Procurement professionals carefully evaluate potential suppliers, negotiate contracts, and establish relationships to ensure a reliable and efficient supply chain. Strategically sourcing materials and services, construction teams can minimise costs, reduce lead times, and enhance project flexibility.

Additionally, procurement practices extend beyond the initial selection of suppliers. They encompass the entire procurement lifecycle, from preparing bid documents and evaluating proposals to managing contracts and supplier relationships (Loosemore et al., 2020). These practices also include managing changes and variations to project scope and ensuring that materials and services are delivered on schedule and within budget. Effective procurement practices are essential for maintaining project control and transparency, mitigating risks, and optimising resource allocation throughout construction. In an industry where efficiency, cost control, and quality are paramount, procurement is a linchpin for successful project delivery, ensuring that construction projects are executed with precision and excellence.

Health and Safety in Construction: Prioritising Well-Being on the Building Site

Health and safety practices are paramount in the construction industry, placing the well-being of workers, stakeholders, and the community at the forefront. The construction sector is characterised by inherent risks, ranging from heavy machinery operation to working at heights and exposure to hazardous materials. Health and safety measures are foundational to creating secure work environments, minimising accidents, and ensuring the safety of all involved in construction projects (Zhang et al., 2020).

Effective health and safety practices in construction begin with a comprehensive understanding of potential risks. This includes rigorous risk assessments to identify and mitigate hazards specific to each project and develop comprehensive safety plans. These plans encompass detailed safety protocols, training requirements, and emergency response procedures. Moreover, implementing personal protective equipment (PPE) is a cornerstone of health and safety practices, with workers equipped with gear such as helmets, safety vests, and protective eyewear to mitigate risks associated with their tasks.

Beyond mitigating risks, health and safety practices foster a culture of safety consciousness within the construction industry. Regular safety training and ongoing communication are critical components of this culture, ensuring that all personnel are educated and informed about safety procedures and potential hazards (Trask & Linderoth, 2023). Construction firms prioritising health and safety also emphasise reporting mechanisms for near misses and incidents, enabling continuous improvement and proactive responses to potential dangers. By valuing health and safety as paramount, the construction industry protects workers' lives and well-being and enhances overall project performance by reducing accidents, delays, and disruptions.

Tendering in Construction: The Gateway to Project Execution

Tendering is a foundational practice in the construction industry, serving as the initial stage of project procurement. It involves inviting bids or proposals from contractors or suppliers to perform specific construction work. Tendering is a crucial practice as it forms the basis for selecting the most suitable contractor or supplier to deliver the required services or materials efficiently, cost-effectively, and in alignment with project requirements (Yu et al., 2021). The tendering process plays a pivotal role in shaping the trajectory of construction projects and ensuring their success.

The first step in the tendering process is the preparation of comprehensive bid documents, which include detailed project specifications, plans, and contract terms. These documents provide prospective bidders with a clear understanding of the project scope, requirements, and expectations. Once the bid documents are prepared, they are released to potential contractors or suppliers, who submit their bids or proposals within a specified timeframe. The bids are evaluated based on various criteria, including cost, qualifications, experience, and adherence to project specifications.

Effective tendering practices are essential for achieving project objectives. They promote transparency, competition, and fairness in selecting contractors or suppliers. Construction project owners can obtain competitive bids through the tendering process, enabling them to select the most qualified and cost-effective parties to execute the work (Signor et al., 2020). Additionally, the tendering process serves as a critical risk management tool, as it allows project owners to assess the qualifications and capabilities of potential contractors or suppliers, ensuring that they are well-equipped to deliver the project successfully. In summary, tendering is the cornerstone of project procurement in construction, guiding the selection of contractors or suppliers and laying the foundation for the successful execution of construction projects.

Partnering in Construction: Fostering Collaboration for Project Success

Partnering is a strategic approach in the construction industry that emphasises collaboration, trust, and shared goals among project stakeholders, including owners, contractors, designers, and other vital participants (Bygballe & Swärd, 2019). This practice seeks to create a collaborative environment where all parties work together to achieve project objectives efficiently, reduce conflicts, and enhance project outcomes. Partnering represents a departure from traditional adversarial relationships in construction and aligns stakeholders in a spirit of mutual benefit.

Establishing a formal partnering agreement or charter is at the core of partnering in construction. This document outlines the project's goals, principles, and the responsibilities of each stakeholder. It also establishes a dispute resolution process to address issues as they arise. Partnering agreements typically emphasise open communication, risk-sharing, and problem-solving commitment. They encourage stakeholders to work together proactively, addressing challenges and conflicts promptly to minimise disruptions and delays.

The benefits of partnering in construction are numerous. It fosters teamwork and collaboration, enhancing project efficiency and productivity. Open communication and shared goals reduce misunderstandings and disputes, resulting in fewer legal conflicts and claims. Moreover, partnering often leads to projects being completed on time and within budget, as stakeholders are collectively focused on achieving the project's success. By emphasising collaboration and trust, partnering practices contribute to a more harmonious and productive construction industry, ultimately benefiting all parties involved in the construction process.

Conclusion

Integrating a comprehensive set of modern construction practices across various stages significantly advances construction processes. This integrated approach, which encompasses practices such as benchmarking, health and safety, tendering, procurement, supply chain management, lean construction, smart construction, sustainable construction, value management, partnering, risk management, and whole life cycle thinking, can be collectively referred to as 'stealth construction.' Stealth construction represents a paradigm shift in the construction industry, where the convergence of these practices forms a cohesive and dynamic framework that transcends conventional project management methodologies. Much like the advanced technologies and strategies employed in stealth aircraft and military operations, stealth construction aims to make construction processes more efficient, effective, and adaptable while remaining 'invisible' to potential challenges and setbacks. This holistic approach emphasises the fusion of technology, data-driven decision-making, sustainability, safety, cost-effectiveness, and collaboration throughout every stage of a construction project. It seeks to optimise project performance, minimise waste, reduce risks, and maximise stakeholder value. In essence, stealth construction represents a bold and transformative direction for the construction industry, one that leverages the best practices of the digital age to create construction projects that are efficient and cost-effective, resilient, sustainable, and attuned to the evolving needs of the modern world.

References

Ahmad, H., Tariq, A., Shehzad, A., Faheem, M. S., Shafiq, M., Rashid, I. A., Afzal, A., Munir, A., Riaz, M., T., Haider, H., T., Afzal, A., Qadir, M., B., Khaliq, Z. (2019). Stealth technology: Methods and composite materials – A review. *Polymer Composites*, *40*(12), 4457–4472.

Albalkhy, W., & Sweis, R. (2021). Barriers to adopting lean construction in the construction industry: A literature review. *International Journal of Lean Six Sigma*, *12*(2), 210–236.

Aslam, M., Gao, Z., & Smith, G. (2020). Exploring factors for implementing lean construction for rapid initial successes in construction. *Journal of Cleaner Production*, *277*, 1–14. https://doi.org/10.1016/j.jclepro.2020.123295

Bennett, K., & Mayouf, M. (2021). Value management integration for whole life cycle: Post covid-19 strategy for the UK construction industry. *Sustainability*, *13*(16), 1–26.

Bygballe, L. E., & Swärd, A. (2019). Collaborative project delivery models and the role of routines in institutionalizing partnering. *Project Management Journal*, *50*(2), 161–176.

Hasik, V., Escott, E., Bates, R., Carlisle, S., Faircloth, B., & Bilec, M. M. (2019). Comparative whole-building life cycle assessment of renovation and new construction. *Building and Environment*, *161*, 1–10. https://doi.org/10.1016/j.buildenv.2019.106218

Kabirifar, K., & Mojtahedi, M. (2019). The impact of engineering, procurement, and construction (EPC) phases on project performance: A case of large-scale residential construction project. *Buildings*, *9*(1), 1–15.

Kim, S. H., Lee, S. Y., Zhang, Y., Park, S. J., & Gu, J. (2023). Carbon-based radar absorbing materials toward stealth technologies. *Advanced Science*, *10*(32), 1–22.

Kim, S. Y., & Nguyen, V. T. (2022). Supply chain management in construction: Critical study of barriers to implementation. *International Journal of Construction Management*, *22*(16), 3148–3157.

Kumar, N., Dixit, A., Kumar, N., & Dixit, A. (2019). Camouflage and stealth technology based on nanomaterials. *Nanotechnology for Defence Applications*, *2*(1), 155–203.

Laurent, J., & Leicht, R. M. (2019). Practices for designing cross-functional teams for integrated project delivery. *Journal of Construction Engineering and Management, 145*(3), 1–11. https://doi.org/10.1061/(ASCE)CO.1943-7862.0001605

Lima, L., Trindade, E., Alencar, L., Alencar, M., & Silva, L. (2021). Sustainability in the construction industry: A systematic review of the literature. *Journal of Cleaner Production, 289*, 1–15. https://doi.org/10.1016/j.jclepro.2020.125730

Loosemore, M., Alkilani, S., & Mathenge, R. (2020). The risks of and barriers to social procurement in construction: A supply chain perspective. *Construction Management and Economics, 38*(6), 552–569.

Madushika, W. H. S., Perera, B. A. K. S., Ekanayake, B. J., & Shen, G. Q. P. (2020). Key performance indicators of value management in the Sri Lankan construction industry. *International Journal of Construction Management, 20*(2), 157–168.

Ramesh, G., Logeshwaran, J., & Rajkumar, K. (2022). The smart construction for image preprocessing of mobile robotic systems using neuro fuzzy logical system approach. *NeuroQuantology, 20*(10), 6354–6367.

Rudyk, A., Semenov, A., Semenova, O., & Kakovkin, S. (2021). Using stealth technologies in mobile robotic complexes and methods of detection of low-sighted objects. *Informatyka, Automatyka, Pomiary w Gospodarce i Ochronie Środowiska, 11*(3), 4–8.

Signor, R., Love, P. E., Belarmino, A. T., & Alfred Olatunji, O. (2020). Detection of collusive tenders in infrastructure projects: Learning from operation car wash. *Journal of Construction Engineering and Management, 146*(1), 1–10. https://doi.org/10.1061/(ASCE)CO.1943-7862.0001737

Trask, C., & Linderoth, H. C. (2023). Digital technologies in construction: A systematic mapping review of evidence for improved occupational health and safety. *Journal of Building Engineering*, 275(4), 1–17.

Westwick, P. (2021). *Lessons from stealth for emerging technologies*. https://cset.georgetown.edu/wp-content/uploads/CSET-Lessons-from-Stealth-for-Emerging-Technologies.pdf

Wolman, I. (2024, May 9). *In first, IDF operates unmanned D9 bulldozers in Gaza*. https://www.ynetnews.com/article/bj5nnk5m0?utm_source=ynetnews.com

Xu, M., Nie, X., Li, H., Cheng, J. C., & Mei, Z. (2022). Smart construction sites: A promising approach to improving on-site HSE management performance. *Journal of Building Engineering, 49*, 1–29. https://doi.org/10.1016/j.jobe.2022.104007

Yu, A. T., Wong, I., Wu, Z., & Poon, C. S. (2021). Strategies for effective waste reduction and management of building construction projects in highly urbanized cities – A case study of Hong Kong. *Buildings, 11*(5), 199–214.

Zhang, M., Shi, R., & Yang, Z. (2020). A critical review of vision-based occupational health and safety monitoring of construction site workers. *Safety Science, 126*, 1–18. https://doi.org/10.1016/j.ssci.2020.104658

Zhao, W., & Yang, Q. (2022). Performance characterization and life cycle assessment of semi-flexible pavement after composite modification of cement-based grouting materials by desulfurization ash, fly ash, and rubber powder. *Construction and Building Materials, 359*(3), 1–14.

Zhu, B., Zhu, C., & Dewancker, B. (2020). A study of development mode in green campus to realize the sustainable development goals. *International Journal of Sustainability in Higher Education, 21*(4), 799–818.

Zohuri, B., & Zohuri, B. (2020). Stealth technology. *Radar Energy Warfare and the Challenges of Stealth Technology, 2*(3), 205–310.

Section 1

Stealth (Pre-construction): Health and Safety, Tendering, and Procurement

Chapter Two

Health and Safety in Stealth Construction

Abstract

This chapter explored health and safety considerations in stealth construction, emphasising the integration of advanced technologies and innovative practices. It commences with a general introduction, followed by a historical overview of safety practices in the construction industry, highlighting the evolution of a safety culture. The chapter examined various health and safety management techniques, including policy formulation, safety training programs, and job safety analysis. Additionally, it discussed current trends such as wearable technology, IoT, VR/AR, and predictive analytics. The unique requirements of stealth construction are addressed, focusing on building cross-section design, visibility, application of radio frequency emission and countermeasures. Finally, it presents a comprehensive approach to achieving stealth construction, emphasising environmental protection, safety, speed, economy, and aesthetics, and provides practical examples to illustrate these concepts.

Keywords: Stealth construction; health and safety; environmental protection; safety; resilience

Introduction

The construction industry plays a crucial role in economic development and urbanisation. Van Hoa et al. (2023) stated that the industry contributes significantly to the labour market of several nations worldwide through its activities, which include rehabilitation, renovation, building or maintenance of transport infrastructures, housing, public buildings and more. However, these activities are associated with significant risks that have necessitated strict health and safety measures (Kheni, 2008). According to the International Labour Organization (ILO), the construction sector is consistently ranked among the most

Stealth Construction: Integrating Practices for Resilience and Sustainability, 21–47

doi:10.1108/978-1-83608-182-120251002

hazardous industries worldwide, with high rates of accidents, injuries, and fatalities (International Labour Organization, 2005). In 2015, the ILO reported that every year, at least 108,000 workers lose their lives on construction sites, accounting for approximately 30% of all occupational fatal injuries. According to data from several industrialised countries, construction workers are 3 to 4 times more likely to die from accidents at work than other workers. The ILO further estimated that the risks associated with construction work may be 3 to 6 times greater in the developing world. Additionally, many workers suffer and die from occupational diseases that arise from past exposure to hazardous substances. Moreover, one significant risk causing construction-related deaths in the global construction industry is falling from heights (The Centre for Construction Research and Training, 2020). The Bureau of Labour Statistics reiterated this by highlighting that in 2020, 49,250 workers were injured from falling to a lower level. In the same year, falling to a lower level caused the death of 645 workers. These incidents have severe human and economic consequences, such as project delays, increased insurance costs, and reputational damage for construction firms. Therefore, the need to prevent fall-related risks has led to developing and implementing safety standards, such as personal protective equipment (PPE), guardrails, and safety nets.

Furthermore, Lingard (2013) opined that it can be challenging to coordinate the activities of different contributors to the design and construction of a facility. Each participant will have individual or organisational business interests and interactions with other project participants. It is important to make sure that the components that make up a facility are compatible with each other and that the design of the parts does not create unexpected hazards or unacceptable risks at their interfaces. The activities of different work crews and trades also need careful management to ensure that the work processes are ordered to reduce the chance of negative impact.

Health and Safety Culture in Construction

In construction organisations, project performance focuses on completing projects within time, budget, and quality standards. However, it is important to note that a weak health and safety culture can significantly hinder attaining these goals. When an organisation fails to prioritise health and safety practices, it increases the likelihood of accidents, injuries, and overall compromised well-being for its employees (Williams et al., 2008).

Health and safety in the construction industry is not only about the immediate physical risks but also about the long-term well-being of workers. Wong et al. (2020) stressed the importance of protecting workers from exposure to hazardous substances common on construction sites, such as asbestos, silica, and lead. These substances can cause serious health problems, such as respiratory diseases and occupational cancers. To prevent this, health and safety measures in construction include using PPE, proper training on handling hazardous materials, and implementing engineering controls to reduce exposure. Several research studies have shown these measures' effectiveness in lowering occupational illnesses among construction workers (Izudi et al., 2017; Sehsah et al., 2020; Tank & Anigbogu, 2012).

According to the definition provided by Tear and Reader (2023) and Glesner et al. (2020), safety culture is a product emerging from an interplay of values, attitudes, competencies, beliefs, and behavioural patterns within an organisational context. Misnan and Mohammed (2007) mentioned that creating a health and safety culture involves ensuring that employees are aware of the hazards present in their workplace, including those they may unintentionally contribute to. This mindset becomes so ingrained that employees naturally take proactive measures to improve safety, leading to a work environment where everyone is committed to recognising and mitigating hazards as part of their daily routine. Williams et al. (2008) further stress that a strong safety culture improves the productivity of construction firms and enhances their competitiveness, which in turn can help position these firms favourably in a globally competitive setting.

Several researchers have categorised health and safety culture within an organisation into two broad perspectives: Functionalist and Interpretative safety culture (Glendon & Stanton, 2000; Olga, 2022; Tappura et al., 2022).

Functionalist Safety Culture: This is a structured, top-down approach to safety management where organisational leaders play a crucial role in shaping and disseminating safety norms (Glendon & Stanton, 2000). In this paradigm, safety practices and protocols are often defined at the managerial level and are expected to be uniformly adhered to by all organisation members. In this model, the managers and practitioners view culture as a variable that can be engineered to suit the prevailing circumstances to affect performance by addressing management system faults, people's safety-related behaviour, risk assessments, and decision-making (Cooper, 2018). According to Guldenmund (2015), functionalists believe that culture is constructed by survival instincts and the need to adapt to a dynamic and sometimes challenging environment. However, this perspective also acknowledges the possibility of manipulation and direction from management. Xenikou (2022) highlighted the significance of leadership commitment in shaping safety culture. The study indicates that management commitment significantly impacts safety performance improvement and that organisational leaders directly impact employees' perceptions of safety and subsequent behaviours. Leadership commitment in this model is demonstrated through allocating resources for safety initiatives, establishing clear safety protocols, and consistently enforcing safety rules.

The functionalist approach to safety culture often involves systematic training programs, audits, and safety performance metrics to monitor and measure compliance. Tappura et al. (2021) opined that structured safety training programs positively influenced safety behaviours and contributed to developing a safety culture in organisations employing a functionalist perspective. However, critics argue that the functionalist model may sometimes lead to a checkbox mentality, where compliance with safety procedures becomes a mere formality rather than a genuine commitment to safety (Guldenmund, 2015; Reason, 1998). Additionally, the top-down nature of this approach can create a disconnect between management and frontline workers, potentially hindering the free flow of information regarding safety concerns from the bottom up.

Interpretative Safety Culture: This is a bottom-up approach that views safety as a socially constructed phenomenon influenced by the collective experiences and interpretations of the workforce (Glendon & Stanton, 2000). This approach recognises that safety is not only a set of prescribed rules but also a dynamic aspect shaped by the daily interactions and shared meanings among employees. The model suggests that frontline workers play an active role in developing safety norms, drawing from their experiences and contextual understanding of the work environment.

Cooper (2018) highlights that in this model, culture is seen as a set of fundamental assumptions created, discovered, or developed by a group as it adapts to external changes and integrates internally. Organisational culture plays a crucial role in ensuring safety in the workplace. The organisational climate, interpersonal relationships between colleagues, and the overall work atmosphere all contribute to developing safety norms. The interpretative model emphasises open communication channels, trust, and shared decision-making regarding safety matters. In this model, employees are encouraged to voice their concerns, share their experiences, and actively participate in identifying and mitigating safety risks. Williams (2008) highlights the importance of employee involvement and participation in safety processes, stating that empowering employees to contribute to safety decision-making positively influences safety culture.

However, some critics argue that this model may face challenges in standardisation and consistency across diverse workgroups. The absence of centralised control may result in variations in safety practices, potentially leading to inconsistencies in safety outcomes (Claxton et al., 2022). Moreover, the functionalist and interpretative safety culture perspectives represent two contrasting yet complementary approaches to fostering a safe working environment. While the functionalist model emphasises top-down control, structured protocols, and leadership commitment, the interpretative model underscores the importance of bottom-up participation, open communication, and the social construction of safety norms (Porter, 2021). The synthesis of these perspectives is crucial for organisations seeking to cultivate a holistic safety culture that integrates the strengths of both approaches, ensuring a balance between organisational directives and the collective wisdom of its workforce (Glendon & Stanton, 2000).

Techniques of Health and Safety Management in Construction

Wilson and Koehn (2000) describe safety management as a process involving systematic control and oversight of safety policies, procedures, and practices specifically designed to ensure safety on project sites. According to Da Silva and Amaral (2019), safety management practices are crucial tools that can be used to improve occupational health and safety performance. These practices are instrumental in mitigating workplace accidents and injuries, contributing to employees' well-being. Li et al. (2015) emphasise the significance of adopting proactive safety measures to reduce the incidence of workplace incidents. Such practices encompass a range of strategies, safety policies, safety training programs, safety planning and implementation, job safety analysis (JSA), and safety inspection and auditing. By systematically implementing these safety management practices, organisations can foster a culture of prevention, minimise risks, and create a secure work environment, aligning with the overarching goal of prioritising the health and safety of the workforce.

Safety Policy

Policy is a foundational technique in health and safety management within the construction industry, providing a structured framework to guide organisational practices and decision-making. Kin and Bonaventura (2006) describe safety policy as a written statement of principles and goals demonstrating top management's commitment to ensuring safe working methods and environment at construction sites. It outlines specific procedures and protocols to achieve this objective, including guidelines for risk assessment, training programs, emergency response, and compliance with relevant regulations. A well-formulated health and safety policy articulates the organisation's commitment to creating a secure work environment and outlines specific responsibilities, procedures, and protocols to achieve this objective (Bakri et al., 2006).

Safety Training Programs

Safety training programs are integral to health and safety management, especially in industries like construction (Boadu et al., 2020). These programs are structured initiatives designed to educate and equip workers with the knowledge, skills, and awareness necessary to navigate potential hazards and contribute to a secure work environment (Lai et al., 2011). Safety training covers various topics, including the proper use of equipment, adherence to safety protocols, and emergency response procedures. Effective programs are interactive, engaging participants in practical scenarios to enhance comprehension and retention. By investing in comprehensive safety training, organisations empower their workforce to make informed decisions, minimise risks, and collectively contribute to safety culture, reducing the likelihood of accidents and injuries on construction sites (Keng & Razak, 2014). Regular updates to training content ensure alignment with evolving industry standards and technological advancements, fostering a continuous commitment to health and safety.

Safety Planning and Implementing

Planning is crucial to ensuring that health and safety efforts are effective. Success in safety management relies on establishing, operating, and maintaining planning systems (Dorji & Hadikusumo, 2006). This technique involves developing and implementing suitable management arrangements to control risks by introducing workplace preventative measures. Organisations can ensure that their safety measures are proportionate and appropriate to the hazards and risks in the organisation. Dorji and Hadikusumo (2006) highlighted the safety planning and implementation process encompassing the following steps.

- Identify health and safety objectives and targets that are achievable and significant.

- Establish performance standards for management and risk control based on a thorough risk assessment and setting minimum acceptable standards that meet legal requirements.
- Address and manage employee risks and other factors that may impact the organisation.
- Document all performance standards to ensure comprehensive record-keeping and traceability.

This pre-emptive process ensures that safety measures are integrated into the project from its inception. Implementing these plans involves the practical execution of safety protocols, training programs, and the utilisation of safety equipment on the construction site (Khalid et al., 2021). Successful safety planning and implementation require a collaborative effort among all stakeholders, from project managers to frontline workers, to create a culture that prioritises and adheres to safety standards.

Job Safety Analysis

Albrechtsen et al. (2019) describe safety analysis as a risk assessment method to identify and evaluate potential hazards associated with a specific work task. The primary objective of JSA is to systematically and incrementally consider all risks related to a particular task of work and to identify ways to mitigate those risks. JSA follows the principles of risk assessment, which Oke et al. (2023) describe as a thorough identification, evaluation, and analysis of potential hazards within a work environment, considering the likelihood and severity of adverse outcomes. Using this technique, construction teams can proactively recognise and address safety concerns before commencing work by breaking down tasks into specific steps and analysing potential hazards related to each step. This process minimises the likelihood of accidents and injuries and provides a comprehensive understanding of the work environment. Choudhar et al. (2018) identified three major steps used in JSA for construction projects.

- ***Identify hazards*:** This step involves identifying the direct and supporting construction activities needed for a domain, defining their procedures, and analysing all possible loss-of-control events that may occur during their execution.
- ***Assess probability:*** In this step, the likelihood of occurrence of each loss-of-control event is evaluated, along with the levels of possible intensifying factors and the likelihood of use of personal safety gear.
- ***Assess severity:*** This step involves associating the possible loss-of-control events with possible accident scenarios and assessing the expected degree of severity for each type of accident scenario.

Safety Inspection and Auditing

Safety inspection and auditing are pivotal techniques in health and safety management within the construction industry. Safety inspections involve systematic assessments of the workplace, equipment, and processes to identify potential hazards and ensure compliance with safety protocols (Keng & Razak, 2014). Safety management systems have created the requirement for safety audits, which are detailed examinations and evaluations of all components of the system to ensure that they comply with prescribed standards (Keng & Razak, 2014). Safety audits include safety inspections, inspection of documents, and interviews (Nikolaos, 2010). These techniques serve as proactive measures, enabling organisations to identify and rectify potential risks before incidents occur. Regular and thorough safety inspections and audits maintain a safe working environment, ensuring safety measures align with evolving industry standards and regulations. The feedback from inspections and audits informs continuous improvement initiatives, fostering a culture of vigilance and commitment to health and safety in construction practices.

Health and Safety for Stealth Construction (At Pre-construction)

In achieving a comprehensive approach to construction management, it is crucial to consider health and safety planning as an intrinsic component of production planning from the project's inception. Pre-construction safety is an essential aspect of construction projects, where potential health and safety risks are identified to prevent accidents and injuries during the construction process. The decisions made at this stage significantly influence the safety of construction workers (Gangolells et al., 2010). This can be achieved by intertwining health and safety planning with overall production strategies. Construction teams can proactively identify potential hazards, implement preventive measures, and seamlessly integrate safety protocols into the project's workflow (Parsamehr et al., 2023). Gangolells et al. (2010) mentioned that this approach prioritises the well-being of construction personnel and fosters a culture of safety that permeates every facet of the construction project, contributing to enhanced efficiency, risk mitigation, and overall project success. Pre-construction health and safety practices are essential when planning construction projects requiring stealth. Stealth construction uses innovative and sustainable practices to ensure health and safety. This construction emphasises integrating safety considerations seamlessly at the project's inception rather than treating safety as a separate or add-on aspect; stealth construction suggests an embedded and integral approach to ensuring the well-being of workers and the project's success. This is achieved by identifying and addressing potential risks before construction activities commence. In other words, safety becomes an inherent and intrinsic aspect of the project rather than a separate entity. The tools for health and safety for stealth construction can be explained explicitly below.

Building Cross-Selection Development

Building cross-section development is a crucial aspect of architectural and construction design that plays a fundamental role in integrating health and safety considerations into stealth construction practices. It involves creating a vertical cut through the structure of a building, which provides a detailed view of its internal components and spatial relationships. A well-optimised building cross-section design prioritises the well-being of workers and occupants by integrating elements such as efficient ventilation systems, proper lighting, and ergonomic layouts.

Optimising building cross-sections in stealth construction is crucial for creating a resilient project with minimal environmental impact. This approach involves more than just meeting safety standards; it also integrates sustainability measures. For instance, using eco-friendly materials and implementing energy-efficient systems contribute to a reduced environmental footprint. Additionally, the resilient nature of stealth construction minimises disruption to neighbouring areas. Ultimately, the health and safety enhancements derived from building cross-section development align with the broader goal of creating construction practices that are environmentally sensitive, socially responsible, and contribute to the overall well-being of communities. It is further discussed below.

- ***The shape of building***

The consideration for the shape of a building in cross-section development plays a crucial role in enhancing health and safety while contributing to the objectives of stealth construction. Keshavarzian et al. (2021) highlighted that by altering the shape of a building, architects can optimise airflow, allowing fresh air to circulate more effectively. For instance, incorporating features like atriums, ventilated courtyards, or wind-catching elements can enhance cross-ventilation and reduce reliance on mechanical systems. Additionally, a thoughtfully developed cross-section incorporates safety features such as easily accessible emergency exits, clear evacuation routes, and structurally sound materials, reducing the likelihood of accidents and injuries during construction and throughout the building's lifecycle (Fu & Liu, 2020). Prioritising this element in the building cross-section development process of a construction project creates a healthier and safer environment for occupants. This approach aligns with the goals of stealth construction, which is to enhance efficiency and overall project sustainability.

- ***Internal construction***

The development of internal construction aspects in building cross sections is crucial for promoting health and safety within structures, thereby contributing to the principles of stealth construction. The thoughtful design of internal spaces involves considerations such as the layout of rooms, placement of structural elements, and incorporation of safety features that are seamlessly integrated. For instance, the strategic placement of support structures, such as columns and beams, not only ensures the structural integrity of the building but also facilitates open, unobstructed pathways that enhance mobility and reduce the risk of accidents (Hossain et al., 2018). Additionally, the proper allocation of spaces for utilities, emergency exits, and safety equipment ensures that these critical elements are discreetly integrated into the elements of the building, minimising their visual impact while maximising their effectiveness in ensuring occupant safety (Dhanakoses et al., 2023).

Internal construction is considered when optimising a building's cross-section, which enhances indoor air quality and environmental comfort. Architects create a healthier indoor environment by thoughtfully addressing natural ventilation, window positioning, and sustainable materials. This approach also reduces reliance on artificial lighting and mechanical ventilation systems, aligning with energy-efficient and environmentally conscious practices. Integrating internal construction elements into the building cross-section fosters health and safety, embodying the principles of stealth construction by harmonising occupant well-being with sustainable design and ensuring the building's resilience.

- ***Material usage***

The materials used in building construction significantly impact the health and safety of occupants. Opting for sustainable and non-toxic materials, such as low-emission paints and formaldehyde-free insulation, can help create indoor environments that are free from harmful emissions, promoting better air quality and safeguarding the health of occupants (Chrobak et al., 2022; Khoshnava et al., 2020). Murtagh et al. (2020) also stated that choosing construction materials can influence a building's resilience to natural disasters and contribute to safety. Incorporating fire-resistant materials, earthquake-resistant structures, and materials with high durability reduces the risk of structural failures and enhances the safety and lifespan of the building.

Incorporating safety measures into the structure of a building is an essential aspect of stealth construction. The use of robust materials for doors, windows, and external walls can enhance the building's resistance to intrusion and external threats, thereby addressing safety concerns without compromising the aesthetic appeal of the architecture (Langar & Kaduskar, 2024). This approach makes resilient material selection a powerful tool for enhancing health and safety. It is an integral part of stealth construction that seeks to mitigate risks and prioritise occupant well-being without overtly drawing attention to safety features.

Visibility

Visibility is a crucial aspect of construction projects that significantly impact health and safety outcomes, making it a key element in stealth construction. It ensures that workers can see and communicate with each other, reducing the risk of accidents and enhancing overall safety on the construction site. Ahmed (2020) mentioned that adequate lighting, transparent barriers, and unobstructed lines of sight contribute to a work environment where potential hazards are easily identifiable and safety protocols can be effectively implemented. Tran et al. (2022) mentioned that prioritising visibility in construction site planning and design aligns with the principles of stealth construction that consider the seamless integration of safety measures into the working environment without overtly disrupting the aesthetics of the construction site.

- ***Colour visibility***

The colour and texture of construction materials and design significantly impact health and safety outcomes in stealth construction. High-visibility colours can be strategically incorporated into safety features such as handrails, signage, and emergency exit, improving their noticeability without compromising the overall aesthetic appeal of the construction. For instance, contrasting colours on stair treads enhance visibility and reduce the risk of slips and falls. Pulver (2021) also added that the use of high-visibility vests by workers while working makes them more visible to equipment operators and more visible in low-light settings, thereby reducing the risk of injury or accidents. These vests are essential for those working on roadways, around heavy machinery, or in close contact with moving vehicles. The fluorescent orange or lime vests with reflective stripes are a common choice, enhancing visibility during the day and at night (Seidu et al., 2023). The seamless integration of colour and texture in construction projects can create an environment where safety features are subtly emphasised, promoting occupant well-being without the need for overt and intrusive safety elements, which aligns with the objectives of stealth construction.

Radio Frequency Emission

As the construction industry moves towards stealth construction, where minimising environmental impact and optimising safety practices are of utmost importance, the role of Radio Frequency (RF) cannot be overemphasised. RF is a spectrum of electromagnetic waves commonly used in communication technologies such as radio, television, and wireless networks. However, its application extends beyond communication, making it an asset in construction. In stealth construction, RF emissions can be utilised for diverse purposes, from monitoring worker safety to optimising project logistics.

- ***Health and safety monitoring***

In stealth construction, RF emission is particularly noteworthy for its ability to address critical aspects of project management. The primary function of real-time worker safety monitoring is facilitated through wearable devices embedded with RF sensors. These devices gather data on vital signs and environmental conditions without impeding the workers' activities. By leveraging RF sensors, these devices can track vital signs such as heart rate, body temperature, and respiratory rate for personalised construction safety monitoring (Awolusi et al., 2018). This comprehensive data collection ensures that potential health risks are identified early. The ability to detect anomalies promptly enables timely intervention to address emerging health concerns, ultimately enhancing worker safety and well-being in various occupational settings.

Therefore, the real-time monitoring system, facilitated by wearable devices with RF sensors, acts as a proactive safeguard for workers. Kanan et al. (2016) highlighted that this RF wearable device comprises several essential components, including a radio transceiver, an RF wake-up sensor, and an alarm actuator. The sensing unit has a processor, radio transceivers, directional antennas, and three ultrasound-based distance sensors. The integrated Wireless Safety Detection System (WSDS) detects potential hazards with construction workers. When a hazard is identified, the WSDS system activates the worker's wearable device, which sends identification information to a visual display unit in the driver's cab. This real-time communication mechanism facilitates immediate awareness for the worker and the equipment operator, enhancing safety by promptly alerting individuals to potential dangers on the construction site. This timely awareness empowers organisations to implement swift and targeted interventions, potentially preventing accidents, injuries, or health issues among workers. Consequently, integrating such advanced monitoring technologies contributes significantly to creating safer and healthier working environments.

Furthermore, Radio Frequency Identification (RFID) technology can also be used on construction sites by attaching RF tags or chips to equipment and materials; with this, teams can track the movement and location of resources in real-time (Mijwil et al., 2023). Rangoli et al. (2022) describe a practical application of Ultra High-Frequency (UHF) RFID tags in construction sites. The system employs an autonomous mechanism to monitor access within the construction site. Electronic devices are attached to the gates of the site's three sections, each equipped with a microcontroller managing a UHF RFID reader and a radio unit for data communication to the gateway. Each worker's helmet is equipped with specific RFID tags, and as they pass beneath the gate, the system identifies and adds them to the in-site personnel list based on the unique tag ID. The gathered data are transmitted to a remote server, providing a real-time overview of access status and records and enhancing overall site security and personnel management. This not only streamlines logistical operations but also enhances overall project efficiency. With the ability to seamlessly monitor the real-time location of assets and personnel, project managers can make informed decisions, allocate resources more effectively, and identify potential bottlenecks in construction workflows (Ikonen et al., 2013). This improves project timelines and resource utilisation, leading to a more efficient and well-organised construction process. This technology integration optimises logistical aspects and underscores a commitment to maintaining a secure and safe working environment, crucial in high-risk industries like construction.

- ***Site-specific hazard detection***

During the preconstruction phase, RF sensors are strategically placed to detect and monitor potential hazards in the construction environment. The deployment of these sensors throughout the site establishes a network capable of continuously monitoring environmental conditions. These RF sensors can detect anomalies indicative of site-specific risks, such as the presence of hazardous gases or the degradation of materials (Geng et al., 2015). This early detection enables project managers to take proactive measures before construction activities commence, mitigating potential risks, and ensuring a safer work environment.

This construction approach, known as stealth construction, emphasises a predictive and discreet strategy to address potential hazards efficiently. When RF technology is optimised to detect site-specific risks, construction teams can operate with a heightened awareness of environmental conditions, allowing for timely intervention and risk mitigation. The data collected by the RF sensors empowers project managers to make informed decisions, implement preventive measures, and uphold safety standards, contributing to a more secure and efficient construction process. Integrating RF sensors for site-specific hazard detection aligns with the industry's commitment to proactive risk management and safety protocols.

- ***Communication***

Effective communication and coordination are crucial for ensuring workers' safety and operations' efficiency in the construction industry. One technology that has significantly contributed to these aspects is RF emission technology. Soltanmohammadlou et al. (2019) elaborated on the Real-Time Location System (RTLS), which enables teams to have real-time insights into the location and movements of personnel and assets, fostering improved communication and negotiation. This heightened environmental awareness facilitates more effective decision-making and coordination of various tasks, ultimately preventing conflicts on construction sites. The result is a streamlined and well-coordinated construction process where teams can communicate proactively, negotiate effectively, and make reliable decisions based on RTLS technology's real-time information, contributing to increased productivity and safety on construction sites.

RF-enabled devices, such as walkie-talkies and smartphones, serve as reliable communication tools on construction sites Ragnoli et al. (2022). These devices operate on radiofrequency, enabling them to maintain communication channels even in areas with limited visibility or challenging environmental conditions. The use of RF technology ensures that construction teams can effectively exchange critical safety-related information, project updates, and emergency alerts in real time. This heightened responsiveness fosters quick decision-making, allowing for swift coordination among team members and reducing the likelihood of accidents by addressing potential issues promptly. In essence, RF emission technology serves as a crucial enabler for effective communication and coordination, contributing significantly to the safety and productivity of construction operations.

Countermeasures

As the construction industry evolves, adopting countermeasures that promote health and safety practices has become increasingly important. These countermeasures include functional construction systems and innovative construction methods, which are pivotal in mitigating potential risks and improving overall safety protocols. Lu and Cai (2019) describe countermeasures in construction safety as the main technologies involved in controlling safety risks, including monitoring, forecasting, communication, information management, intelligent construction equipment, and rescue equipment. These technologies are essential for ensuring that construction sites are safe and secure for workers and the surrounding community. By prioritising safety during the preconstruction phase, stakeholders in the industry can lay the groundwork for a secure and efficient construction process, setting the stage for successful project outcomes and minimising health and safety risks for workers and the broader community.

Additionally, functional construction systems and innovative methods address emerging challenges and create a robust foundation for safety throughout the construction lifecycle.

- ***Functional construction systems***

Functional construction systems are an essential part of countermeasures in stealth construction. These systems are known for adapting and responding dynamically to changing conditions, crucial in mitigating potential health and safety risks. They encompass a range of design, technology, and operational approaches. Some key elements include:

- ***Indoor air quality (IAQ)***

This is critical to creating a healthy living or working environment. One way to ensure optimal IAQ is by implementing heating, ventilation, and air conditioning (HVAC) systems with advanced filtration and ventilation designs (Zhang et al., 2023). These systems are seamlessly integrated into the building design, minimising the visible impact on the aesthetic appeal. By doing so, respiratory issues can be prevented and the overall quality of life can be improved.

- ***Daylighting and lighting control***

This technique maximises natural light through well-designed windows and incorporates lighting control systems that adjust according to occupancy and daylight levels (Seyedolhosseini et al., 2020). This not only enhances occupant well-being but also contributes to energy efficiency. These features are discreetly integrated into stealth construction to create a visually pleasing and safe environment.

- ***Smart building technologies***

This refers to integrating intelligent systems for monitoring and managing building safety. This includes fire detection and suppression systems, security, and emergency response systems (Vijayan et al., 2020). In stealth construction, these technologies are seamlessly integrated into the building infrastructure, minimising their visibility while maximising their effectiveness.

- ***Occupancy monitoring and emergency preparedness***

This technique uses sensors and IoT devices to monitor occupancy levels and promptly respond to emergencies (Khoche et al., 2021). This includes automated evacuation systems and real-time communication platforms. In stealth construction, these features prioritise safety without overtly highlighting their presence.

- ***Seismic resilience***

This refers to integrating structural designs that enhance the building's ability to withstand natural hazards like earthquakes (Asadi et al., 2021). Yin et al. (2022) added that seismic resilience enables a building impacted by an earthquake to recover and return to regular operation after addressing initial damage. This contributes to the resilience of structures and the well-being of occupants. In stealth construction, seismic resilience measures are integrated without compromising the overall aesthetics of the building.

- ***Energy efficiency***

Refers to using energy-efficient technologies and materials to reduce environmental impact and enhance sustainability (Olajiga et al., 2024). This includes green building practices and the use of renewable energy sources. These features are seamlessly integrated into stealth construction to promote a sustainable and safe built environment.

- ***Safe material handling systems***

This refers to implementing advanced material handling systems like robots to reduce the risk of accidents and injuries during construction (Bi et al., 2021). Automation and robotics can contribute to a safer working environment without compromising the visual appeal of the construction site. These systems are designed to move materials with precision and accuracy, reducing the need for manual labour and minimising the risk of injury for workers (Javaid et al., 2021). By automating material handling, workers can avoid repetitive tasks and reduce human errors and downtime (Dabic-Miletic, 2024). This can lead to a safer working environment and increased productivity.

Innovative Methods of Construction

Innovative construction methods that adhere to the principles of stealth construction prioritise the health and safety of workers while minimising the visible impact on the overall aesthetics of the project. Some of these methods include:

- ***Off-site construction methods***

Prefabrication and modular construction offer several advantages for health and safety. Components are assembled off-site and then transported to the construction site for final assembly, reducing on-site construction time and minimising exposure to potential hazards for workers (Ginigaddara et al., 2021). Wang et al. (2020) added that controlled factory conditions also contribute to better safety protocols during the manufacturing process. Additionally, off-site construction aligns with stealth construction principles by reducing on-site activity, noise, and disruptions. The finished components seamlessly integrate into the overall construction without compromising the project's aesthetic appeal.

- ***Building information modelling (BIM)***

BIM is a digital representation of a facility's physical and functional characteristics (Patacas et al., 2020). This tool can be integrated into functional construction systems to plan and visualise projects in a virtual environment. Akram et al. (2022) pointed out that BIM enables project stakeholders to identify and address safety concerns before construction begins. Moreover, Zhang et al. (2013) also established that the technology integrates safety considerations directly into the planning and design phases, providing a proactive approach to hazard detection and mitigation, ensuring a safer construction environment by addressing potential risks before they materialise on the site. By digitally coordinating construction elements, potential clashes and safety concerns can be addressed in the design phase, reducing the need for visible safety features during construction.

- ***Lean construction practices***

Lean construction practices aim to minimise waste, optimise efficiency, and streamline processes (Huda and Berawi (2021). Improved efficiency contributes to a safer and more controlled work environment. Additionally, lean construction practices align with stealth construction by minimising disruptions and improving efficiency. The emphasis on efficiency ensures that construction processes are well-coordinated and that safety measures are implemented seamlessly without drawing unnecessary attention.

- ***Robotics and automation***

Advanced robotics and automation can reduce the risk of injuries for construction workers by performing repetitive or hazardous tasks. Xu and De Soto (2020) stated that autonomous construction equipment and robotic systems reduce the labour force and contribute to a safer work environment, which aligns with the principle of stealth construction of minimising the need for extensive on-site manual labour. The incorporation of these technologies seamlessly integrates safety measures into the construction process without overtly impacting the visual aspects of the project.

- ***Advanced construction materials***

Using innovative and sustainable construction materials can contribute to a safer work environment by reducing exposure to harmful substances (Bradu et al., 2023). Using low-emission materials improves indoor air quality, aligning with stealth construction principles of providing health and safety benefits without requiring visible safety features. Materials contributing to energy efficiency and environmental sustainability can seamlessly integrate into the design.

Health and Safety Towards Stealth Construction

The approach of stealth construction emphasises a proactive and discreet way of integrating health and safety considerations seamlessly into the project's planning. This section explores the various aspects of health and safety in the framework of stealth construction, examining its impact on environmental protection, safety, speed, economy, and aesthetics. The intersection of these elements reflects a holistic construction approach that prioritises workers' well-being and seeks to optimise project outcomes.

Environmental Protection

Stealth construction is a proactive and discreet approach that significantly improves health and safety practices related to environmental protection in the construction industry. By prioritising sustainability, stealth construction fosters a culture where environmentally conscious choices are seamlessly integrated into the construction process. It safeguards worker health and safety and ensures a more responsible and sustainable approach to construction that aligns with broader environmental protection objectives. Some of the extensive ways in which environmental protection is achieved through health and safety practices in stealth construction can be explained below:

- **Sustainable material selection**: This process involves choosing environmentally friendly materials that promote better indoor air quality and occupant health. Maraveas (2020) mentioned that prioritising these practices by replacing conventional materials with waste materials, including rice husk ash (RHA), sugarcane bagasse ash (SCBA), and bamboo leaves ash (BLA), aligns with stealth construction, which actively contributes to reducing the overall environmental impact of a project.
- **Energy-efficient systems**: This refers to implementing energy-efficient technologies and practices that reduce resource consumption and lower emissions (William et al., 2020). This contributes to improved air quality and overall well-being. Additionally, energy-efficient systems, such as smart lighting and HVAC controls, can be integrated discreetly, promoting health and safety while minimising the environmental footprint of the construction (Iddio et al., 2020).
- **Renewable energy integration**: This refers to utilising renewable energy sources to improve air quality and reduce the health risks associated with conventional energy production (Al-Shetwi, 2022). Tan et al. (2021) stated that energy technologies like solar panels, wind turbines, and other renewable energy systems can be incorporated into the construction design, contributing to environmental protection and occupant health.

Safety

Stealth construction is a forward-thinking approach that significantly enhances health and safety practices within the construction industry, specifically concerning safety measures. It involves implementing various measures to ensure the well-being of construction workers and occupants on the construction site. The achievement of safety in stealth construction encompasses several key aspects.

- **Proactive planning and design:** This process involves proactively identifying potential hazards, implementing safety measures in the design, and optimising the layout for efficient workflows to contribute to a safer working environment. Afzal et al. (2021) suggested using Virtual Design Construction (VDC), which allows safety managers to visualise and analyse construction sites virtually to devise proactive safety measures and effective safety training. Rodrigues et al. (2022) also suggested integrating BIM at the design stage, which allows for the simulation of construction processes, identifying potential safety hazards and enabling proactive measures to address them. Additionally, safety measures identified through BIM can be seamlessly implemented, ensuring that safety protocols are discreetly integrated into the construction process without drawing unnecessary attention.

- **Advanced technologies and robotics:** This refers to the adoption of advanced technologies, such as drones, robotics, and automation, that reduce the need for workers to engage in hazardous tasks (Balzan et al., 2020). These technologies enhance safety by minimising exposure to potential risks and accidents, thereby contributing to safety on the construction site.
- **Prefabrication:** This is a process that involves the manufacturing of building components in a controlled environment away from the construction site (Wang et al., 2020). This reduces on-site activities, minimising potential hazards for workers. Additionally, prefabricated components are seamlessly integrated into the construction, reducing the need for extensive on-site work. This aligns with stealth construction principles by minimising visible construction activities and enhancing safety during construction.
- **Innovative PPE:** This enhances worker safety by providing real-time data and alerts about potential dangers on the construction site. Basodan et al. (2021) highlighted that this technology includes smart helmets, vests, lightweight gloves, and protective hearing devices. Rasouli et al. (2024) asserted that these innovative PPEs minimise exposure to hazards, gather data, send notifications, and adapt automatically to various conditions. These advanced PPE solutions are seamlessly integrated into the safety protocols in stealth construction, contributing to worker well-being.

Speed

The approach of stealth construction prioritises optimising project speed without compromising health and safety practices. The implementation of health and safety practices can contribute significantly to achieving speed in construction projects without compromising the well-being of workers. The practices identified below show how speed is achieved in stealth construction through the integration of health and safety measures:

- **Efficient planning and design:** This process involves thorough planning and design to ensure faster construction timelines. Additionally, proactive planning for health and safety ensures that potential hazards are identified and mitigated early in the process. Integrating safety measures into the design minimises disruptions during construction, allowing for a more seamless and speedy execution (Parsamehr et al., 2023).
- **Advanced technologies and automation:** This refers to advanced technologies, such as drones, 3D printing, and robotic construction, that significantly accelerate construction processes (Zhao et al., 2022). Additionally, automation

reduces manual labour requirements and speeds up repetitive tasks. Automated construction methods contribute to worker safety by minimising exposure to hazardous conditions. Speed is achieved without compromising safety, as robots and automated systems handle tasks that could be risky for human workers.

- **Prefabrication:** This is a process that involves the manufacturing of building components in a controlled environment away from the construction site (Wang et al., 2020). This reduces on-site activities, minimising potential hazards for workers. Additionally, prefabricated components are seamlessly integrated into the construction, reducing the need for extensive on-site work. This aligns with stealth construction principles by enhancing speed during the construction process.
- **Lean construction practices:** This process involves eliminating waste, optimising processes, and improving efficiency to achieve faster project delivery (Huda & Berawi, 2021). Additionally, lean practices often lead to safer construction sites by reducing unnecessary movements and enhancing communication (Yilmaz et al., 2022). Integrating safety measures aligns with lean principles, ensuring a balance between speed and safety.
- **BIM for collaborative communication and planning:** BIM allows stakeholders to collaboratively design and plan every aspect of the construction project in a digital environment, fostering efficient communication and decision-making (Kurwi et al., 2021). This collaborative approach ensures that health and safety considerations are proactively incorporated, preventing potential hazards and delays. Real-time information sharing among construction teams, architects, and contractors promotes a cohesive understanding of safety protocols, streamlines workflows, and accelerates project timelines.
- **Supply chain management:** This process involves efficient logistics to minimise delays by ensuring that materials and equipment are available when needed (Okeagu et al., 2021). Additionally, safe logistics practices contribute to worker well-being by reducing the risk of accidents while transporting and handling materials. Integrating safety measures ensures that speed is achieved without compromising worker safety.
- **Preventive maintenance practices:** This process involves regular maintenance and quality assurance practices to prevent delays caused by equipment breakdowns or rework due to errors. Additionally, preventive maintenance ensures that construction equipment operates safely, reducing the risk of accidents and contributing to both speed and overall project integrity (Sarbini et al., 2021).

Economy

Achieving economy in stealth construction involves carefully balancing cost-effectiveness and health and safety. Implementing health and safety measures contributes significantly to economic efficiency in construction projects while aligning with the principles of stealth construction. Below are some of the extensive ways in which economy is achieved:

- **Preventive maintenance practices**: Regular maintenance, when integrated with health and safety measures, can prevent equipment breakdowns and reduce the need for costly repairs. This type of maintenance, known as preventive maintenance, contributes to economic efficiency by minimising downtime and associated costs (Sarbini et al., 2021).
- **Sustainable construction practices**: Sustainable construction practices such as energy-efficient designs and eco-friendly material usage align with health and safety standards and contribute to long-term economic benefits. Lower operational costs and resource-efficient construction practices enhance economic sustainability.
- **Advanced technologies**: Incorporating advanced technologies, such as BIM for safety planning, can enhance collaboration and decision-making. This technology-driven approach streamlines processes, reduces errors, and contributes to economic savings.
- **Supply chain management**: This promotes collaboration and fosters long-term relationships with suppliers, ensuring the health and safety of workers within the supply chain. A stable and well-managed supply chain contributes to cost-effective material procurement and project execution.

Aesthetics

While safety and efficiency are the foundation of stealth construction, aesthetics play an equally important role in shaping the built environment. In stealth construction, aesthetics is achieved by integrating health and safety measures to prioritise the well-being of workers and occupants while maintaining the overall aesthetic appeal. Below is an extensive exploration of how aesthetics is achieved in stealth construction through the implementation of health and safety practices:

- **Integrated design:** In the design phase, aesthetic considerations begin with integrating safety features. This includes incorporating safety elements such as handrails, guardrails, and emergency exits to complement the overall design, making them practically invisible to the observer.
- **Transparent material selection:** Using transparent materials for barriers or partitions maintains an open and airy feel while ensuring safety. Glass barriers, for example, provide both visibility and protection, contributing to a modern and sophisticated appearance.
- **Landscaping and greenery:** Incorporating green spaces and vegetation into the design can enhance aesthetics and contribute to environmental well-being. Living walls, rooftop gardens, or integrated landscaping can soften the built environment while promoting a healthier atmosphere.
- **Colour and texture coordination:** Careful selection of colours and textures can ensure that safety elements blend harmoniously with the overall design. For instance, safety signage and markings can be designed to complement the colour scheme, minimising the visual disruption often associated with construction projects and contributing to a more aesthetically pleasing outcome.

- **Ambient lighting:** Thoughtful lighting design can enhance safety by ensuring well-lit spaces, especially in critical areas. At the same time, strategically placed lighting can create ambience, highlighting architectural features and contributing to an aesthetically pleasing environment.
- **Customised signage:** Safety signage and wayfinding elements can be customised to align with the design aesthetic by choosing fonts, colours, and materials that complement the overall visual language of the construction.
- **Sustainable practices:** Choosing sustainable and recyclable materials can align with aesthetic and environmental goals. Sustainable materials contribute to a modern, eco-friendly appearance while supporting broader health and safety objectives.

Practical Applications of Health and Safety in Stealth Construction

Case Study 1: Urban High-Rise Renovation Project – Minaminagasaki Dormitory, Japan

Overview

Kajima Corporation renovated the Minaminagasaki Dormitory, a high-rise reinforced concrete apartment building in an urban setting in Japan (Liao et al., 2014). This building is considered one of the oldest high-rise reinforced concrete (RC) buildings in Japan. The renovation project aimed to upgrade the facility, enhance energy efficiency, and showcase innovative construction practices, including modular construction. This strategic approach underscored Kajima Corporation's commitment to safety, sustainability, and community-centric development.

Health and Safety Measures

- *Transparent safety barriers:* Transparent barriers made from impact-resistant materials were strategically installed around the construction site. These barriers ensured safety while allowing pedestrians and residents to observe the renovation progress without compromising visibility.
- *Advanced safety technology:* Construction workers were equipped with wearable technology that monitored their well-being in real time. This technology tracked vital signs, provided alerts for potential health issues, and exemplified a proactive approach to worker safety.
- *Noise-reducing strategies:* The project incorporated noise-reduction strategies to minimise disruptions to the surrounding urban environment. Quieter construction equipment was used during specific hours, reducing noise pollution and maintaining positive relations with the neighbouring community.
- *Community engagement:* Kajima Corporation proactively engaged with the local community to communicate project updates, timelines, and safety measures. This included regular community meetings, informational signage, and a 24/7 hotline for residents to address concerns promptly.
- *Sustainable construction practices:* The renovation project prioritised sustainability by incorporating energy-efficient systems, green roofs, and environmentally

friendly materials. These practices aligned with safety and aesthetics, contributing to the overall well-being of the environment and occupants.

- *Modular construction:* The renovation employed modular construction methods to enhance efficiency and minimise on-site disruptions. This was achieved by manufacturing prefabricated components off-site, which reduced construction time and optimised the use of resources. The modular approach ensured precision and quality in the renovation process.

Outcome

This Minaminagasaki Dormitory project showcases how health and safety measures, including modular construction, can be practically applied in renovating a high-rise reinforced concrete apartment building. The strategic combination of safety, sustainability, and innovative construction practices resulted in successful project meeting functional and aesthetic objectives.

Case Study 2: Green Building Construction – Alexander Forbes Building, Sandton, Johannesburg

Overview

The Alexander Forbes Building in Sandton, Johannesburg, is a prime example of sustainable and green building construction (Masia et al., 2020). It was designed by Paragon Architects and constructed through a WBHO-Tiber joint venture. The building has been awarded a 4-Star Green Star office design v1 rating, a testament to its environmentally conscious practices and architectural excellence. The project aimed to construct an energy-efficient and environmentally friendly office space while prioritising the health and safety of construction workers and the surrounding residential community.

- *Transparent safety practices:* The team implemented transparent safety practices during construction to ensure the workforce's well-being. They established clear communication channels, safety signage, and visible safety protocols to minimise risks and accidents.
- *Advanced technologies:* The construction site incorporated advanced safety technologies to monitor and enhance worker safety. Wearable safety equipment and real-time monitoring systems were employed to address potential health and safety concerns promptly.
- *Green concrete usage:* The Alexander Forbes Building used around 43,000 cubic metres of 'green concrete' during construction. This type of concrete typically includes recycled or environmentally friendly materials, contributing to the building's overall green certification.
- *Energy-efficient lighting:* The building implemented motion-sensing activation for lighting throughout the building. This technology ensures that lights are only activated when individuals are present, reducing unnecessary energy consumption and contributing to the overall energy efficiency of the building.

- *Natural daylighting and shading*: The building features an exterior design that mimics the sun's movement with scallops. This design serves dual purposes: it allows for natural daylighting within the building, reduces the need for artificial lighting, and provides natural shading, minimising the reliance on air conditioning for cooling purposes.

Outcome

The Alexander Forbes Building is a sustainable and health-conscious construction model in a residential area with a 4-star Green Star rating. The building's transparent safety practices during construction ensured a secure working environment. Implementing green building features has led to long-term benefits such as energy efficiency, cost savings, and a reduced environmental footprint.

Conclusion

This chapter emphasises the critical role of health and safety practices in the emerging field of stealth construction. Integrating advanced technologies like RF emission, wearable tech, and predictive analytics ensures construction projects achieve unprecedented safety and efficiency levels during the preconstruction phase. The evolution of safety practices, supported by comprehensive training, proactive planning, and cutting-edge innovations, underscores the industry's dedication to worker well-being and project sustainability. Addressing cultural, organisational, and financial challenges is essential for effective implementation. Ultimately, health and safety in stealth construction advocates for a holistic approach that combines technological advancements, regulatory compliance, and sustainable practices to create a safer and more efficient construction environment, setting new industry trends.

References

Afzal, M., Shafiq, M. T., & Al Jassmi, H. (2021). Improving construction safety with virtual-design construction technologies review. *Journal of Information Technology in Construction, 26*, 319–340. https://doi.org/10.36680/j.itcon.2021.018

Ahmed, M. N. (2020). *Effect of improving the visibility of the traffic path on vehicle speeds and safety at work zones* [Master Thesis at University of Alberta, Canada]. https://doi.org/10.7939/r3-04b2-da85

Akram, R., Thaheem, M. J., Khan, S., Nasir, A. R., & Maqsoom, A. (2022). Exploring the role of BIM in construction safety in developing countries: Toward automated hazard analysis. *Sustainability*, *14*(19), 1–23.

Albrechtsen, E., Solberg, I., & Svensli, E. (2019). The application and benefits of job safety analysis. *Safety Science, 113*, 425–437. https://doi.org/10.1016/j.ssci.2018.12.007

Al-Shetwi, A. Q. (2022). Sustainable development of renewable energy integrated power sector: Trends, environmental impacts, and recent challenges. *Science of The Total Environment, 822*, 1–18. https://doi.org/10.1016/j.scitotenv.2022.153645

Asadi, E., Salman, A. M., Li, Y., & Yu, X. (2021). Localized health monitoring for seismic resilience quantification and safety evaluation of smart structures. *Structural Safety, 93*, 1–17. https://doi.org/10.1016/j.strusafe.2021.102127

Awolusi, I., Marks, E., & Hallowell, M. (2018). Wearable technology for personalized construction safety monitoring and trending: Review of applicable devices. *Automation in Construction*, *85*, 96–106. https://doi.org/10.1016/j.autcon.2017.10.010

Bakri, A., Mohamad Zin, R., Misnan, M. S., Mohd Yusof, Z., & Wan Mahmood, W. Y. (2006). *Safety training for construction workers: Malaysian experience*. https://citeseerx.ist.psu.edu/document?repid=rep1&type=pdf&doi=45f4ff90a4d336085bb8d1fb6015d47fcae7119b

Balzan, A., Aparicio, C. C., & Trabucco, D. (2020). Robotics in construction: State-of-art of on-site advanced devices. *International Journal of High-Rise Buildings*, *9*(1), 95–104.

Basodan, R. A., Park, B., & Chung, H. J. (2021). Smart personal protective equipment (PPE): Current PPE needs, opportunities for nanotechnology and e-textiles. *Flexible and Printed Electronics*, *6*(4), 1–15. https://doi.org/10.1088/2058-8585/ac32a9

Bi, Z. M., Luo, C., Miao, Z., Zhang, B., Zhang, W. J., & Wang, L. (2021). Safety assurance mechanisms of collaborative robotic systems in manufacturing. *Robotics and Computer-Integrated Manufacturing*, *67*, 1–10. https://doi.org/10.1016/j.rcim.2020.102022

Boadu, E. F., Wang, C. C., & Sunindijo, R. Y. (2020). Characteristics of the construction industry in developing countries and its implications for health and safety: An exploratory study in Ghana. *International Journal of Environmental Research and Public Health*, *17*(11), 1–20.

Bradu, P., Biswas, A., Nair, C., Sreevalsakumar, S., Patil, M., Kannampuzha, S., Mukherjee, A. G., Wanjari, U. R., Renu K., Vellingiri, K., & Gopalakrishnan, A. V. (2023). Recent advances in green technology and Industrial Revolution 4.0 for a sustainable future. *Environmental Science and Pollution Research*, *30*(60), 124488–124519.

Choudhar, S., Solanki, P., & Gidwani, G. (2018). Job safety analysis (JSA) applied in construction industry. *IJSTE-International Journal of Science Technology & Engineering*, *4*(9), 1–9.

Chrobak, J., Iłowska, J., & Chrobok, A. (2022). Formaldehyde-free resins for the wood-based panel industry: alternatives to formaldehyde and novel hardeners. *Molecules*, *27*(15), 1–16.

Claxton, G., Hosie, P., & Sharma, P. (2022). Toward an effective occupational health and safety culture: A multiple stakeholder perspective. *Journal of Safety Research*, *82*, 57–67. https://doi.org/10.1016/j.jsr.2022.04.006

Cooper, M. D. (2018). The safety culture construct: Theory and practice. In C. Gilbert, J. Benoit, H. Laroche, & C. Bieder (Eds.), *Safety cultures, safety models: Taking stock and moving forward* (pp. 47–61). Springer.

Da Silva, S. L. C., & Amaral, F. G. (2019). Critical factors of success and barriers to the implementation of occupational health and safety management systems: A systematic review of literature. *Safety Science*, *117*, 123–132. https://doi.org/10.1016/j.ssci.2019.03.026

Dabic-Miletic, S. (2024). Benefits and challenges of implementing autonomous technology for sustainable material handling in industrial processes. *Journal of Industrial Intelligence*, *2*(1), 1–13.

Dhanakoses, K., Povatong, P., Wongphyat, W., & Wiriyakraikul, C. (2023). Building adaptations and laboratory safety concerns: A case study of high-rise academic laboratories in a Thai University. *Nakhara: Journal of Environmental Design and Planning*, *22*(2), 312–312.

Dorji, K., & Hadikusumo, B. H. (2006). Safety management practices in the Bhutanese construction industry. *Journal of Construction in Developing Countries*, *11*(2), 53–75.

Fu, M., & Liu, R. (2020). An approach of checking an exit sign system based on navigation graph networks. *Advanced Engineering Informatics*, *46*, 1–11. https://doi.org/10.1016/j.aei.2020.101168

Gangolells, M., Casals, M., Forcada, N., Roca, X., Fuertes, A., Macarulla, M., & Vilella, Q. (2010). Identifying potential health and safety risks at the pre-construction stage. In *Proceedings of the 18th CIB world building congress W099* (pp. 59–73), Salford, United Kingdom.

Geng, Y., Chen, J., Fu, R., Bao, G., & Pahlavan, K. (2015). Enlighten wearable physiological monitoring systems: On-body RF characteristics based human motion classification using a support vector machine. *IEEE Transactions on Mobile Computing, 15*(3), 656–671.

Ginigaddara, B., Perera, S., Feng, Y., & Rahnamayiezekavat, P. (2021). Development of an offsite construction typology: A Delphi study. *Buildings, 12*(1), 1–21.

Glendon, A. I., & Stanton, N. A. (2000). Perspectives on safety culture. *Safety Science, 34*(1–3), 193–214. https://doi.org/10.1016/S0925-7535(00)00013-8

Glesner, C., Van Oudheusden, M., Turcanu, C., & Fallon, C. (2020). Bringing symmetry between and within safety and security cultures in high-risk organizations. *Safety Science, 132*, 1–9. https://doi.org/10.1016/j.ssci.2020.104950

Guldenmund, F. (2015). Organizational safety culture. In *The Wiley Blackwell handbook of the psychology of occupational safety and workplace health* (pp. 437–458). https://doi.org/10.1002/9781118979013.ch19

Hossain, M. A., Abbott, E. L., Chua, D. K., Nguyen, T. Q., & Goh, Y. M. (2018). Design-for-safety knowledge library for BIM-integrated safety risk reviews. *Automation in Construction, 94*, 290–302. https://doi.org/10.1016/j.autcon.2018.07.010

Huda, G., & Berawi, M. A. (2021). Quickly understanding of lean construction to improve project cost performance: A literature review. In *ITB Graduate School Conference, 1*(1), 61–75.

Iddio, E., Wang, L., Thomas, Y., McMorrow, G., & Denzer, A. (2020). Energy efficient operation and modeling for greenhouses: A literature review. *Renewable and Sustainable Energy Reviews, 117*, 1–15. https://doi.org/10.1016/j.rser.2019.109480

Ikonen, J., Knutas, A., Hämäläinen, H., Ihonen, M., Porras, J., & Kallonen, T. (2013). Use of embedded RFID tags in concrete element supply chains. *Journal of Information, Technology and Construction, 18*(7), 119–147.

International Labour Organization (2005). *ILO Convention 155. Number of work-related accidents and illnesses continues to increase*. ILO/WHO Joint Press.

Izudi, J., Ninsiima, V. & Alege, J. B. (2017). Use of personal protective equipment among building construction workers in Kampala, Uganda. *Journal of Environmental and Public Health, 12*(3), 191–199. https://doi.org/10.1155/2017/7930589

Javaid, M., Haleem, A., Singh, R. P., & Suman, R. (2021). Substantial capabilities of robotics in enhancing industry 4.0 implementation. *Cognitive Robotics, 1*, 58–75. https://doi.org/10.1016/j.cogr.2021.06.001

Kanan, R., Elhassan, O., Bensalem, R., & Husein, A. (2016). *A wireless safety detection sensor system. 2016 IEEE SENSORS*, Orlando, Florida, USA, pp. 1–3. https://doi.org/10.1109/ICSENS.2016.7808926

Keng, T. C., & Razak, N. A. (2014). Case studies on the safety management at construction site. *Journal of Sustainability and Management, 9*(2), 90–108.

Keshavarzian, E., Jin, R., Dong, K., & Kwok, K. C. (2021). Effect of building cross-section shape on air pollutant dispersion around buildings. *Building and Environment, 197*, 1–19. https://doi.org/10.1016/j.buildenv.2021.107861

Khalid, U., Sagoo, A., & Benachir, M. (2021). Safety Management System (SMS) framework development–Mitigating the critical safety factors affecting Health and Safety performance in construction projects. *Safety Science, 143*, 1–16. https://doi.org/10.1016/j.ssci.2021.105402

Kheni, N. A. (2008). *Impact of health and safety management on safety performance of small and medium-sized construction businesses in Ghana* [Doctoral dissertation, Loughborough University].

Khoche, S., Chandrasekhar, K. V., Sasirekha, G. V. K., Bapat, J., & Das, D. (2021). Occupancy detection for emergency management of smart building based on indoor localization: A feasibility study. *SN Computer Science*, *2*(419), 1–11. https://doi.org/10.1007/s42979-021-00812-4

Khoshnava, S. M., Rostami, R., Mohamad Zin, R., Štreimikienė, D., Mardani, A., & Ismail, M. (2020). The role of green building materials in reducing environmental and human health impacts. *International Journal of Environmental Research and Public Health*, *17*(7), 2589.

Kurwi, S., Demian, P., Blay, K. B., & Hassan, T. M. (2021). Collaboration through integrated BIM and GIS for the design process in rail projects: Formalising the requirements. *Infrastructures*, *6*(4), 1–19.

Lai, D. N., Liu, M., & Ling, F. Y. (2011). A comparative study on adopting human resource practices for safety management on construction projects in the United States and Singapore. *International Journal of Project Management*, *29*(8), 1018–1032.

Langar, S., & Kaduskar, S. (2024). Resilient materials and technologies for a single-family home: A review. In *Construction research congress 2024* (pp. 197–207).

Li, H., Lu, M., Hsu, S. C., Gray, M., & Huang, T. (2015). Proactive behavior-based safety management for construction safety improvement. *Safety Science*, *75*, 107–117. https://doi.org/10.1016/j.ssci.2015.01.013

Liao, C. T., Chang, H. Y., & Juan, Y. J. (2014). A study on the deterioration condition and renovation work for high-rise reinforced concrete apartment in Japan – The practice project Kajima corporation minaminagasaki dormitory. *Applied Mechanics and Materials*, *479*, 1124–1127.

Lingard, H. (2013). Occupational health and safety in the construction industry. *Construction Management and Economics*, *31*(6), 505–514.

Lu, C., & Cai, C. (2019). Challenges and countermeasures for construction safety during the Sichuan–Tibet railway project. *Engineering*, *5*(5), 833–838.

Maraveas, C. (2020). Production of sustainable construction materials using agro wastes. *Materials*, *13*(2), 1–29.

Masia, T., Kajimo-Shakantu, K., & Opawole, A. (2020). A case study on the implementation of green building construction in Gauteng province, South Africa. *Management of Environmental Quality: An International Journal*, *31*(3), 602–623.

Mijwil, M. M., Hiran, K. K., Doshi, R., & Unogwu, O. J. (2023). Advancing construction with IoT and RFID technology in civil engineering: A technology review. *Al-Salam Journal for Engineering and Technology*, *2*(2), 54–62.

Misnan, M. S., & Mohammed, A. H. (2007). Development of safety culture in the construction industry: A conceptual framework. *In 23rd Annual ARCOM Conference*, September 3rd, Belfast, UK (pp. 13–22).

Murtagh, N., Scott, L., & Fan, J. (2020). Sustainable and resilient construction: Current status and future challenges. *Journal of Cleaner Production*, *268*, 1–10. https://doi.org/10.1016/j.jclepro.2020.122264

Nikolaos. (2010). The measurement of health and safety conditions at work theoretical approaches, tools and techniques a literature review. *International Research Journal of Finance and Economics*, *36*(1), 87–95.

Oke, A. E., Adetoro, P. E., Stephen, S. S., Aigbavboa, C. O., Oyewobi, L. O., & Aghimien, D. O. (2023). *Risk management practices in construction: A global view*. Springer Nature.

Okeagu, C. N., Reed, D. S., Sun, L., Colontonio, M. M., Rezayev, A., Ghaffar, Y. A., Kaye, R. J., Liu, H., Cornett E. M., Fox, C. J., Urman R. D., & Kaye, A. D. (2021). Principles of supply chain management in the time of crisis. *Best Practice & Research Clinical Anaesthesiology*, *35*(3), 369–376.

Olajiga, O. K., Ani, E. C., Sikhakane, Z. Q., & Olatunde, T. M. (2024). A comprehensive review of energy-efficient lighting technologies and trends. *Engineering Science & Technology Journal*, *5*(3), 1097–1111.

Olga, K. (2022). *Safety culture development in dredging and marine contractor companies–three-case study on safety programs* [Master's thesis, uis].
Parsamehr, M., Perera, U. S., Dodanwala, T. C., Perera, P., & Ruparathna, R. (2023). A review of construction management challenges and BIM-based solutions: Perspectives from the schedule, cost, quality, and safety management. *Asian Journal of Civil Engineering*, *24*(1), 353–389.
Patacas, J., Dawood, N., & Kassem, M. (2020). BIM for facilities management: A framework and a common data environment using open standards. *Automation in Construction*, *120*, 1–21. https://doi.org/10.1016/j.autcon.2020.103366
Porter, M. L. (2021). *Preventing injuries and fatalities in inherently dangerous work environments* [Doctoral dissertation, Walden University].
Pulver, S. E. (2021). *The status of construction jobsite emergency alert systems in the construction industry* [Masters Dissertation, Roger Williams University].
Ragnoli, M., Colaiuda, D., Leoni, A., Ferri, G., Barile, G., Rotilio, M., Laurini, E., De Berardinis P., & Stornelli, V. (2022). A LoRaWAN multi-technological architecture for construction site monitoring. *Sensors*, *22*(22), 1–21.
Rasouli, S., Alipouri, Y., & Chamanzad, S. (2024). Smart Personal Protective Equipment (PPE) for construction safety: A literature review. *Safety Science*, *170*, 1–15. https://doi.org/10.1016/j.ssci.2023.106368
Reason, J. (1998). Achieving a safe culture: Theory and practice. *Work & Stress*, *12*(3), 293–306.
Rodrigues, F., Baptista, J. S., & Pinto, D. (2022). BIM approach in construction safety – A case study on preventing falls from height. *Buildings*, *12*(1), 1–17.
Sarbini, N. N., Ibrahim, I. S., Abidin, N. I., Yahaya, F. M., & Azizan, N. Z. N. (2021). Review on maintenance issues toward building maintenance management best practices. *Journal of Building Engineering*, *44*, 1–13. https://doi.org/10.1016/j.jobe.2021.102985
Sehsah, R., El-Gilany, A. H., & Ibrahim, A. M. (2020). Personal protective equipment (PPE) use and its relation to accidents among construction workers. *La Medicina del lavoro*, *111*(4), 285–295.
Seidu, R. K., Sun, L., & Jiang, S. (2023). A systematic review on retro-reflective clothing for night-time visibility and safety. *The Journal of the Textile Institute*, 1–13.
Seyedolhosseini, A., Masoumi, N., Modarressi, M., & Karimian, N. (2020). Daylight adaptive smart indoor lighting control method using artificial neural networks. *Journal of Building Engineering*, *29*, 1–10. https://doi.org/10.1016/j.jobe.2019.101141
Soltanmohammadlou, N., Sadeghi, S., Hon, C. K., & Mokhtarpour-Khanghah, F. (2019). Real-time locating systems and safety in construction sites: A literature review. *Safety Science*, *117*, 229–242. https://doi.org/10.1016/j.ssci.2019.04.025
Tan, K. M., Babu, T. S., Ramachandaramurthy, V. K., Kasinathan, P., Solanki, S. G., & Raveendran, S. K. (2021). Empowering smart grid: A comprehensive review of energy storage technology and application with renewable energy integration. *Journal of Energy Storage*, *39*, 1–22. https://doi.org/10.1016/j.est.2021.102591
Tank, B. L., & Anigbogu, N. A. (2012). The use of personal protective equipment (PPE) on construction sites in Nigeria. In S. Laryea, S. A. Agyepong, R. Leininger, & W. Hughes (Eds.), *Proceedings 4th West Africa built environment research (WABER) conference*, Abu, Nigeria (pp. 1341–1348).
Tappura, S., Jääskeläinen, A., & Pirhonen, J. (2021). Performance implications of safety training. In *Advances in safety management and human performance: Proceedings of the AHFE 2021 virtual conferences on safety management and human factors, and human error, reliability, resilience, and performance*, July 25–29, 2021, USA (pp. 295–301). Springer International Publishing.
Tappura, S., Jääskeläinen, A., & Pirhonen, J. (2022). Creation of satisfactory safety culture by developing its key dimensions. *Safety Science*, *154*, 1–14. https://doi.org/10.1016/j.ssci.2022.105849

Tear, M. J., & Reader, T. W. (2023). Understanding safety culture and safety citizenship through the lens of social identity theory. *Safety Science*, *158*, 1–14. https://doi.org/10.1016/j.ssci.2022.105993

The Centre for Construction Research and Training. (2020). *Highlights 2020*. CDC Stacks. https://stacks.cdc.gov/view/cdc/118256/cdc_118256_DS1.pdf

Tran, S. V. T., Nguyen, T. L., Chi, H. L., Lee, D., & Park, C. (2022). Generative planning for construction safety surveillance camera installation in 4D BIM environment. *Automation in Construction*, *134*, 1–14. https://doi.org/10.1016/j.autcon.2021.104103

Van Hoa, N., Van Thu, P., Dat, N. T., & Loan, L. T. (2023). Analysis of the business environment of the logistics and the construction industry in Vietnam after Covid-19. *International Journal of Advanced Multidisciplinary Research and Studies*, *3*(1), 316–328.

Vijayan, D. S., Rose, A. L., Arvindan, S., Revathy, J., & Amuthadevi, C. (2020). Automation systems in smart buildings: A review. *Journal of Ambient Intelligence and Humanized Computing*, 1–13. https://doi.org/10.1007/s12652-020-02666-9

Wang, M., Wang, C. C., Sepasgozar, S., & Zlatanova, S. (2020). A systematic review of digital technology adoption in off-site construction: Current status and future direction towards industry 4.0. *Buildings*, *10*(11), 1–29.

William, M. A., Elharidi, A. M., Hanafy, A. A., Attia, A., & Elhelw, M. (2020). Energy-efficient retrofitting strategies for healthcare facilities in hot-humid climate: Parametric and economical analysis. *Alexandria Engineering Journal*, *59*(6), 4549–4562.

Wilson, Jr. J. M. & Koehn, E. (2000). Safety management: Problems encountered and recommended solutions. *Journal of Construction Engineering and Management*, *126*(1), 77–79.

Wong, T. K. M., Man, S. S., & Chan, A. H. S. (2020). Critical factors for the use or non-use of personal protective equipment amongst construction workers. *Safety Science*, *126*, 1–12. https://doi.org/10.1016/j.ssci.2020.104663

Xenikou, A. (2022). Leadership and organizational culture. In C. Newton & Knight, R. (Eds.), *Handbook of research methods for organisational culture* (pp. 23–38). Edward Elgar Publishing.

Xu, X., & De Soto, B. G. (2020). On-site autonomous construction robots: A review of research areas, technologies, and suggestions for advancement. In *Proceedings of the 37th International Symposium on Automation and Robotics in Construction (ISARC)* (Vol. 37, pp. 385–392). IAARC Publications.

Yilmaz, A., Dora, M., Hezarkhani, B., & Kumar, M. (2022). Lean and industry 4.0: Mapping determinants and barriers from a social, environmental, and operational perspective. *Technological Forecasting and Social Change*, *175*, 1–16. https://doi.org/10.1016/j.techfore.2021.121320

Yin, C., Kassem, M. M., & Mohamed Nazri, F. (2022). Comprehensive review of community seismic resilience: Concept, frameworks, and case studies. *Advances in Civil Engineering*, *2*, 1–19, https://doi.org/10.1155/2022/7668214

Zhang, J., Chan, C. C., Kwok, H. H., & Cheng, J. C. (2023). Multi-indicator adaptive HVAC control system for low-energy indoor air quality management of heritage building preservation. *Building and Environment*, *246*, 1–17. https://doi.org/10.1016/j.buildenv.2023.110910

Zhang, S., Teizer, J., Lee, J. K., Eastman, C. M., & Venugopal, M. (2013). Building information modelling (BIM) and safety: Automatic safety checking of construction models and schedules. *Automation in Construction*, *29*, 183–195. https://doi.org/10.1016/j.autcon.2012.05.006

Zhao, S., Wang, Q., Fang, X., Liang, W., Cao, Y., Zhao, C., Li, L., Liu, C., & Wang, K. (2022). Application and development of autonomous robots in concrete construction: Challenges and opportunities. *Drones*, *6*(12), 1–26.

Chapter Three

Tendering in Stealth Construction

Abstract

This chapter investigated tendering in stealth construction, emphasising innovative approaches and methodologies that prioritise environmental protection, safety, efficiency, and aesthetics. It began with an overview of the construction industry's tendering processes, followed by an in-depth examination of various tendering types, including competitive and negotiated methods. The study highlighted contemporary trends such as electronic tendering, Building Information Modelling (BIM), green and sustainable procurement, risk management, data analytics, artificial intelligence, lean construction practices, and blockchain technology. Moreover, with a specific focus on stealth construction, the chapter further analysed certain criteria, including building cross-section development, visibility, radio frequency emission, and countermeasures. It explored integrating functional construction systems, including environmental, safety, health, and quality management. Additionally, it discussed methods like green building, modular construction, and low-impact techniques. Lastly, the chapter emphasised the strategies to achieve environmental protection, safety, speed, economy, and aesthetics in tendering for stealth construction.

Keywords: Stealth construction; sustainability; tendering; efficient project delivery; resilient project

Introduction

The construction industry is a multifaceted and dynamic sector that plays a vital role in shaping infrastructure, buildings, and facilities. It encompasses various activities, from small-scale residential projects to large-scale commercial and infrastructure developments. Due to its diverse nature, qualified and capable service providers, including consultants and contractors, are needed to ensure the successful execution of infrastructure construction projects (Kusumarukmi & Adi, 2019).

Stealth Construction: Integrating Practices for Resilience and Sustainability, 49–72

Published under exclusive licence by Emerald Publishing Limited
doi:10.1108/978-1-83608-182-120251003

However, the tendering process has been identified as one of the crucial processes in the preconstruction phase that clients must navigate before commencing any construction project (Kang & Kim, 2018). Several researchers have described tendering as soliciting bids from construction companies for specific services or construction projects (Laryea, 2017; Roy & Kumar, 2017). In construction tendering, qualified contractors or suppliers are invited to submit bids, typically submitted in response to requests for proposals (RFP) issued by government agencies or private clients. The purpose of tendering is to reduce costs and ensure transparency, fairness, and efficiency in selecting the most suitable contractor or supplier for a construction project.

Roy and Kumar (2017) highlighted some of the multifaceted purposes of tendering. Firstly, it aims to create an environment that attracts interest and competitive offers from qualified, experienced construction contractors. This ensures a broad pool of contractors is considered for the project, promoting competitiveness and potentially leading to cost savings. Secondly, tendering seeks a fair price and best value for the construction works, ensuring that the project remains within budget constraints. Additionally, tendering helps establish a clear understanding of the rights and obligations of all parties involved, facilitating smooth project execution and reducing the likelihood of conflicts and disputes. It also allows for resolving general and specific issues related to the tender's bid, ensuring comprehensive coverage of all project aspects. Ultimately, the primary goal of tendering is to secure a construction contractor who can successfully undertake and deliver the project scope within the specified time, cost, and quality parameters.

The construction tendering process commences with developing comprehensive project specifications and requirements, which are documented in a tender document. This document delineates the project's scope, timelines, budgetary constraints, and the criteria for evaluating and selecting the winning bid. Subsequently, the tender document is publicly advertised or shared with a pre-qualified list of contractors or suppliers, inviting them to submit bids. Contractors and suppliers interested in bidding must prepare and submit a detailed tender proposal outlining their project approach, pricing, qualifications, and relevant experience. The client assesses the submitted offers and decides which contractors or suppliers are eligible for negotiations.

Trends of Tendering Practices

The construction tendering process plays a pivotal role in project development. During this stage, contractors submit their bids for construction projects. Over the years, several trends have surfaced in tendering, reflecting technological advancements, evolving regulations, and shifts in industry practices. Below is an extensive overview of key trends in construction tendering:

Electronic Tendering

Electronic tendering, also known as e-tendering, has been identified by construction industry experts as one of the information technology tools capable of changing the industry's culture and improving the tendering processes (Lavelle & Bardon, 2009).

E-tendering systems facilitate the entire tendering process, from creating tender documents to submitting bids, all online. This transition reduces paper usage and eliminates the necessity for physical document storage, resulting in substantial cost and time savings. Abdullahi et al. (2022) asserted that government agencies, service providers, and industry adopt the e-tendering system to enhance efficiency and lower business costs, ultimately delivering more efficient services to the community. Sunmola and Shehu (2020) added that this system facilitates immediate communication among stakeholders, including clients, contractors, and suppliers. This fosters collaboration and ensures transparency throughout the entire tendering process. Stakeholders can retrieve tender documents, submit bids, and stay informed about the tender status in real-time, regardless of location, as long as they have an internet connection. This high level of accessibility significantly boosts efficiency and shortens the time needed for tendering processes, enabling projects to progress swiftly. Additionally, e-tendering systems frequently incorporate automated notifications and reminders, ensuring all stakeholders remain well-informed and meet deadlines (Ahmad et al., 2019).

Building Information Modelling (BIM)

Building Information Modelling (BIM) is also one of the digital revolutionising tools identified in the construction industry (Kensek, 2014). This system enables project stakeholders to enhance their effectiveness through seamless collaboration and coordination. BIM allows for generating detailed 3D models that integrate data related to a project's design, construction, and operation (Ding et al., 2014). During the tendering phase, BIM can be leveraged to improve communication, reduce errors, precise quantity take-offs, and automate time-consuming tasks, contributing to more efficient and reliable tendering processes (Correa & Santos, 2021). Furthermore, BIM enables stakeholders to simulate construction processes and visualise the project's progress, providing a clearer understanding of work sequencing and potential challenges (Abbasnejad et al., 2021). However, Macek (2023) opined that the successful use of BIM in tendering relies on the ability of all participants to meet these high demands and collaborate effectively to create and maintain high-quality BIM models. This entails the architects, engineers, contractors, and other stakeholders ensuring that the BIM models are accurate, detailed, and current. Also, there is a need for a thorough understanding of the software and its capabilities and a commitment to collaboration and data sharing among project team members.

Green and Sustainable Procurement

The need for sustainable and environmentally conscious practices in the construction industry has piqued the interest of both researchers and practitioners, prompting a focus on green and sustainable procurement. Although these trends are still budding, the literature has distinguished between these concepts (Lundberg & Marklund, 2018; Oruezabala & Rico, 2012; Sönnichsen & Clement, 2020). Green procurement has been described as procuring goods, services and works with a reduced environmental impact throughout their life cycle. On the other hand, sustainable procurement encompasses the integration of environmental, economic, and social criteria into the procurement process of goods, services, and works. Numerous tenders now incorporate explicit environmental requirements, urging contractors to embrace green building practices and utilise eco-friendly materials. Sustainable construction standards, including Leadership in Energy and Environmental Design (LEED) and the Building Research Establishment Environmental Assessment Method (BREEAM), are frequently cited in tender documents to steer contractors towards achieving sustainability objectives. This transition towards green procurement positively impacts the environment, fosters innovation, and enhances efficiency in construction processes.

Furthermore, contractors participating in tenders with green and sustainable procurement requirements must demonstrate their commitment to environmental stewardship. This involves sharing sustainability policies, emphasising past green building projects, and detailing strategies for minimising environmental impact during construction. Additionally, collaborating with suppliers offering eco-friendly materials furthers sustainable procurement practices. Integrating green principles into the tender process ensures construction projects reduce carbon footprints, preserve natural resources, and foster healthier and sustainably built environments.

Risk Management

The construction process is exposed to all types of risk, including economic, socio-political, technical and business risks within the environment where they are operating, making it very difficult for the contractor to meet the project quality, operational requirements and the planned delivery date of the project (Mhetre et al., 2016). The growing focus on risk management during the tendering process aligns with a broader industry trend of proactive risk mitigation in construction.

However, the advent of risk management in the construction industry enables contractors, consultants, clients, and suppliers to achieve specific project goals while mitigating adverse effects on quality, schedule, and expenses (Oke et al., 2023). Contractors are now expected to provide detailed risk management plans as part of the tender submission process. These plans entail identifying project-related risks, evaluating their potential consequences, and suggesting mitigation measures to reduce or eliminate them. During the tender evaluation, contractors' handling of these risk factors is meticulously assessed, emphasising the practicality and efficacy of their proposed risk management strategies. De Lima (2020) stated that this will ensure clients make better-informed choices when selecting project contractors. This approach ensures early identification and resolution of potential risks, minimising the chances of expensive delays or disputes. Incorporating risk management into the tendering process is a proactive step that enhances the successful execution of construction projects.

Data Analytics and Artificial Intelligence (AI)

In construction, the tendering process involves various manual and time-consuming tasks. These tasks include sifting through unstructured data within tender documents, copying and pasting requirements, and writing tender descriptions. Unfortunately, these manual processes can be inefficient and lead to missed opportunities. However, integrating data-driven automated technologies like AI can revolutionise the tendering process (Flechsig et al., 2022). Contractors can utilise these technologies to analyse extensive historical data, including past project costs, market trends, and supplier performance. By harnessing AI algorithms, contractors can achieve more accurate project cost predictions, identify potential risks, and optimise their bidding strategies. This data-driven approach enhances tendering process efficiency and elevates the overall quality of project planning and management (Regona et al., 2022).

Moreover, using data analytics and artificial intelligence (AI) tools empowers contractors to streamline operations and enhance project results (Jahani et al., 2021). By analysing data from diverse sources, such as project schedules, resource allocation, and material costs, contractors can pinpoint inefficiencies and areas for improvement. This enables them to optimise their workflows, minimise waste, and boost productivity. Elmousalami (2020) added that AI-driven tools can automate routine tasks, like generating cost estimates or creating schedules, freeing contractors' time to focus on strategic aspects of their projects. Integrating data analytics and AI in the tendering process fosters innovation and efficiency within the construction industry, improving project outcomes and greater client satisfaction.

Lean Construction

Initially inspired by the manufacturing industry, lean construction principles are now being applied in the construction tendering process. These principles focus on waste reduction, efficiency optimisation, and enhancing overall project outcomes (Ahmed & Wong, 2018). A crucial element of lean construction during tendering involves minimising unnecessary paperwork and administrative tasks, streamlining the process, and saving valuable time. Eliminating waste, such as excess documentation or redundant procedures, ensures contractors concentrate on value-added activities directly contributing to project success.

Another key lean construction principle in tendering is emphasising collaboration and communication among project stakeholders through BIM (Nowotarski et al., 2016). Lean practices encourage the early involvement of all parties, including contractors, architects, and suppliers, to ensure that project requirements are clearly understood and met. This collaborative approach helps to identify potential issues early in the tendering process, leading to more efficient problem-solving and decision-making. Applying lean construction principles in tendering improves efficiency, communication, and project outcomes, ultimately benefiting contractors and clients.

Blockchain Technology

Blockchain technology is gaining significant interest due to its potential to transform the tendering process. The adoption of blockchain by stakeholders can establish immutable records for bid submissions, evaluations, and contract awards (Ahmadisheykhsarmast et al., 2023). This ensures that all transactions are securely recorded and cannot be modified, resulting in high transparency throughout the tendering process. Furthermore, blockchain streamlines verification by consolidating relevant information in a single accessible ledger, reducing the risk of errors and fraud.

Furthermore, blockchain technology holds promise for simplifying payment processes in construction projects. Smart contracts, self-executing agreements with terms directly encoded in code, can automate payments based on predefined conditions. This reduces administrative overhead related to payment processing and mitigates payment disputes. Overall, blockchain technology can potentially transform the tendering process in construction by improving transparency, security, and efficiency.

Tendering for Stealth Construction

Tendering is a necessary process that sets the foundation for a successful and efficient project. During the tendering process, a critical factor is choosing contractors and suppliers who align with the principles of stealth construction. This entails providing tender documents that detail the project's sustainability objectives, environmental protection criteria, and safety standards. Contractors must showcase their comprehension of these requirements and propose effective strategies for meeting them in their bids.

Cheaitou et al. (2019) highlighted that in the preconstruction stage, tendering involves two critical aspects: risk management and cost optimisation. Contractors must articulate their risk management strategies, detailing how to identify, assess, and mitigate risks throughout the project. This proactive approach helps minimise the chances of delays, disputes, and cost overruns. Lines and Kumar (2018) asserted that contractors must provide comprehensive cost estimates and proposals that showcase their ability to execute the project within budget constraints. The rigorous process of evaluating bids and selecting contractors who offer the best value for money ensures project owners have a project delivered within budgeted cost and time.

Moreover, Ciribini et al. (2015) suggested that tendering can leverage innovative technologies and practices like BIM, commonly used to create detailed project models. BIM enhances the precision of cost estimations and improves overall project understanding. Additionally, contractors may be obligated to incorporate green building practices and sustainable construction standards into their proposals, aligning with the project's sustainability goals. Preconstruction tendering sets the stage for a successful project by carefully selecting contractors and suppliers capable of delivering high-quality, sustainable, cost-effective solutions. However, certain important criteria must be considered to achieve stealth construction during the tendering process.

Building Cross-Section Development

In stealth construction, the development of building cross-sections plays a pivotal role, especially during the tendering phase. A building's cross-section represents a vertical slice through its structure, revealing the arrangement and interplay of key components such as walls, floors, roofs, and openings. However, in stealth construction, the cross-section transcends mere technical drawings; it becomes a strategic tool. Contractors utilise cross-sections to integrate sustainability, energy efficiency, and resilience features into the design. To achieve the desired project outcomes, contractors must interpret and effectively implement these cross-sections during the tendering process.

Moreover, the development of building cross-sections requires meticulous attention to various factors. Contractors must grasp the design intent and ensure the cross-sections harmonise with the project's sustainability objectives. These objectives may include maximising natural light, optimising thermal performance, and incorporating renewable energy systems. Additionally, cross-sections should seamlessly integrate resilient features such as blast-resistant materials while preserving the building's aesthetics and functionality. Contractors bidding on stealth construction projects must showcase their proficiency in interpreting and executing these complex cross-sections.

The development of building cross-sections also goes beyond the physical structure. It encompasses the seamless integration of smart technologies and advanced materials. Contractors must be proficient in using BIM to create and analyse digital representations of the building's cross-sections. BIM enables virtual simulations and optimisations, improving efficiency, reducing waste, and enhancing overall building performance. By incorporating these cutting-edge technologies into their bids, contractors demonstrate their commitment to innovation and ability to deliver high-quality, sustainable, and secure buildings aligned with the principles of stealth construction.

- ***Shape of the building***

The building's shape is crucial during cross-section development, particularly during the tendering phase. Pacheco et al. (2012) emphasised that the shape of a building not only influences aesthetics but also significantly impacts sustainability, energy efficiency, and resilience. Contractors vying for stealth construction projects must meticulously evaluate the building's form and its effects on these aspects. For instance, buildings with streamlined and aerodynamic shapes can decrease energy consumption by minimising wind resistance and optimising natural ventilation. Jafari and Alipour (2021) highlighted that this approach can result in cost savings and environmental benefits, making it an essential consideration during the tendering process. Achieving aesthetics and efficiency in building design involves integrating certain technologies and strategies, such as photovoltaic cells, natural ventilation, heat recovery, external shading, and efficient lighting controls (Chandrasekar, 2023; Shafaghat & Keyvanfar, 2022). As the global floor area continues to expand, prioritising both efficiency and flexibility becomes crucial for the sustainability of our energy system. Through the adoption of more efficient building designs and stringent energy performance standards for appliances, we can mitigate the increase in peak electricity demand driven by electrification and renewables growth.

Moreover, the architectural shape of a building significantly impacts its resilience against external threats (D'Ascanio et al., 2016). Buildings with irregular or complex designs may be more susceptible to damage during natural disasters. When bidding for stealth construction projects, contractors must showcase their expertise in designing and constructing buildings with resilient shapes. This involves integrating symmetrical designs in buildings, which perform better during disasters. Buildings with predictable shapes are more accessible to analyse and reinforce, while irregular shapes may exacerbate the effects of natural forces (e.g., wind and earthquakes). Addressing these factors in their proposals shows that contractors demonstrate their grasp of building shape's role in stealth construction. They also highlight their ability to deliver structures that align with project requirements for sustainability, energy efficiency, and resilience.

- ***Internal construction***

Internal construction encompasses the arrangement and layout of the building's components, like walls, floors, ceilings, and utilities (Sully, 2015). In stealth construction, these internal elements must fulfil functional requirements and integrate features that enhance sustainability, energy efficiency, and innovative construction. Contractors competing for stealth construction projects should meticulously evaluate the internal construction of the building and propose inventive solutions that align with these fundamental principles.

- ***Material usage***

The materials for constructing a building profoundly impact its sustainability, energy efficiency, and resilience. Contractors competing for stealth construction projects must thoughtfully evaluate the types of materials they propose to use and understand how these materials contribute to overall building performance. Miller and Doh (2015) highlighted that contractors might recommend incorporating sustainable materials and construction methods that minimise energy consumption and environmental impact. This could involve utilising recycled materials, energy-efficient insulation, and passive design approaches to enhance natural light and ventilation. Additionally, contractors may propose integrating smart technologies and advanced building management systems for efficient energy monitoring and optimisation. By emphasising these internal construction aspects in their proposals, contractors showcase their dedication to sustainability and energy efficiency, which are critical factors during the tendering process for stealth construction projects.

Golz (2016) also identified that the choice of materials significantly influences the resilience of a building. Contractors may recommend using resilient materials that withstand impact, fire, and other external threats. This could involve integrating blast-resistant materials, reinforced concrete, or ballistic glass in areas susceptible to natural disasters. Emphasising these material considerations in their proposals shows that contractors demonstrate their grasp of the importance of material selection in stealth construction. They also showcase their capability to deliver buildings that align with project requirements for sustainability, energy efficiency, and resilience.

Energy Transmission

Energy transmission in the context of stealth construction involves effectively transferring and distributing energy within a building, primarily focusing on minimising waste and optimising efficiency. During the tendering process, contractors vying for stealth construction projects must meticulously assess how energy transmission will impact the overall performance and sustainability of the building. This assessment includes evaluating the design and construction of heating, cooling, lighting, and electrical systems to ensure they minimise energy loss and maximise energy efficiency. Furthermore, contractors are expected to propose innovative solutions for energy transmission that align with the project's sustainability objectives, ultimately leading to reduced energy consumption and lower operating costs over the long term.

Furthermore, energy transmission in stealth construction goes beyond conventional systems to encompass the integration of renewable energy sources and smart technologies. Elcock (2007) highlighted nanotechnologies as one of the inventions that offer the potential to directly generate electricity from renewable sources such as solar, wind, and geothermal energy. It was further mentioned that we can enable distributed electricity production by harnessing this energy at or near its source. This approach minimises transmission losses and reduces the reliance on traditional right-of-way (ROW)-based transmission systems for electricity. Integrating nanotechnology-driven energy solutions into building designs contributes to sustainability and resilience (Roco et al., 2011). Additionally, Abdulraheem et al. (2024) suggested that smart metres and sensors can also be employed to monitor and optimise energy usage. Therefore, contractors must carefully consider how these elements can be seamlessly integrated into the building's design and construction to enhance its energy efficiency and sustainability. Addressing the issue of energy transmission in their project proposals shows contractors' dedication to sustainability and innovation in stealth construction, positioning themselves as industry leaders and offering clients state-of-the-art solutions for energy-efficient buildings.

Countermeasures

Countermeasures in stealth construction involve the tactics and technologies employed to mitigate risks and ensure the building remains resilient in a natural disaster. During the tendering process, contractors competing for stealth construction projects must present effective countermeasures that align with the project's resilient goals. These measures may involve implementing management systems or resilient features to safeguard the building against potential threats and vulnerabilities. Contractors must demonstrate their grasp of the associated risks and propose comprehensive construction techniques that balance resilience while preserving the building's functionality and aesthetic appeal.

Furthermore, countermeasures in stealth construction extend beyond resilient features to include innovative solutions that leverage technology and sustainable practices. Contractors must propose technologies that are not only effective but also sustainable and environmentally friendly. This may include using renewable energy sources, energy-efficient systems, smart building management systems and BIM to enhance resilience, sustainability, and efficiency. Addressing the issues of countermeasures in their bids enables contractors to demonstrate their commitment to delivering safe, resilient, and sustainable buildings that meet the project's requirements and exceed client expectations. Some of these Management systems and technologies are explained explicitly below:

- ***Environmental management system (EMS)***

Campos et al. (2016) describe the EMS as an integral component of an organisation's management system that focuses on managing the environmental aspects of the organisation's activities, products, and services. Lam et al. (2011) further corroborated that this system can also function as a critical element in defining green specifications for buildings. EMS is essential for seamlessly integrating environmental factors throughout the construction process, beginning with the tendering phase. When contractors integrate an EMS into their tendering procedures, they showcase their dedication to reducing the environmental footprint of their projects. This involves implementing strategies to minimise waste generation, advocating for sustainable material sourcing, and ensuring compliance with pertinent environmental regulations. Furthermore, integrating an EMS into the tendering process enables contractors to identify and mitigate potential environmental risks before they escalate proactively. This proactive approach leads to cost savings and prevents delays by addressing environmental issues early in the project lifecycle. Additionally, an EMS empowers contractors to enhance their environmental performance and improve their reputation as environmentally responsible organisations. Integrating environmental considerations into the tendering process through EMS, contractors can effectively demonstrate their unwavering commitment to sustainability and set themselves apart in competitive bidding.

- ***Occupational safety and health (OSH) management system***

Adequate safety and health management systems are critical in the construction industry, especially during tendering. Jaafar et al. (2017) emphasised that an efficient OSH management system is vital to tackle the elevated fatality rates in construction, which surpass those in other industries. This system ensures that workers operate safely, leading to enhanced productivity and reduced costs related to workplace injuries and illnesses (Shamsuddin et al., 2015). Integrating an OSH system during the tendering stage demonstrates contractors' commitment to worker well-being and project success. At this stage, the system should incorporate provisions for regular safety inspections and audits to identify and address potential risks (Yiu et al., 2019). Additionally, contractors must ensure that workers have access to personal protective equipment (PPE) and that all safety procedures are communicated and understood (Wong et al., 2020). This proactive approach contributes to a safer and more productive construction industry.

- ***Quality management system (QMS)***

QMS encompasses a systematic approach, documentation, guidance, and audits integral to every project management process, from project initiation through the final steps of project closure (Aized, 2012). Mane and Patil (2015) added that this system also aims to uphold the standard of construction work, ensure customer satisfaction, and contribute to companies' long-term competitiveness and survival. Implementing a robust QMS helps contractors demonstrate their commitment to delivering high-quality work in the tendering process. This includes establishing quality control measures that cover all aspects of the construction process, from materials sourcing to project completion. Additionally, a robust QMS during tendering helps contractors deliver quality work and builds trust with clients, which shows the capacity of the contractor to deliver a project that benefits all stakeholders (Pater, 2019). It also allows contractors to identify potential quality issues early and proactively address them (Loushine et al., 2006). This approach significantly reduces the chances of costly rework and delays, benefiting all stakeholders involved.

- ***Supply chain management system***

Construction supply chain management (CSCM) involves strategically planning, coordinating, and optimising all activities related to procuring, managing, and delivering materials needed for a stealth construction project (Irizarry et al., 2013). This system is a crucial component of the tendering process and serves as a strategy to ensure efficiency, cost-effectiveness, and timely delivery of materials and services. Contractors must carefully manage their supply chains to secure competitive pricing, reliable delivery schedules, and high-quality materials (Benton & McHenry, 2010). This can be achieved by building supplier relationships, contract negotiation, and logistics coordination to ensure timely availability of materials. Furthermore, this system assists contractors in risk mitigation and handling disruptions (DuHadway et al., 2019). Through vetting suppliers, contractors ensure reliable partners who meet quality and delivery needs. These systems also optimise inventory management, reducing waste and enhancing efficiency (Benton & McHenry, 2010). Therefore, this system is essential for contractors looking to improve competitiveness, reduce risks, and deliver successful construction projects.

- ***Technology integration***

Technology integration has been identified as one of the important processes of stealth construction, especially at the tendering stage, where it enhances efficiency and communication among project stakeholders (Lu et al., 2015). These processes include tools and platforms that enhance online collaboration, document management, and real-time communication. They empower contractors to streamline the tendering process, reduce paperwork, minimise errors, and make faster decisions. Furthermore, integrating technology into the tendering process allows contractors to use innovative tools like BIM and immersive technologies, which include Virtual Reality (VR), Augmented Reality (AR), and Mixed Reality (MR). Khan et al. (2021) highlighted integrating this BIM and immersive technologies, which helps visualise project designs, identify issues early, create detailed models, analyse construction sequences, and optimise plans. Ultimately, this leads to more accurate cost estimates and better project outcomes, showcasing contractors as industry leaders in innovation and efficiency.

Tendering Towards a Sustainable Construction

Tendering towards stealth construction refers to a strategic approach that combines innovative building design, careful material selection, and advanced technological tools to achieve successful project outcomes while prioritising sustainability. Integrating these elements in construction ensures projects operate more efficiently and minimise environmental impact. This construction process prioritises resilience, sustainability, and efficiency. Johnson and Klassen (2022) identified that stakeholders, including contractors, clients, and consultants, leverage technology and a collaborative approach to tendering to ensure efficient communication, streamlined processes, and successful project outcomes. Ciribini et al. (2015) highlighted that using BIM for tendering is based on model checking, which reduces costs, time, and risks. Additionally, potential claims, disputes, or conflicts decrease because inconsistencies become more detectable. As a result, honest bidding is encouraged, and bidders are motivated to submit higher-quality offers, competing based on their ability to execute design and construction work.

In stealth construction, there is a significant need to balance various important principles, including environmental protection, safety, speed, economy, and aesthetic considerations, which are the core principles that guide the construction industry towards more efficient and sustainable practices.

Environmental Protection

During the early stage of construction, which involves the tendering process, it is important to consider strategies that minimise the environmental impacts of the projects. This involves the various aspects of the construction process, from design and material selection to construction methods and waste management. Choden and Soratana (2020) asserted that contractors seeking to define environmental criteria should outline these features in tender notices or contract clauses to receive the most economically advantageous tender. Palmujoki et al. (2010) further added that specifying green products precisely in tender documents promotes sustainable choices. Certain principles highlight how environmental protection can be achieved in stealth construction during the tendering process; some of these are explained below;

- ***Energy-efficient technologies***

This is one of the principles adopted to ensure construction projects meet sustainability objectives and reduce environmental impact. Lee et al. (2015) agreed that contractors must prioritise high-efficiency heating, ventilation, and air conditioning (HVAC) systems, energy-efficient lighting, and renewable energy sources like solar panels or wind turbines when bidding for projects. These technologies significantly reduce energy consumption, operating costs, and greenhouse gas emissions. Specifying these energy-efficient technologies during the tendering process shows the contractors' commitment to sustainability and provides clients with environmentally friendly and cost-effective solutions. Additionally, these technologies offer long-term benefits for building owners and occupants. Such buildings are more comfortable, healthier, and have lower energy bills than traditional structures (Ismaeil & Sobaih, 2023).

- ***Sustainable material specifications***

Integrating sustainable material specifications in the tendering process is crucial for promoting environmentally friendly practices in stealth construction. Oyegoke et al. (2009) explained that the instruction to tender in the tendering process usually encourages contractors bidding for projects to stipulate requirements for using recycled or locally sourced materials and those with low environmental impact throughout their life cycle. These materials reduce the project's overall carbon footprint and minimise waste.

- ***Waste management plan (WMP)***

In promoting sustainable practices in stealth construction, it is essential to integrate WMPs into the tendering process. Contractors bidding for projects are expected to submit detailed waste reduction and recycling plans and environmental protection measures as part of their bids (Hu, 2011). This plan serves as a strategic document which outlines how waste will be managed within a specific context, including details about the types of waste, their estimated volumes (determined based on historical data), and the final destinations for waste disposal or treatment (Oladiran, 2009). A comprehensive WMP proposes waste minimisation strategies to reduce waste generation and improve overall management practices. Furthermore, integrating WMPs into construction projects offers substantial environmental advantages. Saad et al. (2022) identified that reusing and recycling materials can help decrease the amount of waste going into landfills and preserve natural resources. Mawed (2020) added that reusing and recycling materials can save costs. Recycled materials are usually less expensive than new ones, and by minimising waste, there is no need for unnecessary purchases. Including waste reduction and recycling plans during the tendering process is a proactive measure to promote sustainable construction practices and minimise the environmental impact of projects.

Safety

Ensuring safety in construction is crucial, and achieving a high level of safety during stealth construction involves meticulous planning and strict adherence to best practices. Boadu et al. (2021) stated that in the tendering stage, stakeholders specify health and safety (H&S) requirements in tender documents by clearly defining objectives and planning. This enables contractors to allocate resources for H&S, leading to better contractor selection and safer outcomes.

There are several ways in which safety can be achieved at this stage, some of which include:

- ***Safety plans***

Integrating safety plans in stealth construction at the tendering stage involves providing a detailed document that lays out procedures and protocols to safeguard the health and well-being of workers on-site. This plan is vital in preventing accidents, injuries, and fatalities by including risk assessments, hazard identification, emergency procedures, and PPE use guidelines. Contractors participating in the bidding process are expected to provide comprehensive safety plans that detail strategies for identifying, assessing, and mitigating safety hazards (Raza et al., 2022). Getuli et al. (2020) also suggested that these plans should highlight the significance of regular safety meetings and effective communication to inform workers about potential risks.

- ***Advanced technologies***

The use of advanced technologies for safety is one of the core principles in stealth construction, and it offers innovative solutions to enhance workers' safety or mitigate risks. Afzal et al. (2021) suggested using technology-driven applications, including BIM and immersive technologies, enabling visualisation and simulation of design and construction data. Unlike traditional methods, these technologies enhance hazard recognition and improve safety planning and management. Patel et al. (2022) also added that contractors can embrace wearable technologies like smart helmets or vests for real-time monitoring of workers' health and safety. These devices track vital signs, detect fatigue or dehydration, and alert workers and supervisors to potential hazards, ensuring a safer work environment. Furthermore, to ensure workers comply with safety regulations, drones can conduct site inspections and identify potential hazards without putting workers at risk (Umar, 2021).

- ***Safety training and certifications***

Safety training and certifications are significant in stealth construction to ensure a safe working environment. Tender documents are expected to specify mandatory comprehensive safety training for all workers to minimise potential risks. Also, contractors must provide evidence of certifications and ongoing training programs to ensure their workforce is well-prepared to handle safety hazards. For instance, the Construction Skills Certification Scheme (CSCS) provides cards that validate individuals' training and qualifications for construction work (Khan et al., 2022). Additionally, certifications like the Construction Health and Safety Technician (CHST) and Safety Trained Supervisor Construction (STSC) are widely recognised. These certifications often require specific hours of safety training and relevant work experience (Tariqa, 2021). Contractors can leverage such programs to demonstrate their commitment to safety.

Speed

One of the fundamental principles of stealth construction at the tendering stage involves prioritising speed in project delivery. This approach emphasises efficiency and aims to minimise disruptions. Some of the effective strategies for achieving speed in stealth construction during the tendering process are discussed below:

- ***Lean construction practices***

Lean construction focuses on creating sustainable customer satisfaction by minimising waste across construction processes (Ramani & KSD, 2021). It involves waste reduction from material storage to structure acceptance. Lalmi et al. (2021) stated that the lean concept integrates principles like Just in Time (JIT), Total Quality Management (TQM), and supply chain management, drawing inspiration from TOYOTA's renowned experience. For instance, implementing lean practices, such as just-in-time delivery of materials and minimising downtime and delays, enables contractors to streamline processes and reduce costs. Xing et al. (2021) added that this principle emphasises the importance of collaboration and communication among project stakeholders, fostering a team-based approach to project delivery. Therefore, integrating lean construction practices into the tendering process is crucial for contractors aiming to optimise efficiency and achieve successful construction projects.

- ***Prefabrication and modular construction***

In stealth construction, highlighting the possible implementation of prefabrication and modular construction methods during the tendering process yields substantial benefits. Prefabrication involves manufacturing building components off-site, allowing for concurrent on-site and off-site work (Rocha et al., 2022). This approach significantly reduces construction time and enhances project efficiency. Similarly, modular construction, where pre-built modules are assembled on-site, streamlines the process and minimises on-site labour. Both methods are implemented for repetitive project elements like walls, floors, and ceilings, where standardised components can be swiftly assembled on-site. Additionally, the speed of construction afforded by prefabrication and modular construction can result in earlier project completion and reduced financing costs. According to Thai et al. (2020), modular construction can increase construction time by up to 50%.

- ***Advanced project management tools***

Implementing advanced project management tools and software during tendering can significantly enhance construction project outcomes. Rahimian et al. (2020) suggested a hybrid application of BIM and machine learning as an advanced tool capable of streamlining communication, monitoring progress in real time, and facilitating quick decision-making, leading to improved project efficiency. Pérez et al. (2024) also stated that these tools benefit all stakeholders by enhancing coordination among project stakeholders, reducing delays, and minimising cost overruns. Additionally, these tools enable better tracking of project timelines and budgets, allowing for more effective project planning and resource allocation (Awofadeju et al., 2024; Odejide & Edunjobi, 2024).

Economy

In stealth construction, prioritising resilience, efficiency, and resource conservation is crucial. Moreover, achieving cost-effectiveness without compromising quality or safety is paramount. During the tendering process for this approach, meticulous planning and specific strategies are essential to achieve an economically viable project.

- ***Value engineering***

Promoting value engineering during the tendering process can result in substantial cost savings and enhanced project efficiency. Venkataraman and Pinto (2023) described value engineering as a methodological approach that systematically reviews project needs and scope to identify cost-saving opportunities without compromising performance or quality. Gunduz et al. (2022) highlighted that value engineering and sustainability can effectively reduce the overall cost of sustainable projects. Contractors achieve this by analysing each project component and proposing alternative materials or methods that offer better value. Baarimah et al. (2021) revealed that value engineering can lead to innovative solutions that improve project outcomes by challenging traditional assumptions and exploring new approaches; contractors can find creative ways to reduce costs and enhance project efficiency.

- ***Life cycle cost analysis***

Integrating life cycle cost analysis into the tendering process is crucial for making informed decisions that contribute to the overall project economy of stealth construction. This approach goes beyond the initial construction costs and considers the long-term costs associated with maintenance, operation, and energy consumption (Larsen et al., 2022). Conducting a thorough life cycle cost analysis will enable contractors to identify solutions that offer the best value over the project's life. This may involve opting for materials, equipment, or technologies with higher upfront costs but lower long-term costs (Kumari et al., 2022). Rad et al.

(2021) suggested that BIM plugins can use cost and resiliency factors to predict the project lifecycle. This allows stakeholders to select a cost-efficient and resilient design option for stealth construction.

- ***Risk management strategies***

Developing proactive risk management strategies at the tendering stage is essential for the successful performance of stealth construction. This involves a systematic approach to identifying potential risks, assessing their possible impact on project costs, and developing comprehensive mitigation plans (Smith & Merritt, 2020). By proactively identifying risks such as supply chain disruptions, labour shortages, or unforeseen site conditions, contractors can anticipate challenges and prepare detail-specific strategies to mitigate the identified risks (Boateng et al., 2022). These strategies may include contingency planning, diversifying suppliers, or integrating flexible scheduling.

Aesthetics

In stealth construction, focusing on achieving aesthetics during the early stage of the project involves integrating design principles that enhance the project's visual appeal while maintaining resilience. Here are seven extensive ways in which aesthetics can be achieved in stealth construction through the tendering process:

- ***Integrated design architecture***

Making this principle a priority in the tendering process is significant for ensuring that stealth construction projects blend seamlessly with their surrounding environments. This involves carefully selecting materials, colours, and forms harmonising with the existing environment or urban context. Ozturk and Soygazi (2024) asserted that generative AI can enhance design processes, predictive modelling, operation and maintenance of a building. By thoughtfully integrating this element, contractors and designers can create structures that enhance visual coherence rather than disrupt it. This approach improves aesthetics and fosters community acceptance and socio-cultural appreciation (Kosorić et al., 2021). Therefore, by prioritising integrated design architecture, tender documents can set the standard for projects that are not only environmentally conscious but also visually appealing.

- ***Innovative façade system***

Implementing innovative facade solutions is crucial for achieving aesthetic appeal and functional benefits in stealth construction projects. Johnsen and Winther (2015) illustrated that dynamic facades can adapt to environmental conditions, enhancing resilience and energy efficiency by regulating heat gain and loss. These facades may incorporate features like movable shading devices, smart glass, or responsive cladding systems that adjust based on weather patterns or user preferences. Additionally, Collins et al. (2017) identified green walls as another innovative facade solution that enhances visual appeal and offers environmental advantages such as improved air quality, reduced urban heat island effect, and natural insulation. Therefore, prioritising innovative façade solutions in the tendering process ensures contractors can offer projects that are not only environmentally responsible but also visually striking and aligned with stealth construction principles.

- ***Low-impact development practices***

Low Impact Development (LID) practices, as conceptualised by Anne (2008), involve design and development processes that harmonise with nature. LID aligns with the objectives of stealth construction, which prioritise the sustainable management of natural water, vegetation, and artificial landscape elements to create environmentally friendly and visually appealing urban spaces. Dhurve (2024) added that integrating natural materials such as wood and stone can result in visually appealing buildings that harmonise effortlessly with their surroundings. Therefore, contractors can prioritise environmental preservation and visual harmony at the tendering stage by proposing these construction techniques. This approach allows projects to achieve both sustainability and aesthetic excellence, aligning with the concept of stealth construction.

Tendering in the Pre-construction Stage

Tendering is crucial at pre-construction because it sets the foundation for the entire project, ensuring transparency, competitiveness, and cost-effectiveness. By conducting the tendering process early, project owners can evaluate multiple bids from different contractors, allowing them to select the most qualified and cost-efficient option. This competitive bidding process helps prevent cost overruns and ensures the project begins with a clear financial plan. Additionally, early tendering allows for thorough vetting of contractors, ensuring that only those with the requisite experience, capabilities, and resources are considered. This reduces the risk of project delays or substandard work due to an inadequately prepared contractor.

Tendering at the pre-construction stage also facilitates better planning and coordination. When contractors are selected before construction begins, they can collaborate with architects, engineers, and project managers during the design phase. This collaboration allows for integrating contractor expertise into the planning process, potentially identifying cost-saving opportunities and optimising construction methods. Moreover, early contractor involvement can help anticipate and mitigate potential risks, enhancing the overall project timeline and quality. By contrast, tendering during or after construction could lead to significant disruptions, increased costs, and logistical challenges, as changes would need to be managed mid-project. Thus, initiating tendering at the pre-construction stage is essential for ensuring a smooth, efficient, and well-coordinated project execution.

Conclusion

The exploration of the various tendering types and contemporary trends (including electronic tendering, BIM, green procurement, and blockchain technology) highlights how these innovations streamline processes and enhance project efficiency. Integrating functional construction systems and low-impact construction techniques within tender specifications is pivotal for aligning projects with stealth construction principles. Furthermore, the study underscores the importance of environmental protection, safety, speed, economy, and aesthetics. Specific strategies like sustainable design, comprehensive safety plans, and advanced project management tools demonstrate the various benefits of adopting these approaches. Ultimately, tendering in stealth construction advocates for a holistic tendering framework that promotes sustainable and resilient construction practices, ensuring that future developments are well-integrated and revolutionary.

References

Abbasnejad, B., Nepal, M. P., Ahankoob, A., Nasirian, A., & Drogemuller, R. (2021). Building information modelling (BIM) adoption and implementation enablers in AEC firms: A systematic literature review. *Architectural Engineering and Design Management*, *17*(5–6), 411–433.

Abdullahi, B., Ibrahim, Y. M., Ibrahim, A. D., & Bala, K. (2022). Development of web-based e-Tendering system for Nigerian public procuring entities. *International Journal of Construction Management*, *22*(2), 278–291.

Abdulraheem, M. I., Moshood, A. Y., Chen, Y., Chen, H., Zhang, H., & Hu, J. (2024). Advancements in designing smart and intelligent nanocoatings. *Sustainable Approach to Protective Nanocoatings*, *1*(1), 57–87.

Afzal, M., Shafiq, M. T., & Al Jassmi, H. (2021). Improving construction safety with virtual-design construction technologies review. *Journal of Information Technology in Construction*, *26*, 319–340. https://doi.org/10.36680/j.itcon.2021.018

Ahmad, T., Aljafari, R., & Venkatesh, V. (2019). The government of Jamaica's electronic procurement system: Experiences and lessons learned. *Internet Research*, *29*(6), 1571–1588.

Ahmadisheykhsarmast, S., Senji, S. G., & Sonmez, R. (2023). Decentralised tendering of construction projects using blockchain-based smart contracts and storage systems. *Automation in Construction*, *151*, 1–18. https://doi.org/10.1016/j.autcon.2023.104900

Ahmed, M. E. A. M., & Wong, L. S. (2018). Assessment of lean construction practice at selected construction sites in Klang Valley. *International Journal of Engineering & Technology*, *7*(4.35), 125–130.

Aized, T. (2012). *Total quality management and six sigma*. BoD–Books on Demand.

Anne, G. (2008). *Achieving sustainable site through low impact development practices*. http://www.rcinef.org/Achieving-Sustainable-Site-Design-through-LID-Practices.pdf

Awofadeju, M. O., Ademiju, T. A., & Adzande, O. A. (2024). Implementing project management information system and project risk management in developing economies: A case study of Nigeria. *International Journal of Humanities Social Science and Management (IJHSSM)*, *4*(2), 09–17.

Baarimah, A. O., Alaloul, W. S., Liew, M. S., Al-Aidrous, A. H. M., Alawag, A. M., & Musarat, M. A. (2021, November). Integration of building information modelling

(BIM) and value engineering in construction projects: A bibliometric analysis. In *2021 Third International Sustainability and Resilience Conference: Climate Change*, Sakheer, Bahrain, November 15 (pp. 362–367). https://doi.org/10.1109/IEEECONF53624.2021.9668045

Benton, W. C., & McHenry, L. F. (2010). *Construction purchasing and supply chain management*. McGraw-Hill.

Boateng, A., Ameyaw, C., & Mensah, S. (2022). Assessment of systematic risk management practices on building construction projects in Ghana. *International Journal of Construction Management, 22*(16), 3128–3136.

Boadu, E. F., Sunindijo, R. Y., & Wang, C. C. (2021). Health and safety consideration in the procurement of public construction projects in Ghana. *Buildings, 11*(3), 1–20.

Campos, L. M., Trierweiller, A. C., de Carvalho, D. N., & Šelih, J. (2016). Environmental management systems in the construction industry: A review. *Environmental Engineering and Management Journal, 15*(2), 453–460.

Chandrasekar, M. (2023). Building-integrated solar photovoltaic thermal (BIPVT) technology: A review on the design innovations, aesthetic values, performance limits, storage options and policies. *Advances in Building Energy Research, 17*(2), 223–254.

Cheaitou, A., Larbi, R., & Al Housani, B. (2019). Decision making framework for tender evaluation and contractor selection in public organizations with risk considerations. *Socio-Economic Planning Sciences, 68*, 1–12. https://doi.org/10.1016/j.seps.2018.02.007

Choden, T., & Soratana, K. (2020). *Incorporation of environmental criteria in the tendering process: Moving towards green public procurement in Bhutan* [Doctoral dissertation, Naresuan University].

Ciribini, A. L. C., Bolpagni, M., & Oliveri, E. (2015). An innovative approach to e-public tendering based on model checking. *Procedia Economics and Finance, 21*, 32–39. https://doi.org/10.1016/S2212-5671(15)00147-1

Collins, R., Schaafsm, M., & Hudson, M., D. (2017). The value of green walls to urban biodiversity. *Land Use Policy, 64,* 114–123.

Correa, S. L. M., & Santos, E. T. (2021). BIM support in the tendering phase of infrastructure projects. In E. Toledo Santos & S. Scheer (Eds.), *Proceedings of the 18th international conference on computing in civil and building engineering. ICCCBE 2020. Lecture Notes in Civil Engineering* (Vol. 98). Springer. https://doi.org/10.1007/978-3-030-51295-8_27

D'Ascanio, F., Di Ludovico, D., & Di Lodovico, L. (2016). Design and urban shape for a resilient city. *Procedia-Social and Behavioral Sciences, 223*, 764–769. https://doi.org/10.1016/j.sbspro.2016.05.265

De Lima, K. C. P. (2020). *Evaluation of risk management practices in the tendering process within the construction industry in Mozambique* [Doctoral dissertation, Dublin, National College of Ireland].

Dhurve, S. (2024). The challenges and opportunities of transitioning from conventional materials to natural and sustainable materials in commercial projects. *Journal of Research in Infrastructure Designing*. 7(1), 28–46.

Ding, L., Zhou, Y., & Akinci, B. (2014). Building information modeling (BIM) application framework: The process of expanding from 3D to computable nD. *Automation in Construction, 46*, 82–93. https://doi.org/10.1016/j.autcon.2014.04.009

DuHadway, S., Carnovale, S., & Hazen, B. (2019). Understanding risk management for intentional supply chain disruptions: Risk detection, risk mitigation, and risk recovery. *Annals of Operations Research, 283*, 179–198. https://doi.org/10.1007/s10479-017-2452-0

Elcock, D. (2007). *Potential impacts of nanotechnology on energy transmission applications and needs (No. ANL/EVS/TM/08-3)*. Argonne National Laboratory (ANL), Argonne, IL, United States. https://doi.org/10.2172/924389

Elmousalami, H. H. (2020). Artificial intelligence and parametric construction cost estimate modeling: State-of-the-art review. *Journal of Construction Engineering and Management, 146*(1), 1–15. https://doi.org/10.1061/(ASCE)CO.1943-7862.0001678

Flechsig, C., Anslinger, F., & Lasch, R. (2022). Robotic process automation in purchasing and supply management: A multiple case study on potentials, barriers, and implementation. *Journal of Purchasing and Supply Management*, *28*(1), 1–21. https://doi.org/10.1016/j.pursup.2021.100718

Getuli, V., Capone, P., Bruttini, A., & Isaac, S. (2020). BIM-based immersive Virtual Reality for construction workspace planning: A safety-oriented approach. *Automation in Construction*, *114*, 1–20. https://doi.org/10.1016/j.autcon.2020.103160

Golz, S. (2016). Resilience in the built environment: How to evaluate the impacts of flood resilient building technologies? In *3rd European Conference on Flood Risk Management (FLOODrisk 2016), E3S Web of Conferences* (Vol. 7, p. 13001). EDP Sciences, October 20, Dresden, Germany. https://doi.org/10.1051/e3sconf/20160713001

Gunduz, M., Aly, A. A., & El Mekkawy, T. (2022). Value engineering factors with an impact on design management performance of construction projects. *Journal of Management in Engineering*, *38*(3), 1–12.

Hu, Y. (2011). Minimization management of construction waste. In *International symposium on water resource and environmental protection*, Xi'an, China, May 20 (pp. 2769–2772). https://doi.org/10.1109/ISWREP.2011.5893453

Irizarry, J., Karan, E. P., & Jalaei, F. (2013). Integrating BIM and GIS to improve the visual monitoring of construction supply chain management. *Automation in Construction*, *31*, 241–254. https://doi.org/10.1016/j.autcon.2012.12.005

Ismaeil, E. M., & Sobaih, A. E. E. (2023). High-performance glazing for enhancing sustainable environment in arid region's healthcare projects. *Buildings*, *13*(5), 1–20.

Jaafar, M. H., Arifin, K., Aiyub, K., Razman, M. R., Ishak, M. I. S., & Samsurijan, M. S. (2018). Occupational safety and health management in the construction industry: A review. *International Journal of Occupational Safety and Ergonomics*, *24*(4), 493–506.

Jafari, M., & Alipour, A. (2021). Aerodynamic shape optimization of rectangular and elliptical double-skin façades to mitigate wind-induced effects on tall buildings. *Journal of Wind Engineering and Industrial Aerodynamics*, *213*, 1–25. https://doi.org/10.1016/j.jweia.2021.104586

Jahani, N., Sepehri, A., Vandchali, H. R., & Tirkolaee, E. B. (2021). Application of industry 4.0 in the procurement processes of supply chains: A systematic literature review. *Sustainability*, *13*(14), 1–25.

Johnsen, K., & Winther, F. V. (2015). Dynamic facades, the smart way of meeting the energy requirements. *Energy Procedia*, *78*, 1568–1573. https://doi.org/10.1016/j.egypro.2015.11.210

Johnson, P. F., & Klassen, R. D. (2022). New directions for research in green public procurement: The challenge of inter-stakeholder tensions. *Cleaner Logistics and Supply Chain*, *3*, 1–7. https://doi.org/10.1016/j.clscn.2021.100017

Kang, H. W., & Kim, Y. S. (2018). A model for risk cost and bidding price prediction based on risk information in plant construction projects. *KSCE Journal of Civil Engineering*, *22*, 4215–4229. https://doi.org/10.1007/s12205-018-0587-4

Kensek, K. (2014). *Building information modeling*. Routledge.

Khan, M., Khalid, R., Anjum, S., Khan, N., Cho, S., & Park, C. (2022). Tag and IoT based safety hook monitoring for prevention of falls from height. *Automation in Construction*, *136*, 1–13. https://doi.org/10.1016/j.autcon.2022.104153

Khan, A., Sepasgozar, S., Liu, T., & Yu, R. (2021). Integration of BIM and immersive technologies for AEC: A scientometric-SWOT analysis and critical content review. *Buildings*, *11*(3), 1–34.

Kosorić, V., Lau, S. K., Tablada, A., Bieri, M., & M. Nobre, A. (2021). A holistic strategy for successful photovoltaic (PV) implementation into Singapore's built environment. *Sustainability*, *13*(11), 1–35.

Kumari, M., Gupta, P., & Deshwal, S. S. (2022). Integrated life cycle cost comparison and environment impact analysis of the Concrete and Asphalt Roads. *Materials Today: Proceedings*, *60*, 345–350. https://doi.org/10.1016/j.matpr.2022.01.240

Kusumarukmi, E. I., & Adi, T. J. W. (2019). Public tendering process for construction projects: Problem identifications, analysis, and proposed solutions. *International Conference on Sustainable Civil Engineering Structures and Construction Materials (SCESCM 2018)*. In *MATEC Web of Conferences* (Vol. 258, pp. 1–8). https://doi.org/10.1051/matecconf/201925802013

Lalmi, A., Fernandes, G., & Souad, S. B. (2021). A conceptual hybrid project management model for construction projects. *Procedia Computer Science*, *181*, 921–930. https://doi.org/10.1016/j.procs.2021.01.248

Lam, P. T. I., Chan, E. H. W., Chau, C. K., & Poon, C. S. (2011), Environmental management system vs green specifications: How do they complement each other in the construction industry? *Journal of Environmental Management*, 92, 788–795. https://doi.org/10.1016/j.jenvman.2010.10.030

Larsen, V. G., Tollin, N., Sattrup, P. A., Birkved, M., & Holmboe, T. (2022). What are the challenges in assessing circular economy for the built environment? A literature review on integrating LCA, LCC and S-LCA in life cycle sustainability assessment, LCSA. *Journal of Building Engineering*, *50*, 1–16. https://doi.org/10.1016/j.jobe.2022.104203

Laryea, S. (2017). Impact of tendering procedure on price formation in construction contracts: Case study of the competitive negotiation procedure. In *Procs 7th West Africa Built Environment Research (WABER) Conference*, August 16–18, Accra, Ghana (pp. 853–869).

Lavelle, D., & Bardon, A. (2009). E-tendering in construction: Time for a change? *Northumbria Working Paper Series: Interdisciplinary Studies in the Built and Virtual Environment*, *2*(2), 104–112.

Lee, P., Lam, P. T. I., & Lee, W. L. (2015). Risks in energy performance contracting (EPC) projects. *Energy and Buildings*, *92*, 116–127. https://doi.org/10.1016/j.enbuild.2015.01.054

Lines, B. C., & Kumar, G. G. R. (2018). Developing more competitive proposals: Relationship between contractor qualifications-based proposal content and owner evaluation scores. *Journal of Construction, Engineering, and Management*, *144*, 1–12. https://doi.org/10.1061/(ASCE)CO.1943-7862.0001479

Loushine, T. W., Hoonakker, P. L., Carayon, P., & Smith, M. J. (2006). Quality and safety management in construction. *Total Quality Management and Business Excellence*, *17*(9), 1171–1212.

Lu, Y., Li, Y., Skibniewski, M., Wu, Z., Wang, R., & Le, Y. (2015). Information and communication technology applications in architecture, engineering, and construction organizations: A 15-year review. *Journal of Management in Engineering*, *31*(1), 1–19. https://doi.org/10.1061/(ASCE)ME.1943-5479.0000319

Lundberg, S. & Marklund, P. O. (2018). Green public procurement and multiple environmental objectives. *Journal of Industrial Business and Economics*, 45 (1), 37–53. https://doi.org/10.1007/s40812-017-0085-6

Macek, D. (2023). Use of BIM as a support for tendering of facility management services. *Buildings*, *13*(3), 1–19.

Mane, P. P., & Patil, J. R. (2015). Quality management system at construction project: A questionnaire survey. *International Journal of Engineering Research and Applications*, *5*(3), 126–130.

Mawed, M. (2020). Construction and demolition waste management in the UAE: Application and obstacles. *GEOMATE Journal*, *18*(70), 235–245.

Mhetre, K., Konnur, B. A., & Landage, A. B. (2016). Risk management in construction industry. *International Journal of Engineering Research*, *5*(1), 153–155.

Miller, D., & Doh, J. H. (2015). Incorporating sustainable development principles into building design: A review from a structural perspective including case study. *The Structural Design of Tall and Special Buildings*, *24*(6), 421–439.

Mostashari-Rad, F., Ghasemi-Mobtaker, H., Taki, M., Ghahderijani, M., Kaab, A., Chau, K. W., & Nabavi-Pelesaraei, A. (2021). Exergoenvironmental damages assessment of horticultural crops using ReCiPe2016 and cumulative exergy demand frameworks. *Journal of Cleaner Production*, *278*, 1–18. https://doi.org/10.1016/j.jclepro.2020.123788

Nowotarski, P., Pastawaski, J. & Matyja, J. (2016). Improving construction processes using lean management methodologies- cost case study. *Procedia Engineering*, *16*, 1037–1042. https://doi.org/10.1016/j.proeng.2016.08.845

Odejide, O. A., & Edunjobi, T. E. (2024). AI in project management: Exploring theoretical models for decision-making and risk management. *Engineering Science & Technology Journal*, *5*(3), 1072–1085.

Oke, A. E., Adetoro, P. E., Stephen, S. S., Aigbavboa, C. O., Oyewobi, L. O., & Aghimien, D. O. (2023). *Risk management practices in construction: A global view*. Springer Nature.

Oladiran, O. J. (2009). Innovative waste management through the use of waste management plans on construction projects in Nigeria. *Architectural Engineering and Design Management*, *5*(3), 165–176.

Oruezabala, G., & Rico, J.-C. (2012). The impact of sustainable public procurement in supplier management – The case of French public hospitals. *Industrial Marketing Management*, *41*, 573–580. https://doi.org/10.1016/j.indmarman.2012.04.004

Oyegoke, A. S., McDermott, P., & Abbott, C. (2009). Achieving sustainability in construction through the specialist task organisation procurement approach. *International Journal of Procurement Management*, *2*(3), 288–313.

Ozturk, G. B., & Soygazi, F. (2024). Generative AI use in the construction industry. In *applications of generative AI* (pp. 161–187). Springer International Publishing.

Pacheco, R., Ordóñez, J., & Martínez, G. (2012). Energy efficient design of building: A review. *Renewable and Sustainable Energy Reviews*, *16*(6), 3559–3573.

Palmujoki, A., Parikka-Alhola, K., & Ekroos, A. (2010). Green public procurement: Analysis on the use of environmental criteria in contracts. *Review of European Community & International Environmental Law*, *19*(2), 250–262.

Patel, V., Chesmore, A., Legner, C. M., & Pandey, S. (2022). Trends in workplace wearable technologies and connected-worker solutions for next-generation occupational safety, health, and productivity. *Advanced Intelligent Systems*, *4*(1), 1–30. https://doi.org/10.1002/aisy.202100099

Pater, T. (2019). *Quality management for integrated contracts: Can the client improve it?* [A Master Dissertation, Department of Construction, Management and Engineering, Eindhoven University of Technology].

Pérez, Y., Ávila, J., & Sánchez, O. (2024). Influence of BIM and lean on mitigating delay factors in building projects. *Results in Engineering*, *1*(1), 1–39. https://doi.org/10.1016/j.rineng.2024.102236

Rahimian, F. P., Seyedzadeh, S., Oliver, S., Rodriguez, S., & Dawood, N. (2020). On-demand monitoring of construction projects through a game-like hybrid application of BIM and machine learning. *Automation in Construction*, *110*, 1–14. https://doi.org/10.1016/j.autcon.2019.103012

Ramani, P. V., & KSD, L. K. L. (2021). Application of lean in construction using value stream mapping. *Engineering, Construction and Architectural Management*, *28*(1), 216–228.

Raza, M. S., Tayeh, B. A., & Ali, T. H. (2022). Owner's obligations in promoting occupational health and safety in preconstruction of projects: A literature viewpoint. *Results in Engineering*, *16*, 1–12. https://doi.org/10.1016/j.rineng.2022.100779

Regona, M., Yigitcanlar, T., Xia, B., & Li, R. Y. M. (2022). Opportunities and adoption challenges of AI in the construction industry: A PRISMA review. *Journal of Open Innovation: Technology, Market, and Complexity*, *8*(1), 1–31.

Rocha, P. F., Ferreira, N. O., Pimenta, F., & Pereira, N. B. (2022). Impacts of prefabrication in the building construction industry. *Encyclopedia*, *3*(1), 28–45.

Roco, M. C., Hersam, M. C., Mirkin, C. A., Brinker, C. J., & Ginger, D. (2011). Nanotechnology for sustainability: Energy conversion, storage, and conservation. *Nanotechnology research directions for societal needs in 2020: Retrospective and outlook*, *1*, 261–303. https://doi.org/10.1007/978-94-007-1168-6_7

Roy, B. H., & Kumar, M. R. M. (2017). Study on different types of tenders, tender qualification/processing, quantitative techniques for tendering, planning and monitoring-a review. *International Journal of Engineering Research & Technology*, *6*(11), 552–555.

Saad, A., Bal, M., & Khatib, J. (2022). The need for a proper waste management plan for the construction industry: A case study in Lebanon. *Sustainability*, *14*(19), 1–17.

Shafaghat, A., & Keyvanfar, A. (2022). Dynamic façades design typologies, technologies, measurement techniques, and physical performances across thermal, optical, ventilation, and electricity generation outlooks. *Renewable and Sustainable Energy Reviews*, *167*, 1–22. https://doi.org/10.1016/j.rser.2022.112647

Shamsuddin, K. A., Ani, M. N. C., Ismail, A. K., & Ibrahim, M. R. (2015). Investigation the safety, health and environment (SHE) protection in construction area. *International Research Journal of Engineering and Technology*, *2*(6), 624–636.

Smith, P. G., & Merritt, G. M. (2020). *Proactive risk management: Controlling uncertainty in product development* (1st ed.). Taylor and Francis. https://doi.org/10.4324/9780367807542

Sönnichsen, S. D., & Clement, J. (2020). Review of green and sustainable public procurement: Towards circular public procurement. *Journal of Cleaner Production*, *245*, 1–18. https://doi.org/10.1016/j.jclepro.2019.118901

Sully, A. (2015). *Interior design: Conceptual basis*. Springer.

Sunmola, F. T., & Shehu, Y. U. (2020). A case study on performance features of electronic tendering systems. *Procedia Manufacturing*, *51*, 1586–1591. https://doi.org/10.1016/j.promfg.2020.10.221

Tariqa, M. (2021). Temporary works forum guidance: Advice to the young engineer joining the temporary works sector. In *1st International Conference on Recent Advances in Civil and Earthquake Engineering (ICCEE-2021)*, October 08, Peshawar, Pakistan (pp. 147–149).

Thai, H. T., Ngo, T., & Uy, B. (2020). A review on modular construction for high-rise buildings. *Structures*, *28*, 1265–1290. https://doi.org/10.1016/j.istruc.2020.09.070

Umar, T. (2021). Applications of drones for safety inspection in the Gulf Cooperation Council construction. *Engineering, Construction and Architectural Management*, *28*(9), 2337–2360.

Venkataraman, R. R., & Pinto, J. K. (2023). *Cost and value management in projects*. John Wiley & Sons.

Wong, T. K. M., Man, S. S., & Chan, A. H. S. (2020). Critical factors for the use or non-use of personal protective equipment amongst construction workers. *Safety Science*, *126*, 1–12. https://doi.org/10.1016/j.ssci.2020.104663

Xing, W., Hao, J. L., Qian, L., Tam, V. W., & Sikora, K. S. (2021). Implementing lean construction techniques and management methods in Chinese projects: A case study in Suzhou, China. *Journal of Cleaner Production*, *286*, 1–13. https://doi.org/10.1016/j.jclepro.2020.124944

Yiu, N. S., Chan, D. W., Shan, M., & Sze, N. N. (2019). Implementation of safety management system in managing construction projects: Benefits and obstacles. *Safety Science*, *117*, 23–32. https://doi.org/10.1016/j.ssci.2019.03.027

Chapter Four

Procurement in Stealth Construction

Abstract

The chapter highlighted the key themes explored on procurement in construction, emphasising its significance in driving sustainability, efficiency, and innovation within the industry. It began with an introduction to the topic, followed by an exploration of the importance of procurement planning and the conceptual framework of procurement. The study then delved into sustainable procurement practices in construction, highlighting the role of technology, innovation, and stakeholder engagement in advancing procurement strategies. It also examined the concept of procurement in stealth construction, focusing on critical aspects such as the cross-section of the building, energy transmission, and countermeasures. The chapter summarised the principles guiding procurement towards stealth construction, emphasising the importance of environmental protection, safety, project efficiency, cost-effectiveness, and aesthetics in driving sustainable construction practices.

Keywords: Stealth construction; sustainable procurement; environmental protection; efficient project delivery; efficient procurement

Introduction

Alhammadi et al. (2023) stated that procurement in construction stands as a cornerstone process, orchestrating the acquisition of goods, services, and resources vital for the realisation of construction projects. It involves a multifaceted approach encompassing sourcing, negotiation, and contract management, ensuring the timely and cost-effective provision of materials, equipment, and expertise necessary for project execution. This pivotal function demands thorough planning and strategic decision-making to navigate the complexities inherent in the construction industry. Giuffrida and Rovigatti (2022) corroborate that procurement drives project success, from selecting suppliers to managing risks and ensuring

Stealth Construction: Integrating Practices for Resilience and Sustainability, 73–95

Published under exclusive licence by Emerald Publishing Limited
doi:10.1108/978-1-83608-182-120251004

compliance. Effective procurement strategies can optimise resource utilisation, minimise delays, and enhance project quality. Consequently, understanding the variation of procurement in construction is paramount for stakeholders across the industry spectrum, from developers and contractors to suppliers and subcontractors, as it directly influences project outcomes and stakeholders' bottom lines.

Moreover, Murphy and Eadie (2019) added that procurement in construction extends beyond mere transactional processes; it encompasses broader objectives such as sustainability, innovation, and stakeholder engagement. Sustainable procurement practices aim to minimise environmental impact, promote social responsibility, and foster economic development within local communities. When sustainable criteria are integrated into procurement decisions, construction projects can mitigate environmental risks, enhance resource efficiency, and contribute to long-term sustainability goals. Althabatah et al. (2023) also highlighted that fostering innovation through strategic supplier partnerships and technology adoption can drive continuous improvement, streamline processes, and enhance project outcomes. Furthermore, effective stakeholder engagement throughout the procurement lifecycle fosters transparency, trust, and collaboration, mitigating conflicts, optimising communication channels, and ultimately facilitating project success. In essence, procurement in construction transcends transactional activities to become a strategic enabler of sustainable development, innovation, and stakeholder satisfaction within the built environment.

The Importance of Procurement Planning

Procurement planning stands as a cornerstone in the construction industry, dictating the success trajectory of projects from conception to completion. At its core, Watermeyer (2023) and Ogunsanya et al. (2022) illustrated that effective procurement planning involves meticulously analysing project requirements, identifying key stakeholders, and formulating strategies to acquire necessary resources promptly and cost-effectively, as summarised in Fig. 2. When procurement needs and objectives are outlined early in the project lifecycle, stakeholders can proactively address potential challenges, mitigate risks, and capitalise on opportunities. Furthermore, robust procurement planning facilitates alignment between project goals and strategies, ensuring resource allocation aligns with project priorities. This strategic alignment optimises resource utilisation and enhances project efficiency and overall performance. Consequently, investing time and effort into comprehensive procurement planning sets the stage for smoother project execution, reduced costs, and improved project outcomes.

In addition, procurement planning is a keystone for fostering collaboration and communication among project stakeholders (Bohari et al., 2020). Organisations can leverage diverse perspectives, harness collective expertise, and build consensus around procurement strategies by engaging key stakeholders – such as project managers, designers, contractors, and suppliers – early in the procurement planning process. This collaborative approach enhances transparency and accountability and fosters a sense of ownership and commitment among stakeholders. Additionally, Can-Saglam et al. (2021) and Manners-Bell (2017) mentioned that effective procurement planning enables proactive risk management by

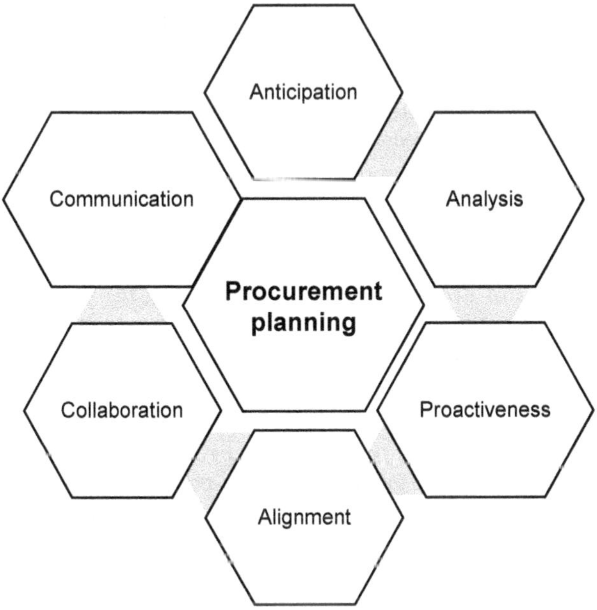

Fig. 2. Importance of Procurement Planning.

identifying and addressing potential bottlenecks, supply chain disruptions, and regulatory compliance issues before they escalate. Furthermore, by anticipating and mitigating risks through strategic procurement planning, construction projects can minimise delays, avoid costly disputes, and maintain project momentum. Ultimately, the importance of procurement planning lies in its ability to lay a solid foundation for successful project outcomes by aligning resources, mitigating risks, and fostering collaboration among stakeholders.

Concepts of Procurement

Methods: Various procurement methods play a pivotal role in shaping the dynamics of construction projects, offering stakeholders diverse avenues to acquire goods, services, and expertise (Plunkett, 2021). Ghadamsi and Braimah (2012) and Jimoh et al. (2015) illustrated that from traditional procurement, where design and construction phases are separate, to more collaborative approaches like design and build, procurement methods significantly influence project outcomes. Design and build, for instance, integrates design and construction responsibilities under a single entity, streamlining communication, reducing project duration, and potentially lowering costs. On the other hand, construction management allows for early contractor involvement, fostering collaboration and innovation throughout the project lifecycle. Each procurement method carries distinct advantages and challenges, necessitating careful consideration based on project goals, complexity, and stakeholder preferences. Thus, selecting the most appropriate procurement method is crucial for optimising project performance and achieving desired outcomes in the dynamic landscape of construction projects.

Process steps: Procurement encompasses several vital steps to acquiring goods and services in construction projects (Ershadi et al., 2021; Morledge et al., 2021). Beginning with needs identification, stakeholders analyse project requirements and define the scope of procurement activities. Subsequently, supplier identification and prequalification involve researching potential vendors, assessing their capabilities, and establishing a pool of qualified suppliers. Tendering and bidding then come into play, where invitations are extended to qualified suppliers to submit proposals or bids for the project. Following a thorough evaluation, the contract award marks the formal selection of the supplier or contractor, signalling the commencement of project execution. Each step in the procurement process demands meticulous planning, clear communication, and adherence to legal and ethical standards to ensure the efficient and effective procurement of resources essential for project success.

Key considerations: According to Morledge et al. (2021) and Lysons and Farrington (2020), this serves as a guide for decision-making processes and ultimately shapes the success of construction projects. Among these considerations, cost is paramount, as it directly impacts project budgets and financial viability. Balancing cost considerations with quality requirements ensures project outcomes meet stakeholders' expectations without compromising on budget constraints. Time is another critical factor, as delays in procurement can cascade into project delays, leading to increased costs and stakeholder dissatisfaction. Effective risk management is also essential, necessitating identifying and mitigating potential risks associated with procurement activities to safeguard project timelines and budgets. Additionally, ethical sourcing practices and compliance with legal regulations ensure transparency, integrity, and accountability throughout the procurement process, fostering stakeholder trust and mitigating reputational risks.

Sustainable Procurement Practices in Construction

As mentioned by Marcelline et al. (2022) and Yu et al. (2020), sustainable procurement practices in construction have emerged as a fundamental strategy for mitigating environmental impact, promoting social responsibility, and fostering economic development within local communities. Sustainable procurement involves integrating environmental, social, and economic criteria into procurement decisions to minimise adverse environmental and societal effects while maximising long-term value. Russell (2020) reiterated that this approach encompasses various initiatives, such as sourcing environmentally friendly materials, prioritising suppliers with strong sustainability credentials, and implementing green building standards like Leadership in Energy and Environmental Design (LEED) or Building Research Establishment Environmental Assessment Method (BREEAM). By prioritising sustainability in procurement decisions, construction projects can reduce carbon emissions, minimise waste generation, and conserve natural resources, contributing to broader sustainability goals and mitigating the industry's environmental footprint.

Furthermore, Sönnichsen and Clement (2020) added that sustainable procurement practices in construction extend beyond environmental considerations to encompass social and economic dimensions. Socially responsible procurement initiatives prioritise fair labour practices, diversity, and inclusion within the supply chain, ensuring workers are treated ethically and provided with safe working conditions. Additionally, supporting local businesses and engaging with diverse suppliers foster economic development within communities, strengthens social ties, and promotes resilience. When procurement practices align with social sustainability goals, construction projects can enhance social equity, create opportunities for marginalised groups, and contribute to inclusive economic growth, ultimately fostering more robust, resilient communities.

Moreover, Amarah and Langston (2017) and Goh et al. (2023) state that sustainable procurement practices in construction yield numerous tangible benefits for project stakeholders, ranging from cost savings to enhanced reputation and stakeholder satisfaction. While initial investments in sustainable procurement may require upfront costs, they often yield long-term financial savings through reduced energy consumption, operational efficiency improvements, and lower waste disposal expenses. Furthermore, demonstrating a commitment to sustainability enhances a company's reputation as a responsible corporate citizen, attracting environmentally conscious clients, investors, and employees. By integrating sustainable procurement practices into their operations, construction firms can differentiate themselves in the market, gain a competitive edge, and build stronger relationships with stakeholders, positioning themselves for long-term success in a rapidly evolving business landscape.

Advancing Procurement Through Technology, Innovation, and Stakeholder Engagement

Technology and innovation are revolutionising procurement practices in the construction industry, streamlining processes, enhancing efficiency, and driving cost savings (Martínez-Peláez et al., 2023). One significant advancement is the adoption of procurement management software, which enables stakeholders to automate routine tasks, track procurement activities in real-time, and analyse data for informed decision-making. Additionally, artificial intelligence (AI) and machine learning (ML) facilitate predictive analytics, allowing organisations to forecast demand, optimise inventory levels, and identify cost-saving opportunities (Kalasani, 2023). Moreover, blockchain technology offers transparent and secure transactional records, mitigating risks associated with fraud and ensuring integrity throughout the procurement process. Furthermore, the emergence of digital marketplaces and e-procurement platforms enables seamless collaboration between buyers and suppliers, expanding market reach and fostering competition. When technology and innovation are harnessed, procurement in construction can become more agile, transparent, and responsive to evolving project needs and market dynamics.

Furthermore, Elsayegh and El-adaway (2021) and Amirtash et al. (2021) pointed out that stakeholders' engagement and collaboration are integral components of effective procurement practices in construction, facilitating communication, alignment of goals, and mutual understanding among project participants. Early stakeholder involvement allows for identifying diverse perspectives, requirements, and expectations, enabling procurement strategies to be tailored to meet project-specific needs. Furthermore, collaboration between project owners, designers, contractors, and suppliers fosters innovation and value engineering, identifying cost-effective solutions and optimising project outcomes. Moreover, stakeholder engagement extends beyond traditional project participants to include local communities, regulatory authorities, and other relevant stakeholders. When stakeholders are engaged throughout the procurement process, construction projects can address concerns, mitigate conflicts, and build trust and consensus, enhancing project credibility and social licence to operate. Effective stakeholder engagement and collaboration improve project outcomes and contribute to long-term relationships and reputation management within the construction industry.

Moreover, stakeholder engagement and collaboration in procurement are essential for fostering sustainability and corporate social responsibility (CSR) within the construction industry (Montalbán-Domingo et al., 2022). Engaging with suppliers and subcontractors committed to sustainable practices ensures that environmental and social criteria are integrated into procurement decisions, thereby reducing environmental footprint, promoting ethical labour standards, and supporting local communities. Additionally, collaboration with regulatory authorities and industry organisations enables compliance with relevant regulations, codes, and standards, ensuring that projects meet legal requirements and industry best practices. Furthermore, stakeholder engagement enhances transparency and accountability, allowing stakeholders to participate in decision-making processes, provide feedback, and hold project participants accountable for their actions. By embracing stakeholder engagement and collaboration, construction projects can demonstrate their commitment to sustainability, CSR, and ethical business practices, enhancing their reputation and competitive advantage in the marketplace.

Furthermore, Loosemore et al. (2022) reiterated that stakeholder engagement and collaboration in procurement play a crucial role in managing risks and building resilience within the construction industry. Projects can proactively address potential challenges by involving stakeholders in risk identification and mitigation strategies, such as supply chain disruptions, regulatory changes, and unforeseen events. Additionally, collaboration with stakeholders enables the development of contingency plans and alternative sourcing strategies to mitigate risks and ensure project continuity. Moreover, engaging with local communities and other stakeholders affected by the project fosters resilience by building social capital, enhancing community support, and facilitating effective crisis management. Construction projects can improve their adaptability, responsiveness, and long-term sustainability in an increasingly volatile and uncertain environment by actively involving stakeholders in risk management and resilience-building efforts.

Procurement in Stealth Construction

Procurement in stealth construction represents a cutting-edge approach that integrates advanced technologies throughout the construction process, revolutionising traditional procurement practices. Sawhney et al. (2020) and Rahimian et al. (2021) have previously asserted that the implementation of this innovative methodology by leveraging a combination of digital tools, such as Building Information Modelling (BIM), drones, and augmented reality (AR), streamline procurement activities from planning to execution. Through BIM, stakeholders can collaboratively visualise project designs, identify potential clashes, and optimise material quantities, leading to more accurate procurement specifications and reduced waste. Drones play a crucial role in surveying and monitoring construction sites, providing real-time data on progress, inventory levels, and site conditions, which informs procurement decisions and facilitates inventory management. Moreover, AR technology enhances stakeholders' engagement by enabling immersive experiences, allowing clients and contractors to visualise design concepts and make informed decisions during procurement.

Furthermore, procurement in stealth construction extends beyond adopting technology to encompass innovative construction practices prioritising sustainability, efficiency, and resilience. Okogwu et al. (2023) illustrated that sustainable procurement principles guide the selection of materials and suppliers, emphasising environmentally friendly products, ethical sourcing practices, and energy-efficient solutions. Additionally, off-site manufacturing and modular construction techniques optimise resource utilisation, reduce construction waste, and accelerate project timelines, offering a more sustainable alternative to traditional construction methods. Furthermore, lean construction principles, such as just-in-time (JIT) delivery and value stream mapping, streamline procurement processes, minimise inventory holding costs, and enhance project productivity.

Moreover, procurement in stealth construction emphasises stakeholder collaboration and engagement as fundamental drivers of project success, as Oke et al. (2022) facilitated, where new technology practice drives sustainable building projects. Early involvement of stakeholders, including clients, architects, contractors, and suppliers, fosters a culture of collaboration, innovation, and shared ownership throughout the procurement process. Ma et al. (2023) and Nwajei et al. (2022) identified that collaborative procurement approaches, such as integrated project delivery (IPD) and collaborative contracting, facilitate open communication, risk-sharing, and joint decision-making, aligning project objectives and optimising resource allocation. Additionally, stakeholder engagement extends beyond traditional project participants to include local communities, regulatory agencies, and other relevant stakeholders. When stakeholders are involved early in procurement activities, stealth construction projects can address diverse perspectives, anticipate challenges, and build trust and consensus, enhancing project credibility and social licence to operate. Ultimately, procurement in stealth construction embodies a holistic approach that integrates technology, innovation, and stakeholder collaboration to deliver sustainable, efficient, and resilient projects in an ever-evolving construction landscape.

As the construction industry drives towards resilience, aspects like cross-section of the building (*shape of the building, internal construction, and material usage*), energy transmission (*renewable energy, climate action planning, and energy efficiency*), and countermeasures (*functional construction systems and preventive measures*) have been considered a drive towards stealth construction.

Cross-section of the Building

In stealth construction, the cross-section of the building, encompassing its shape, internal construction, and material usage, as stated by Pan et al. (2024), plays a pivotal role in shaping the project's outcomes and aligning with the principles of efficiency, sustainability, and innovation inherent in this methodology. To start with, the shape of the building influences various aspects of construction procurement, including material selection, structural design, and energy efficiency (Allen & Iano, 2019; Rückert & Shahriari, 2014). Stealth construction often favours streamlined and aerodynamic building forms that optimise structural performance and reduce material consumption. This preference for efficient shapes minimises construction costs and enhances the building's energy efficiency by reducing heat loss and improving airflow, aligning with sustainability goals. Procurement decisions regarding materials and construction methods are guided by the building's shape, emphasising lightweight and durable materials that complement the architectural design while minimising environmental impact.

Furthermore, the internal construction of the building is a critical consideration in stealth construction procurement, impacting spatial layout, functionality, and occupant comfort (Finnigan, 2020; Null, 2013). Innovative techniques, such as prefabrication and modular construction, are often employed to optimise procurement efficiency and accelerate project timelines. By prefabricating building components off-site, construction procurement can be streamlined, reducing lead times, minimising waste, and improving cost predictability. Additionally, modular construction allows for greater flexibility in design and adaptation to changing project requirements, facilitating agile procurement practices. Furthermore, the internal construction of the building influences material usage and resource allocation, focusing on efficient space utilisation, sustainable building practices, and occupant well-being. Procurement decisions related to internal construction prioritise materials that enhance indoor air quality, promote natural lighting, and support flexible use of space, aligning with the principles of stealth construction.

Lastly, material usage in stealth construction procurement is guided by a commitment to sustainability, resilience, and performance (Saad et al., 2024). Sustainable procurement practices prioritise selecting environmentally friendly materials, such as recycled content, low-emission products, and renewable resources, to minimise the building's carbon footprint and promote resource conservation. Additionally, resilient materials that withstand environmental stresses, such as extreme weather events and seismic activity, are prioritised to enhance the building's longevity and reduce lifecycle costs. Stealth construction embraces innovative materials and construction techniques, such as advanced composites, carbon fibre, and engineered timber, which offer superior strength-to-weight ratios, durability, and environmental performance. Procurement decisions regarding material usage are informed by lifecycle assessments, performance specifications, and sustainability certifications, ensuring that materials meet quality standards while aligning with project objectives. Overall, the cross-section of the building influences procurement decisions in stealth construction by shaping architectural design, construction methods, and material selection to achieve efficient, sustainable, and resilient built environments.

Energy Transmission

In construction procurement, energy transmission, encompassing renewable energy integration, climate action planning, energy management, and energy efficiency, emerges as a critical consideration driving stealth construction (Alvarez-Alvarado et al., 2024; Halgamuge, 2024). Firstly, procuring renewable energy technologies is a cornerstone of energy transmission in stealth construction projects. Stakeholders procure solar panels, wind turbines, or other renewable energy systems through strategic sourcing and partnerships to power construction sites or integrate them into building designs. Moreover, climate action planning is tightly woven into the fabric of energy transmission in stealth construction procurement. Through collaborative efforts among project stakeholders, comprehensive climate action plans are developed to assess risks, set emission reduction targets, and implement strategies for resilience. This involves procuring climate modelling services, conducting environmental impact assessments, and engaging with experts to develop robust mitigation and adaptation strategies.

Furthermore, Irmak et al. (2023) and Halgamuge (2024) illustrated that energy management practices are integral to optimising energy transmission in stealth construction procurement endeavours. Procurement decisions are informed by energy management principles, guiding the selection of energy-efficient equipment, materials, and technologies. Through strategic partnerships with energy service providers, stakeholders procure advanced energy management systems, smart metres, and real-time monitoring solutions to track energy usage, identify inefficiencies, and optimise performance. Through embedding energy management considerations into procurement strategies, construction projects reduce operational costs, enhance resource efficiency, and promote environmental stewardship.

Additionally, Murtagh et al. (2021) stated that energy efficiency measures are prioritised in construction procurement to enhance energy transmission within stealth construction projects. Procurement decisions favour energy-efficient building materials, systems, and technologies that minimise energy consumption and maximise performance. This may involve procuring high-performance insulation, energy-efficient lighting fixtures, or smart thermostats to build energy efficiency.

In summary, energy transmission within construction procurement in stealth construction projects encompasses a finite approach integrating renewable energy, climate action planning, energy management, and energy efficiency measures. When these considerations are incorporated into procurement strategies, construction projects can achieve significant environmental, social, and economic benefits, advancing towards a more sustainable and resilient built environment. Through strategic partnerships, innovative solutions, and proactive procurement practices, stealth construction projects pave the way for a greener, more sustainable future.

Countermeasures

In construction procurement, countermeasures, encompassing functional construction systems and preventive measures, are pivotal components that contribute to stealth construction. To begin with, functional construction systems are procured strategically to mitigate potential risks and enhance project resilience (Malik et al., 2022). Procurement decisions prioritise selecting robust and adaptable construction systems that can withstand environmental stresses, such as extreme weather events, seismic activity, or technological disruptions. This may involve sourcing durable building materials, advanced structural components, or resilient infrastructure systems that contribute to the longevity and reliability of construction projects.

Moreover, Rane (2023) stated that preventive measures are crucial in mitigating risks and safeguarding construction projects against potential threats. Through proactive procurement practices, stakeholders procure preventive measures, such as safety equipment, risk management services, and insurance coverage, to protect against accidents, liabilities, and unforeseen events. This involves partnering with reputable suppliers, contractors, and insurance providers to ensure compliance with safety regulations, mitigate legal risks, and protect project assets. Additionally, procurement decisions may include acquiring advanced technologies, such as sensors, surveillance systems, and cybersecurity solutions, to prevent security breaches, data loss, or unauthorised access to construction sites.

Furthermore, Lubis et al. (2023) illustrated that countermeasures in construction procurement extend beyond risk mitigation to encompass measures that optimise project performance and efficiency. Strategic procurement decisions prioritise acquiring innovative technologies, automation solutions, and lean construction practices that streamline processes, minimise waste, and optimise resource utilisation. This may involve partnering with technology providers, research institutions, or industry experts to access cutting-edge solutions that improve productivity, reduce costs, and accelerate project timelines. Additionally, procurement strategies may include implementing quality management systems, performance monitoring tools, and continuous improvement processes to ensure compliance with project specifications and stakeholder expectations. When countermeasures are integrated into procurement practices, stealth construction projects achieve higher efficiency, reliability, and performance levels, enhancing project outcomes and stakeholder satisfaction.

Summarising it all, countermeasures in construction procurement are critical in enhancing stealth construction projects' resilience, efficiency, and performance. Through strategic sourcing, proactive risk management, and investment in innovative solutions, construction stakeholders can mitigate potential threats, optimise resource utilisation, and foster a culture of continuous improvement. Stealth construction projects navigate challenges, seize opportunities, and achieve sustainable success in an ever-evolving construction landscape by integrating functional construction systems and preventive measures into procurement strategies.

Procurement Towards Stealth Construction

Procurement towards stealth construction embodies a progressive approach that integrates innovative strategies, advanced technologies, and collaborative partnerships to optimise project outcomes and drive sustainable development within the construction industry. This strategic procurement prioritises efficiency, sustainability, and resilience, guided by lean construction principles and green building practices. Aldherwi (2021) and Bechtsis et al. (2022) illustrated that stakeholders leverage data-driven insights, digital solutions, and collaborative procurement approaches to streamline processes, minimise waste, and maximise value throughout the project lifecycle. When technology-enabled procurement platforms like BIM and e-procurement systems are embraced, stakeholders automate tasks, enhance transparency, and empower data-driven decision-making. Collaborative partnerships among project stakeholders foster open communication, trust, and mutual respect, enabling stakeholders to address complex challenges, leverage diverse expertise, and unlock collaboration that drives project success. Additionally, partnerships with sustainable suppliers and vendors facilitate access to cutting-edge solutions, reduce environmental impact, and enhance social responsibility throughout the supply chain, laying the foundation for resilient, future-ready projects.

Furthermore, procurement towards stealth construction fosters a culture of continuous improvement, innovation, and adaptability in the construction industry. By embracing emerging technologies such as AI, Internet of Things (IoT), and virtual reality (VR), stakeholders enhance project efficiency, accelerate timelines, and optimise resource allocation. Engebø et al. (2020) and Elghaish et al. (2020) added that through collaborative procurement approaches such as IPD and alliance contracting, stakeholders align project goals, share risks and rewards, and optimise project delivery processes. This collaborative atmosphere extends beyond traditional project boundaries to include engagement with local communities, regulatory authorities, and other stakeholders, fostering an approach to project development. Procurement towards stealth construction empowers stakeholders to navigate challenges, seize opportunities, and build resilient, future-ready projects that meet the evolving needs of society while leaving a positive legacy for future generations.

In aiming for stealth construction through procurement, we have pinpointed five vital sustainable principles: protecting the environment, ensuring safety, speeding up project delivery, being cost-effective, and focusing on aesthetics. These principles highlight the importance of using stealth construction technologies at every stage of the construction process, from the beginning to the end and throughout the project's lifetime.

Protecting the Environment

Environmental protection in stealth construction is a fundamental objective achieved through strategic procurement practices prioritising sustainability, resource efficiency, and environmental stewardship. Ogunmakinde et al. (2022) and Birkeland (2022) mentioned that through proactive measures and innovative solutions, stakeholders aim to minimise environmental impact, conserve natural resources, and promote biodiversity while delivering construction projects that meet the highest environmental standards. One key aspect of environmental protection in stealth construction procurement is the selection of sustainable materials and construction methods. Procurement decisions prioritise materials with low environmental impact, such as recycled, reclaimed, or locally sourced materials, to reduce carbon footprint and minimise waste. Additionally, stakeholders seek to incorporate eco-friendly construction techniques, such as modular construction, prefabrication, and green building systems, which optimise resource utilisation, reduce energy consumption, and minimise construction waste.

Furthermore, environmental protection in stealth construction procurement extends to energy management and renewable energy integration. Hafez et al. (2023) and Dion et al. (2023) illustrated that stakeholders prioritise procuring energy-efficient technologies, such as high-performance insulation, energy-efficient lighting, and heating, ventilation, and air conditioning (HVAC) systems, to minimise energy consumption and reduce greenhouse gas emissions. Additionally, renewable energy sources, such as solar panels, wind turbines, and geothermal systems, are procured to power construction sites and integrated into building designs, reducing reliance on fossil fuels and mitigating environmental impact. Stealth construction projects reduce their carbon footprint, lower operational costs, and contribute to the transition to a sustainable energy future by investing in energy-efficient technologies and renewable energy solutions.

Moreover, environmental protection in stealth construction procurement involves waste management and recycling initiatives. Stakeholders implement procurement strategies prioritising suppliers and contractors committed to waste reduction, recycling, and responsible disposal practices (Chidiobi et al., 2023). Shooshtarian et al. (2022) added that through partnerships with waste management companies and recycling facilities, construction projects establish waste diversion programs, segregate recyclable materials, and minimise landfill disposal. Additionally, stakeholders procure recycled-content materials and products with extended producer responsibility (EPR) schemes, promoting circular economy principles and reducing the environmental footprint of construction projects.

Taking all into summary, environmental protection in stealth construction is achieved through strategic procurement practices that prioritise sustainability, resource efficiency, and environmental stewardship. By selecting sustainable materials, incorporating eco-friendly construction methods, integrating energy-efficient technologies, and implementing waste management and recycling initiatives, stakeholders minimise environmental impact, conserve natural resources, and promote biodiversity in construction projects. Through proactive procurement strategies, stealth construction projects demonstrate a commitment to environmental sustainability and contribute to advancing green building practices within the construction industry.

Ensuring Safety

Ensuring safety in stealth construction through procurement involves comprehensive measures and strategic partnerships to minimise risks, enhance workplace safety, and protect the well-being of workers, stakeholders, and the public. Cooney (2016) and Koc and Gurgun (2021) ascertained that in proactive procurement practices, stakeholders prioritise safety considerations at every stage of the construction process, from project planning to execution, fostering a culture of safety and resilience within the construction industry. One crucial aspect of ensuring safety in stealth construction procurement is the selection of reputable suppliers and contractors with robust safety records and demonstrated commitment to safety standards. Procurement decisions prioritise safety-conscious vendors who adhere to strict safety regulations, provide comprehensive employee safety training, and implement effective safety protocols on construction sites.

Moreover, ensuring safety in stealth construction procurement involves procuring safety equipment, personal protective gear, and emergency response resources. Kasperson (2019) and Tukesiga (2022) highlighted that stakeholders prioritise procuring high-quality safety equipment, such as helmets, gloves, safety harnesses, and fall protection systems, to protect workers from common construction hazards, such as falls, slips, and impact injuries. Additionally, emergency response resources, such as first aid kits, fire extinguishers, and emergency evacuation plans, are procured to prepare for unforeseen incidents and ensure prompt and effective response in case of emergencies. Furthermore, ensuring safety in stealth construction procurement involves the implementation of stringent safety standards, policies, and procedures throughout the supply chain. Also, El-Wafa and Mosly (2024) added that stakeholders establish clear safety guidelines, protocols, and performance metrics for suppliers, subcontractors, and vendors, ensuring alignment with industry best practices and regulatory requirements. Additionally, procurement contracts include provisions for safety compliance, risk management, and liability insurance, holding suppliers accountable for maintaining safe working conditions and adhering to safety standards.

Summarising it all, ensuring safety in stealth construction through procurement involves a multifaceted approach encompassing supplier selection, safety equipment procurement, emergency preparedness, and supply chain management. When partnering with reputable suppliers, procuring high-quality safety equipment, and implementing stringent safety standards and protocols, construction projects minimise risks, prevent accidents, and protect the health and well-being of workers and stakeholders. Through proactive procurement practices and strategic partnerships, stealth construction projects demonstrate a commitment to safety excellence and set a benchmark for industry-wide safety standards and practices.

Speeding Up Project Delivery

Speeding up project delivery in stealth construction through procurement involves implementing efficient processes, leveraging technology, and fostering collaborative partnerships to streamline project timelines and accelerate construction schedules. Ajayi et al. (2017) and Mahi (2024) asserted that when procurement strategies are optimised, stakeholders can minimise delays, reduce lead times, and enhance overall project efficiency, ensuring timely project completion and maximising stakeholder satisfaction.

Furthermore, as stated in their studies, Dara et al. (2024) and Yang and Pan (2021) pointed out that one key aspect of speeding up project delivery in stealth construction procurement is the adoption of JIT procurement practices. Rather than stockpiling materials in advance, JIT procurement involves procuring materials and equipment as needed, minimising inventory holding costs and reducing storage space requirements. Through strategic partnerships with suppliers and vendors, stakeholders can establish reliable supply chains and implement JIT procurement strategies to ensure timely delivery of materials and equipment to the construction site. Additionally, JIT procurement enables stakeholders to adapt quickly to changing project requirements and minimise the risk of material shortages or overstocking, thereby accelerating project delivery schedules and optimising resource utilisation (Antwi, 2024). Moreover, speeding up project delivery in stealth construction procurement involves leveraging digital technologies and procurement platforms to automate processes, streamline communication, and enhance collaboration among project stakeholders. Digital procurement platforms, such as e-procurement systems and supply chain management software, enable stakeholders to centralise procurement activities, track procurement status in real-time, and automate routine tasks, such as sourcing, ordering, and invoicing.

In addition, speeding up project delivery in stealth construction procurement entails fostering collaborative partnerships and engaging with stakeholders early in the project planning phase. Through collaborative procurement approaches, such as IPD and design-build contracts, stakeholders work together to align project goals, share risks and rewards, and optimise project delivery processes (Adamtey, 2021; Jobidon et al., 2021; Othman & Youssef, 2021). From the outset, construction projects can streamline decision-making, reduce conflicts, and accelerate project timelines by involving key stakeholders, including owners, designers, contractors, and suppliers. Additionally, collaborative partnerships enable stakeholders to leverage each other's expertise, resources, and insights, fostering innovation and driving continuous improvement throughout the project lifecycle.

In summary, speeding up project delivery in stealth construction through procurement involves implementing efficient procurement strategies, leveraging technology, and fostering collaborative partnerships to streamline processes and optimise project schedules. By adopting JIT procurement practices, leveraging digital technologies, and fostering collaborative partnerships, construction projects can minimise delays, reduce lead times, and accelerate project delivery schedules, ensuring timely project completion and maximising stakeholder satisfaction. Through proactive procurement practices and strategic partnerships, stealth construction projects demonstrate a commitment to efficiency, innovation, and excellence in project delivery.

Being Cost-Effective

Achieving cost-effectiveness in stealth construction through procurement entails strategic sourcing, value engineering, and efficient resource allocation to optimise project budgets and deliver maximum value for stakeholders. Olanrewaju et al. (2022) mentioned that an essential aspect is the strategic selection of suppliers and subcontractors based on their ability to provide high-quality materials and services at competitive prices. Through rigorous vendor evaluations and negotiations, stakeholders can secure favourable pricing, discounts, and incentives, reducing procurement costs without compromising quality. Additionally, value engineering principles are applied to identify cost-saving opportunities and optimise project designs, specifications, and materials while maintaining performance and functionality (Othman & Abdelrahim, 2020). Wuni and Shen (2022) and Chen et al. (2022) showed that collaboration with designers, engineers, suppliers and stakeholders can identify alternative materials, construction methods, and design solutions that reduce costs while meeting project requirements. Furthermore, efficient resource allocation ensures that resources, such as labour, equipment, and materials, are utilised optimally to minimise waste and maximise productivity. When adopting lean construction principles and optimising supply chain logistics, stakeholders can reduce project costs, improve project efficiency, and enhance overall cost-effectiveness in stealth construction procurement.

Moreover, achieving cost-effectiveness in stealth construction through procurement involves leveraging technology and innovation to streamline processes, enhance efficiency, and reduce overhead costs. Cox (2022) added that digital procurement platforms and e-procurement systems enable stakeholders to automate routine tasks, such as sourcing, ordering, and invoicing, reducing administrative overhead and minimising errors. Additionally, Radman et al. (2022) illustrated that advanced technologies such as BIM, drones, and real-time monitoring systems provide stakeholders valuable insights and data-driven decision-making tools to optimise project performance and resource utilisation. Leveraging technology-enabled solutions, stakeholders can improve project efficiency, minimise project costs, and enhance overall cost-effectiveness in stealth construction procurement. Furthermore, Walker and Rowlinson (2020) and Viana et al. (2020) showed that strategic partnerships and collaborative procurement approaches, such as IPD and alliance contracting, enable stakeholders to share risks and rewards, align project goals and optimise resource allocation, further enhancing cost-effectiveness and value creation. Through proactive procurement practices and strategic partnerships, stealth construction projects achieve cost-effectiveness while delivering high-quality projects on time and within budget, maximising value for all stakeholders.

Focusing on Aesthetics

In stealth construction, focusing on aesthetics through procurement involves strategically selecting materials, finishes, and design elements to enhance the built environment's visual appeal and architectural integrity. Iacovidou et al. (2021) discussed that procuring high-quality materials and finishes that complement the project's design vision and aesthetic objectives is vital. Lucchi et al. (2023) mentioned that stakeholders can prioritise sourcing materials with appealing textures, colours, and patterns that contribute to the overall aesthetic harmony of the project. Through partnerships with reputable suppliers and manufacturers, stakeholders ensure access to a diverse range of aesthetically pleasing materials, such as natural stone, wood, glass, and metal, that meet design specifications and quality standards. Additionally, procurement decisions may involve custom fabrication or speciality finishes to achieve unique architectural features and design accents that enhance the project's visual impact and distinguish it from conventional construction projects.

Moreover, achieving aesthetic excellence in stealth construction through procurement involves collaboration with architects, designers, and design experts to integrate artistic elements and creative expression into the built environment, as identified by Shen et al. (2010), Cicmil and Marshall (2005), and Dokter et al. (2021). Dokter et al. (2021) reiterated that stakeholders engage in a collaborative design process to explore innovative design concepts, materials, and techniques that reflect the project's aesthetic vision and cultural context. Foreman (2014) and Amusan et al. (2021) discussed that by partnering with artists, craftsmen, and artisans, stakeholders could incorporate custom-designed elements, such as sculptures, murals, and installations, that add artistic value and enrich the architectural experience for occupants and visitors alike. Furthermore, procurement decisions may include sourcing sustainable and environmentally friendly materials that align with the project's aesthetic goals while promoting responsible stewardship of natural resources. When aesthetics is prioritised in procurement practices, stealth construction projects create visually stunning and emotionally engaging spaces that inspire, delight, and leave a lasting impression on users and communities. Through proactive procurement strategies and creative collaborations, stealth construction projects achieve aesthetic excellence while delivering projects that transcend functionality and elevate the human experience.

Procurement for Stealth Construction from the Pre-construction Stage

In the pre-construction stage of stealth construction, procurement is crucial in setting the foundation for project success by laying the groundwork for efficient resource allocation, strategic planning, and collaborative partnerships. To begin with, during the pre-construction phase, procurement activities focus on sourcing and procuring materials, equipment, and services essential for project execution. This involves conducting market research, soliciting bids from suppliers and subcontractors, and negotiating contracts to secure favourable terms and pricing. When proactively engaging with vendors and establishing procurement agreements early in the project lifecycle, stakeholders can mitigate risks, optimise project budgets, and ensure timely delivery of materials and services, setting the stage for seamless project execution.

Furthermore, in the pre-construction stage, procurement activities extend beyond traditional sourcing and contracting to include value engineering, risk assessment, and sustainability planning. Stakeholders collaborate with designers, engineers, and suppliers to identify cost-saving opportunities, optimise project designs, and select materials and construction methods that enhance project efficiency and sustainability. Additionally, procurement decisions in the pre-construction phase involve assessing and mitigating risks associated with supply chain disruptions, market fluctuations, and regulatory compliance. When conducting thorough risk assessments and implementing risk management strategies, stakeholders can anticipate potential challenges, develop contingency plans, and safeguard project timelines and budgets.

Moreover, procurement in the pre-construction stage of stealth construction facilitates collaboration and stakeholder engagement, laying the groundwork for successful project delivery. Stakeholders leverage procurement processes to establish collaborative partnerships, align project goals, and foster a culture of transparency, accountability, and trust among project participants. When key stakeholders are involved, including owners, designers, contractors, and suppliers, early in the procurement process, construction projects can leverage diverse expertise, insights, and resources to optimise project outcomes and mitigate risks. Additionally, procurement in the pre-construction phase enables stakeholders to address regulatory requirements, permitting, and environmental considerations upfront, minimising delays and disruptions during later stages of project execution.

Procurement for stealth construction aligns well with the pre-construction stage rather than the construction or post-construction stages due to its proactive and preparatory nature. During the pre-construction phase, stakeholders can lay the groundwork for project success by conducting thorough planning, procurement, and risk assessment activities. Engaging in procurement activities early in the project lifecycle, stakeholders can address critical issues such as material sourcing, cost estimation, and supply chain management before construction. This proactive approach enables stakeholders to anticipate challenges, mitigate risks, and optimise project budgets and schedules, setting the stage for efficient project execution and successful project delivery.

Furthermore, procurement in the pre-construction stage allows stakeholders to establish collaborative partnerships, align project goals, and foster a culture of teamwork and innovation among project participants. By involving key stakeholders, including owners, designers, contractors, and suppliers, early in the procurement process, construction projects can leverage diverse expertise, insights, and resources to optimise project outcomes and mitigate risks. Additionally, procurement in the pre-construction phase enables stakeholders to address regulatory requirements, permitting, and environmental considerations upfront, minimising delays and disruptions during later stages of project execution. Overall, by positioning procurement activities within the pre-construction stage, stealth construction projects can maximise efficiency, minimise risks, and lay the foundation for successful project delivery.

Conclusion

The exploration of procurement in stealth construction has underscored its pivotal role in driving efficiency, sustainability, and innovation within the construction industry. Through proactive procurement practices, stakeholders can optimise project outcomes, mitigate risks, and foster collaborative partnerships that enhance project delivery from inception to completion. Focusing on fundamental principles such as environmental protection, safety, cost-effectiveness, and aesthetics, procurement in stealth construction facilitates the integration of advanced technologies, sustainable practices, and creative design solutions that elevate the built environment while minimising environmental impact. Moreover, by positioning procurement activities within the pre-construction stage, stakeholders can lay the groundwork for successful project execution, setting the stage for seamless construction processes and optimal project outcomes. Ultimately, procurement in stealth construction represents a transformative approach that empowers stakeholders to navigate challenges, seize opportunities, and build resilient, future-ready projects that meet the evolving needs of society while leaving a positive legacy for future generations.

References

Adamtey, S. A. (2021). A case study performance analysis of design-build and integrated project delivery methods. *International Journal of Construction Education and Research*, *17*(1), 68–84.

Ajayi, S. O., Oyedele, L. O., Akinade, O. O., Bilal, M., Alaka, H. A., & Owolabi, H. A. (2017). Optimising material procurement for construction waste minimization: An exploration of success factors. *Sustainable Materials and Technologies*, *11*(1), 38–46.

Aldherwi, A. (2021). *Conceptualising a procurement 4.0 model for a truly data driven procurement*. https://www.diva-portal.org/smash/record.jsf?

Alhammadi, A., Soar, J., Yusaf, T., Ali, B. M., & Kadirgama, K. (2023). Redefining procurement paradigms: A critical review of buyer-supplier dynamics in the global petroleum and natural gas industry. *The Extractive Industries and Society*, *16*(1), 1–11.

Allen, E., & Iano, J. (2019). *Fundamentals of building construction: Materials and methods*. John Wiley & Sons.

Althabatah, A., Yaqot, M., Menezes, B., & Kerbache, L. (2023). Transformative procurement trends: Integrating industry 4.0 technologies for enhanced procurement processes. *Logistics*, *7*(3), 1–40. https://doi.org/10.3390/logistics7030063

Alvarez-Alvarado, M. S., Apolo-Tinoco, C., Ramirez-Prado, M. J., Alban-Chacón, F. E., Pico, N., Aviles-Cedeno, J., Recalde A. A., Moncayo-Rea, F., Velasquez, W., & Rengifo, J. (2024). Cyber-physical power systems: A comprehensive review about technologies drivers, standards, and future perspectives. *Computers and Electrical Engineering*, *116*, 1–27. https://doi.org/10.1016/j.compeleceng.2024.109149

Amarah, B., & Langston, C. (2017). Development of a triple bottom line stakeholder satisfaction model. *Journal of Corporate Real Estate*, *19*(1), 17–35.

Amirtash, P., Parchami Jalal, M., & Jelodar, M. B. (2021). Integration of project management services for international engineering, procurement and construction projects. *Built Environment Project and Asset Management*, *11*(2), 330–349.

Amusan, L. M., Oluwatobi, O., Dalshe, C., Ezenduka, J., Emetere, M., Owolabi, J. D., & Tunji- Olayeni, P. F. (2021, February). Towards improving artisan and craftsmen productivity. *IOP Conference Series: Earth and Environmental Science, Volume 655, 4th International Conference on Science and Sustainable Development (ICSSD 2020), "Advances in Sciences and Technology for Sustainable Development"*, 3–5 August 2020, Center for Research, Innovation and Discovery, Covenant University, Ota, Nigeria.

Antwi, K. B. (2024). *Applicable lean methods and the implementation in a building refurbishment project* [Doctoral dissertation, Hochschule Rhein-Waal].

Bechtsis, D., Tsolakis, N., Iakovou, E., & Vlachos, D. (2022). Data-driven secure, resilient and sustainable supply chains: Gaps, opportunities, and a new generalised data sharing and data monetisation framework. *International Journal of Production Research*, *60*(14), 4397–4417.

Birkeland, J. (2022). Nature positive: Interrogating sustainable design frameworks for their potential to deliver eco-positive outcomes. *Urban Science*, *6*(2), 1–29.

Bohari, A. A. M., Skitmore, M., Xia, B., Teo, M., & Khalil, N. (2020). Key stakeholder values in encouraging green orientation of construction procurement. *Journal of Cleaner Production*, *270*, 1–11. https://doi.org/10.1016/j.jclepro.2020.122246

Can-Saglam, Y., Yildiz Çankaya, S., & Sezen, B. (2021). Proactive risk mitigation strategies and supply chain risk management performance: An empirical analysis for manufacturing firms in Turkey. *Journal of Manufacturing Technology Management*, *32*(6), 1224–1244.

Chen, Q., Adey, B. T., Haas, C. T., & Hall, D. M. (2022). Exploiting digitalization for the coordination of required changes to improve engineer-to-order materials flow management. *Construction Innovation*, *22*(1), 76–100.

Chidiobi, C. C., Booth, C. A., & Lamond, J. E. (2023). Prioritising materials procurement and construction waste management attributes. *Proceedings of the Institution of Civil Engineers-Waste and Resource Management*, *177*(1), 29–43.

Cicmil, S., & Marshall, D. (2005). Insights into collaboration at the project level. Complexity, social interaction and procurement mechanisms. *Building Research & Information*, *33*(6), 523–535.

Cooney, J. P. (2016). *Health and safety in the construction industry: A review of procurement, monitoring, cost effectiveness and strategy*. https://www.proquest.com/docview/2570357853?pq-

Cox, K. (2022). Digital purchasing and procurement systems: Evolution and current state. *The Digital Supply Chain*, *1*, 181–197. https://doi.org/10.1016/B978-0-323-91614-1.00011-3

Dara, H. M., Raut, A., Adamu, M., Ibrahim, Y. E., & Ingle, P. (2024). Reducing non-value adding activities through lean tools for the precast industry. *Heliyon*, *10*, 1–18. https://doi.org/10.1016/j.heliyon.2024.e2914

Dion, H., Evans, M., & Farrell, P. (2023). Hospitals management transformative initiatives; Towards energy efficiency and environmental sustainability in healthcare facilities. *Journal of Engineering, Design and Technology*, *21*(2), 552–584.

Dokter, G., Thuvander, L., & Rahe, U. (2021). How circular is current design practice? Investigating perspectives across industrial design and architecture in the transition towards a circular economy. *Sustainable Production and Consumption*, *26*, 692–708. https://doi.org/10.1016/j.spc.2020.12.032

Elghaish, F., Hosseini, M. R., Talebi, S., Abrishami, S., Martek, I., & Kagioglou, M. (2020). Factors driving success of cost management practices in integrated project delivery (IPD). *Sustainability*, *12*(22), 1–14.

Elsayegh, A., & El-adaway, I. H. (2021). Holistic study and analysis of factors affecting collaborative planning in construction. *Journal of Construction Engineering and Management*, *147*(4), 1–15.

El-Wafa, M. M. A., & Mosly, I. (2024). An extensive examination of the barriers faced by contractors leading to project delays. *Global Journal of Engineering and Technology Advances*, *18*(3), 152–167.

Engebø, A., Lædre, O., Young, B., Larssen, P. F., Lohne, J., & Klakegg, O. J. (2020). Collaborative project delivery methods: A scoping review. *Journal of Civil Engineering and Management*, *26*(3), 278–303.

Ershadi, M., Jefferies, M., Davis, P., & Mojtahedi, M. (2021). Achieving sustainable procurement in construction projects: The pivotal role of a project management office. *Construction Economics and Building*, *21*(1), 45–64.

Finnigan, S. J. (2020). *Human-centred smart buildings: Reframing smartness through the lens of human-building interaction* [Doctoral dissertation, Newcastle University].

Foreman, A. (2014). *A micro-ethnographic study of creative behavior of title 1 urban art students: How do context, collaboration and content play a role in the development of creativity?* https://www.proquest.com/docview/1540514040?pq-

Ghadamsi, A., & Braimah, N. (2012). The influence of procurement methods on project performance: A conceptual framework. *Brunel University*, *1*(1), 860–871.

Giuffrida, L. M., & Rovigatti, G. (2022). Supplier selection and contract enforcement: Evidence from performance bonding. *Journal of Economics & Management Strategy*, *31*(4), 980–1019.

Goh, C. S., Su, F., & Rowlinson, S. (2023). Exploring economic impacts of sustainable construction projects on stakeholders: The role of integrated project delivery. *Journal of Legal Affairs and Dispute Resolution in Engineering and Construction*, *15*(3), 1–9. https://doi.org/10.1061/JLADAH.LADR-963

Hafez, F. S., Sa'di, B., Safa-Gamal, M., Taufiq-Yap, Y. H., Alrifaey, M., Seyedmahmoudian, M., Stojcevski, A., Horan, B., & Mekhilef, S. (2023). Energy efficiency in sustainable buildings: A systematic review with taxonomy, challenges, motivations, methodological aspects, recommendations, and pathways for future research. *Energy Strategy Reviews*, *45*, 1–30. https://doi.org/10.1016/j.esr.2022.101013

Halgamuge, M. N. (2024). Leveraging deep learning to strengthen the cyber-resilience of renewable energy supply chains: A survey. *IEEE Communications Surveys and Tutorials*, *1*(1), 1–30. https://doi.org/10.1109/COMST.2024.3365076

Iacovidou, E., Purnell, P., Tsavdaridis, K. D., & Poologanathan, K. (2021). Digitally enabled modular construction for promoting modular components reuse: A UK view. *Journal of Building Engineering*, *42*, 1–9. https://doi.org/10.1016/j.jobe.2021.102820

Irmak, E., Kabalci, E., & Kabalci, Y. (2023). Digital transformation of microgrids: A review of design, operation, optimization, and cybersecurity. *Energies*, *16*(12), 1–58.

Jimoh, R. A., Oyewobi, L. O., & Aliu, N. (2015). *Impact of procurement methods and project types on construction projects performance*. http://repository.futminna.edu.ng:8080/jspui/handle/123456789/492

Jobidon, G., Lemieux, P., & Beauregard, R. (2021). Building information modeling in Quebec's procurement for public infrastructure: A case for integrated project delivery. *Laws*, *10*(2), 1–16.

Kalasani, R. R. (2023). *An exploratory study of the impacts of artificial intelligence and machine learning technologies in the supply chain and operations field* [Doctoral dissertation, University of the Cumberlands].

Kasperson, R. E. (2019). *Corporate management of health and safety hazards: A comparison of current practice*. Routledge.

Koc, K., & Gurgun, A. P. (2021). Stakeholder-associated life cycle risks in construction supply chain. *Journal of Management in Engineering*, *37*(1), 1–27. https://doi.org/10.1061/(ASCE)ME.1943-5479.0000881

Loosemore, M., Keast, R., Barraket, J., Denny-Smith, G., & Alkilani, S. (2022). The risks and opportunities of social procurement in construction projects: A cross-sector collaboration perspective. *International Journal of Managing Projects in Business*, *15*(5), 793–815.

Lubis, A. M., Widyarini, M., & Sunardi, O. (2023). Risk mitigation in supply chain disruption during pandemic covid-19 in EPC project, study case Pt Xyz. *Jurnal Ekonomi*, *12*(4), 1404–1416.

Lucchi, E., Adami, J., & Stawinoga, A. E. (2023). Social acceptance of photovoltaic systems in heritage buildings and landscapes: Exploring barriers, benefits, drivers, and challenges for technical stakeholders in northern Italy. *Sustainable Energy Technologies and Assessments*, *60*, 1–14. https://doi.org/10.1016/j.seta.2023.103544

Lysons, K., & Farrington, B. (2020). *Procurement and supply chain management*. Pearson UK.

Ma, Q., Cheung, S. O., & Li, S. (2023). Optimum risk/reward sharing framework to incentivize integrated project delivery adoption. *Construction Management and Economics, 41*(6), 519–535.

Mahi, R. (2024). Optimizing supply chain efficiency in the manufacturing sector through ai- powered analytics. *International Journal of Management Information Systems and Data Science, 1*(1), 41–50.

Malik, A., Khan, K. I. A., Qayyum, S., Ullah, F., & Maqsoom, A. (2022). Resilient capabilities to tackle supply chain risks: Managing integration complexities in construction projects. *Buildings, 12*(9), 1–26.

Manners-Bell, J. (2017). *Supply chain risk management: Understanding emerging threats to global supply chains*. Kogan Page Publishers.

Marcelline, S. R., Chengang, Y., Ralison Ny Avotra, A. A., Hussain, Z., Zonia, J. E., & Nawaz, A. (2022). Impact of green construction procurement on achieving sustainable economic growth influencing green logistic services management and innovation practices. *Frontiers in Environmental Science, 9*(1), 1–14. https://doi.org/10.3389/fenvs.2021.815928

Martínez-Peláez, R., Ochoa-Brust, A., Rivera, S., Félix, V. G., Ostos, R., Brito, H., Félix R. A., & Mena, L. J. (2023). Role of digital transformation for achieving sustainability: Mediated role of stakeholders, key capabilities, and technology. *Sustainability, 15*(14), 1–27. https://doi.org/10.3390/su151411221

Montalbán-Domingo, L., García-Segura, T., Sanz-Benlloch, A., Pellicer, E., Torres-Machi, C., & Molenaar, K. (2022). Assessing social performance of construction companies in public-works procurement: Data envelopment analysis based on the benefit of the doubt approach. *Environmental Impact Assessment Review, 96*, 1–15. https://doi.org/10.1016/j.eiar.2022.106844

Morledge, R., Smith, A. J., & Appiah, S. Y. (2021). *Building procurement*. John Wiley & Sons.

Murphy, M., & Eadie, R. (2019). Socially responsible procurement: A service innovation for generating employment in construction. *Built Environment Project and Asset Management, 9*(1), 138–152.

Murtagh, N., Owen, A. M., & Simpson, K. (2021). What motivates building repair- maintenance practitioners to include or avoid energy efficiency measures? Evidence from three studies in the United Kingdom. *Energy Research & Social Science, 73*, 1–11. https://doi.org/10.1016/j.erss.2021.101943

Null, R. (Ed.). (2013). *Universal design: Principles and models*. CRC Press.

Nwajei, U. O. K., Bølviken, T., & Hellström, M. M. (2022). Overcoming the principal-agent problem: The need for alignment of tools and methods in collaborative project delivery. *International Journal of Project Management, 40*(7), 750–762.

Ogunmakinde, O. E., Egbelakin, T., & Sher, W. (2022). Contributions of the circular economy to the UN sustainable development goals through sustainable construction. *Resources, Conservation and Recycling, 178*, 1–13. https://doi.org/10.1016/j.resconrec.2021.106023

Ogunsanya, O. A., Aigbavboa, C. O., Thwala, D. W., & Edwards, D. J. (2022). Barriers to sustainable procurement in the Nigerian construction industry: An exploratory factor analysis. *International Journal of Construction Management, 22*(5), 861–872.

Oke, A. E., Kineber, A. F., Ekundayo, D., Tunji-Olayeni, P., & Edwards, D. J. (2022). Exploring the cyber technology critical success factors for sustainable building projects: A stationary analysis approach. *Sustainability, 14*(22), 1–16.

Okogwu, C., Agho, M. O., Adeyinka, M. A., Odulaja, B. A., Eyo-Udo, N. L., Daraojimba, C., & Banso, A. A. (2023). Exploring the integration of sustainable materials in supply chain management for environmental impact. *Engineering Science & Technology Journal, 4*(3), 49–65.

Olanrewaju, A., Bong, M. Z. X., & Preece, C. (2022). Establishment of pre-qualification criteria for the selection of subcontractors by the prime constructors for building projects. *Journal of Building Engineering*, *45*, 1–19. https://doi.org/10.1016/j.jobe.2021.103644

Othman, A. A. E., & Abdelrahim, S. M. (2020). Achieving sustainability through reducing construction waste during the design process: A value management perspective. *Journal of Engineering, Design and Technology*, *18*(2), 362–377.

Othman, A. A. E., & Youssef, L. Y. W. (2021). A framework for implementing integrated project delivery in architecture design firms in Egypt. *Journal of Engineering, Design and Technology*, *19*(3), 721–757.

Pan, Y., Zhong, W., Zheng, X., Xu, H., & Zhang, T. (2024). Natural ventilation in vernacular architecture: A systematic review of bioclimatic ventilation design and its performance evaluation. *Building and Environment*, *253*, 1–23. https://doi.org/10.1016/j.buildenv.2024.111317

Plunkett, S. (2021). Introduction to procurement methods in construction. In A. Speaight (Ed.), *Architect's legal handbook* (pp. 175–180). Routledge. https://doi.org/10.4324/9780429279546

Radman, K., Jelodar, M. B., Lovreglio, R., Ghazizadeh, E., & Wilkinson, S. (2022). Digital technologies and data-driven delay management process for construction projects. *Frontiers in Built Environment*, *8*, 1–20. https://doi.org/10.3389/fbuil.2022.1029586

Rahimian, F. P., Goulding, J. S., Abrishami, S., Seyedzadeh, S., & Elghaish, F. (2021). *Industry 4.0 solutions for building design and construction: A paradigm of new opportunities*. Routledge.

Rane, N. (2023). Integrating Building Information Modelling (BIM) and Artificial Intelligence (AI) for smart construction schedule, cost, quality, and safety management: Challenges and opportunities. *Social Science Research Network*, *1*(1), 1–20. https://doi.org/10.2139/ssrn.4616055

Rückert, K., & Shahriari, E. (2014). *Guideline for sustainable, energy efficient architecture and construction* (Vol. 10). Universitätsverlag der TU Berlin.

Russell, E. F. (2020). *Leading role or bit player? Main contractors, supply chain and sustainable construction* [Doctoral dissertation, University of Surrey].

Saad, S., Rasheed, K., Ammad, S., Ali, Z., & Zaland, A. (2024). The conclusive role of AI in industry 4.0 construction materials. In S. Saad, S. Ammad, & Rasheed, K. (Eds.), *AI in material science* (pp. 262–270). CRC Press.

Sawhney, A., Riley, M., Irizarry, J., & Riley, M. (2020). *Construction 4.0*. Routledge publisher.

Shen, W., Hao, Q., Mak, H., Neelamkavil, J., Xie, H., Dickinson, J., Thomas, J. R., Pardasani, A., & Xue, H. (2010). Systems integration and collaboration in architecture, engineering, construction, and facilities management: A review. *Advanced Engineering Informatics*, *24*(2), 196–207.

Shooshtarian, S., Maqsood, T., Wong, P. S., & Bettini, L. (2022). Application of sustainable procurement policy to improve the circularity of construction and demolition waste resources in Australia. *Materials Circular Economy*, *4*(1), 1–22.

Sönnichsen, S. D., & Clement, J. (2020). Review of green and sustainable public procurement: Towards circular public procurement. *Journal of Cleaner Production*, *245*, 1–18. https://doi.org/10.1016/j.jclepro.2019.118901

Tukesiga, P. (2022). *Health and safety performance on construction sites in Rwanda* [Doctoral dissertation, JKUAT-COHRED].

Viana, M. L., Hadikusumo, B. H., Mohammad, M. Z., & Kahvandi, Z. (2020). Integrated Project Delivery (IPD): An updated review and analysis case study. *Journal of Engineering, Project & Production Management*, *10*(2), 147–161.

Walker, D. H., & Rowlinson, S. M. (Eds.). (2020). *Routledge handbook of integrated project delivery*. Routledge.

Watermeyer, R. (2023). Project procurement management in developing countries. In G. Ofori (Ed.), *Building a body of knowledge in project management in developing countries* (pp. 355–386). https://doi.org/10.1142/11950

Wuni, I. Y., & Shen, G. Q. (2022). Developing critical success factors for integrating circular economy into modular construction projects in Hong Kong. *Sustainable Production and Consumption*, *29*, 574–587. https://doi.org/10.1016/j.spc.2021.11.010

Yang, Y., & Pan, W. (2021). Automated guided vehicles in modular integrated construction: Potentials and future directions. *Construction Innovation*, *21*(1), 85–104.

Yu, A. T. W., Yevu, S. K., & Nani, G. (2020). Towards an integration framework for promoting electronic procurement and sustainable procurement in the construction industry: A systematic literature review. *Journal of Cleaner Production*, *250*, 1–18. https://doi.org/10.1016/j.jclepro.2019.119493

Section 2

Stealth (Construction): Supply Chain Management, Lean Construction, and Smart Construction

Chapter Five

Supply Chain Management for Stealth Construction

Abstract

The chapter explored the critical components, challenges, and technological advancements in construction supply chain management (CSCM), focusing on stealth construction (STC). It delved into STC encompassing nature, highlighting its unique challenges in its supply chain management and the necessity for adaptive technologies. It further discussed the benefits of tailoring supply chain management specifically for STC, emphasising the importance of developing the building's cross-section, managing visibility, controlling energy transmission, and implementing countermeasures. Practical applications of CSCM in STC are also examined. This chapter sheds light on the complexities of managing supply chains in STC contexts and offers insights into strategies and technologies to address these challenges effectively.

Keywords: Construction supply chain management; stealth construction; technological advancements; adaptive construction technologies; sustainable construction

Introduction

Kim and Nguyen (2022) stated that construction supply chain management (CSCM) is a critical aspect of the construction industry, encompassing the planning, coordination, and optimisation of the flow of materials, information, and finances throughout the construction process. It is a dynamic and complex system that involves various stakeholders, from suppliers and manufacturers to contractors and end-users. The effective management of the construction supply chain is vital for ensuring timely project completion, cost control, and overall project success. This exploration delves into the extensive facets of CSCM, elucidating its

Stealth Construction: Integrating Practices for Resilience and Sustainability, 99–127

Published under exclusive licence by Emerald Publishing Limited
doi:10.1108/978-1-83608-182-120251005

key components, challenges, technological advancements, and its crucial role in the contemporary construction scenery.

Critical Components of CSCM

CSCM seamlessly integrates numerous components to facilitate a smooth and efficient construction process. These components include procurement, logistics, inventory management, demand forecasting, supplier relationships, and distribution (Cigolini et al., 2022; Golpîra, 2020; Le et al., 2021). Procurement strategies involve sourcing and acquiring materials, equipment, and services. Logistics focuses on transporting and storing these materials, ensuring they are available at the construction site when needed. Inventory management aims to optimise the balance between having enough materials to meet demand without excess and minimising costs and waste. Demand forecasting involves predicting material needs based on project timelines and progress. Supplier relationships are crucial for maintaining a reliable and responsive network of partners, and distribution ensures materials reach the right place at the right time.

Technological Advancements in CSCM

In recent years, technological advancements have significantly transformed CSCM. Fatorachian and Kazemi (2021) illustrated that integrating digital tools, data analytics, and automation has enhanced the efficiency and transparency of supply chain operations. Building Information Modelling (BIM) allows for the digital representation of the entire construction project, visualising and coordinating the supply chain. Real-time tracking systems using Internet of Things (IoT) devices provide visibility into the movement of materials, enabling better decision-making and reducing the risk of delays (Blanchard, 2021). Advanced analytics and artificial intelligence help in demand forecasting, inventory optimisation, and risk management. Cloud-based collaboration platforms facilitate seamless communication and coordination among stakeholders, fostering a more connected and responsive construction supply chain.

Role in Project Cost and Time Management

Harris et al. (2021) asserted that CSCM is crucial in project cost and time management. Efficient supply chain practices contribute to cost reduction by minimising waste, optimising inventory, and negotiating favourable procurement terms. Furthermore, timely delivery of materials is paramount to avoiding delays in construction schedules. CSCM ensures that suitable materials are available at the construction site when needed, preventing costly interruptions (Hussein et al., 2021). By streamlining processes, improving communication, and leveraging technology, construction projects can achieve cost savings and adhere to project timelines, ultimately enhancing overall project success.

Sustainability in CSCM

Hossain et al. (2020) and Lăzăroiu et al. (2020) stated that sustainability has become a focal point in the contemporary construction prospect, and CSCM is

Table 2. Summary of Sustainability in Supply Chain Management.

Authors	**Sustainability in Supply Chain Management**
Hossain et al. (2020) and Lăzăroiu et al. (2020)	Sustainable construction supply chain management selection of environmentally friendly materials optimises transportation routes to reduce carbon emissions by adopting circular economy principles.
Hossain (2023)	It considers ethical sourcing, ensuring suppliers adhere to responsible and fair labour practices.
Zhou et al. (2024)	It minimises the ecological footprint of construction activities by promoting eco-friendly materials, reducing energy consumption, and implementing waste reduction strategies.
Kedir and Hall (2021) and Shooshtarian et al. (2022)	It encompasses practices such as recycling, reusing materials, and minimising waste, reducing the need for new resource extraction, lowering disposal costs, and enhancing overall operational efficiency.
Pham and Pham (2021), Huq and Stevenson (2020), and Modak et al. (2020)	Incorporating sustainability into construction supply chain management improves ethical sourcing in materials procurement and social responsibility.
Zimon et al. (2020)	Sustainability helps keep up to date with evolving regulatory frameworks and standards to promote environmentally responsible and socially conscious construction practices.

integral to achieving sustainable practices, as summarised in Table 2. This involves selecting environmentally friendly materials, optimising transportation routes to reduce carbon emissions, and adopting circular economy principles. Sustainable CSCM also considers ethical sourcing, ensuring suppliers adhere to responsible and fair labour practices (Hossain, 2023). As the construction industry increasingly embraces sustainable development goals (SDGs), CSCM becomes a vehicle for aligning supply chain practices with broader environmental and social responsibilities.

According to Zhou et al. (2024), one of the primary benefits of incorporating sustainability into CSCM is the positive impact on the environment. Sustainable CSCM focuses on minimising the ecological footprint of construction activities by promoting eco-friendly materials, reducing energy consumption, and implementing waste reduction strategies. By sourcing materials responsibly, optimising transportation routes to reduce emissions, and adopting green building practices, construction projects contribute to biodiversity preservation, reduction in greenhouse gas emissions, and overall conservation of natural resources. The environmental benefits extend beyond the construction phase, aligning with global efforts to mitigate climate change and foster a more sustainable built environment.

Furthermore, Kedir and Hall (2021) and Shooshtarian et al. (2022) highlighted that this sustainability often goes hand in hand with resource efficiency, leading to substantial cost savings over the project lifecycle. When implementing practices such as recycling, reusing materials, and minimising waste, construction projects can reduce the need for new resource extraction, lower disposal costs, and enhance overall operational efficiency. Sustainable CSCM also encourages using energy-efficient technologies, such as renewable energy sources and advanced building systems, which contribute to environmental goals and result in long-term cost savings through reduced energy consumption and operational expenses.

In addition, a significant correlation was found in the studies between Pham and Pham (2021) and Modak et al. (2020), where it was stated that integrating sustainability into CSCM extends beyond environmental considerations to encompass ethical sourcing and social responsibility. Furthermore, Huq and Stevenson (2020) added that construction projects prioritising sustainability often engage in responsible procurement practices, ensuring that materials are sourced from suppliers adhering to fair labour practices and ethical standards. This commitment to social responsibility fosters positive relationships with local communities and contributes to the overall well-being of workers and stakeholders involved in the construction supply chain. Sustainable CSCM becomes a driving force for promoting social equity, fair labour practices, and community engagement.

CSCM for Stealth Construction (STC)

STC represents a paradigm shift in the traditional construction view, seamlessly integrating various construction practices and technologies across stages: before, during, and post-construction. This approach necessitates a specialised CSCM strategy to ensure the discreet and strategic execution of projects.

CSCM takes on a unique dimension when applied to STC. This approach involves the correlation of materials, information, and resources throughout the construction process, aligning with the projects' discreet and strategic nature across construction stages. This section discusses the challenges, strategic considerations, benefits, and technological integration necessary for an effective CSCM tailored specifically for STC projects.

The Encompassing Nature of STC

The encompassing nature of STC is characterised by the seamless integration of construction practices throughout the entire project whole lifecycle. In the pre-construction phase, the emphasis is on strategic planning and design considerations that lay the foundation for the project's discreet execution. This involves rapt attention to detail in site selection, design elements, and logistical planning to ensure that the construction process aligns with the project's covert objectives. The pre-construction stage sets the stage for the subsequent phases, establishing the framework for a construction project that seamlessly integrates into its surroundings, as opinionated in the study by Biersteker et al. (2021).

During the construction phase of STC, a carefully orchestrated flow of materials, technologies, and labour becomes paramount to maintaining secrecy, as mentioned by Bontempi et al. (2021) and Omer and Noguchi (2020) that materials and technologies are paramount towards achieving SDGs 7, 8, 9, 11, 12, and 13 stating affordable and clean energy, decent work and economic growth, industry, innovation and infrastructure, sustainable cities and communities, responsible consumption and production, and climate action. This stage involves the physical construction of the structure and the strategic coordination of resources to minimise visibility and potential exposure. The selection of suppliers and subcontractors, adherence to tight schedules, and efficient logistics planning are critical components. This stage is where CSCM for STC plays a pivotal role in ensuring the discreet execution of the project, emphasising the need for a tailored approach that aligns with the unique demands of covert construction.

Post-construction marks the continuation of the encompassing approach in STC, focusing on ongoing maintenance and sustainability practices. Beyond completing the physical structure, sustaining the inconspicuous integration of the project into its environment becomes crucial. This involves adopting environmentally friendly maintenance practices, ensuring the long-term durability of the structure, and contributing to the overall sustainability of the built environment (Munaro et al., 2020). The post-construction phase completes the lifecycle of STC, emphasising the need for continued strategic planning and adaptation of construction practices to maintain the covert nature of the structure over time.

Challenges in CSCM for STC

STC, emphasising discretion and secrecy, presents a distinctive set of challenges in CSCM. First and foremost is the challenge of maintaining confidentiality throughout the supply chain. Limited information dissemination is crucial to the success of stealth projects, yet this conflicts with the conventional transparency required in supply chain operations. The restricted flow of information poses difficulties in supplier and subcontractor selection, requiring a delicate balance between sharing essential details and safeguarding project confidentiality (Montecchi et al., 2021). Moreover, the fragmented nature of the construction supply chain can be intensified in STC, where various elements must converge seamlessly to maintain the covert nature of the project. Overcoming these challenges demands a meticulous approach to communication, coordination, and a heightened level of trust among stakeholders to ensure the secure and discreet progression of the project.

Furthermore, the unpredictable nature of STC projects introduces uncertainty and risk management challenges in CSCM. The discreet execution often means operating within unpredictable environments and timelines, as unexpected events can arise without warning. This uncertainty extends to material requirements, project timelines, and potential disruptions due to external factors. Addressing these challenges necessitates agile supply chain practices, proactive risk management strategies, and the flexibility to adapt to unforeseen circumstances (Arowosegbe et al., 2024). The unique challenges of uncertainty in STC further underscore the need for a customised CSCM approach to navigate these intricacies precisely.

In addition, the security of the supply chain becomes a paramount challenge in STC. The risk of unauthorised access or exposure increases significantly with heightened sensitivity to information confidentiality. Ensuring the security of the supply chain involves implementing robust cybersecurity measures, secure communication channels, and strict access controls. The challenge lies in safeguarding physical materials during transit and protecting the digital aspects of the supply chain, including project plans, designs, and communication channels (Stodt et al., 2024). Overcoming these security challenges requires a comprehensive strategy that combines technological solutions with rigorous procedural controls.

Finally, integrating advanced technologies introduces challenges in ensuring that these technologies align with the discreet nature of STC. Implementing BIM, real-time tracking systems, and encryption technologies requires careful consideration to prevent unintended exposure of project details. The challenge lies in adopting and customising technologies that enhance CSCM without compromising the security and confidentiality crucial to STC projects.

Technologies for Adaptive CSCM in STC

STC, often associated with advanced technologies and efficient management practices, can benefit significantly from integrating various technologies into the construction supply chain. Below are some technologies that could improve supply chain management in STC.

Blockchain Technology (Ullah et al., 2022): Blockchain ensures a secure and transparent ledger system for transactions and data sharing. STC can create a decentralised and tamper-proof record of every transaction, from procurement to payment. This enhances trust among stakeholders and minimises the risk of fraud.

IoT (Mishra et al., 2022): IoT devices, such as sensors and actuators, provide real-time data on the location, condition, and usage of construction materials and equipment. This data is invaluable for optimising inventory levels, preventing theft, and ensuring that resources are used efficiently. For example, IoT-enabled tracking devices can notify the supply chain manager if materials are delayed or equipment requires maintenance.

Radio-Frequency Identification (RFID) Technology (Bacchetta et al., 2021): RFID tags offer a more granular tracking level than traditional barcodes. RFID can be applied to construction materials and equipment for automated and accurate tracking in STC. This technology expedites inventory management and authentication processes, reducing the likelihood of errors associated with manual tracking methods.

Predictive Analytics (Olaniyi et al., 2023): Predictive analytics leverages historical and real-time data to forecast future trends and potential issues. This technology helps anticipate demand fluctuations, identify potential delays, and optimise resource allocation in STC supply chain management. By proactively addressing challenges, construction projects can maintain timelines and reduce the risk of disruptions.

Drones and Aerial Imaging (Powers et al., 2022): Drones with cameras and sensors offer a bird's-eye view of construction sites. This technology aids in monitoring the progress of construction, verifying the quality of work, and assessing the condition of materials. Aerial imaging can create accurate 3D models of construction sites, providing valuable insights for inventory management and project planning.

3D Printing (George & George, 2024): 3D printing in STC enables on-site fabrication of components, reducing the reliance on centralised manufacturing and transportation. This technology allows for customising construction elements, minimises waste, and accelerates construction timelines. It is particularly beneficial for projects that require unique components.

Augmented Reality (AR) and Virtual Reality (VR) (Einizinab et al., 2023): AR and VR technologies enhance visualisation and communication in STC projects. Design plans can be overlaid onto physical spaces, facilitating accurate on-site construction. VR simulations aid in training workers and testing construction sequences before implementation. This reduces errors, enhances collaboration, and improves overall project efficiency.

Autonomous Vehicles (Melenbrink et al., 2020): Autonomous vehicles like drones and robotic vehicles can autonomously transport materials within construction sites. This reduces the need for manual labour in material handling, improves safety, and ensures a more efficient flow of materials. These vehicles can be programmed to follow predefined routes and schedules, optimising the supply chain.

Machine Learning for Inventory Optimisation (Oyewole et al., 2024): Machine learning algorithms analyse historical data to predict future consumption patterns, supplier performance, and external factors affecting the supply chain. By continuously learning from new data, these algorithms optimise inventory levels, ensuring that materials are available when needed while minimising excess stock. This prevents both shortages and overstock situations.

Supply Chain Management Software (Wuni & Shen, 2023): Specialised CSCM software integrates various supply chain functions, providing a centralised platform for communication and coordination. This software streamlines procurement, logistics, and inventory management processes, offering real-time visibility into the supply chain. It facilitates data-driven decision-making, improves stakeholder collaboration, and enhances supply chain efficiency in STC projects.

Benefits of Tailoring CSCM for STC

Tailoring CSCM for STC presents many benefits that directly contribute to the success and seamless execution of discreet and strategic projects. One significant advantage lies in the optimised flow of materials and resources throughout the construction process (Golpîra, 2020). Customising the supply chain to align with the discreet nature of the construction will allow materials to be sourced, transported, and utilised with precision, minimising the risk of exposure or unauthorised access. This optimisation enhances operational efficiency, ensuring suitable materials arrive at the construction site discreetly and on schedule. The streamlined material flow contributes to the overall success of the STC project, allowing for a covert integration of the structure into its environment.

Efficient logistics planning is another critical benefit that can be derived from tailoring CSCM for STC, as stated by Dutta et al. (2020). The discreet execution of STC often hinges on avoiding unnecessary attention during material transportation. Tailoring logistics strategies involves coordinating deliveries precisely, utilising inconspicuous transportation modes, and operating during low-visibility periods. This strategic approach to logistics planning contributes to maintaining the covert nature of the project, minimising the likelihood of disruptions that could compromise its secrecy. The result is a supply chain that operates seamlessly, with materials arriving discreetly, enhancing the project's overall success.

Furthermore, Koctas-Cotur et al. (2024) mentioned that the strategic selection of suppliers and subcontractors, a critical component of tailored CSCM, fosters a collaborative environment conducive to the unique demands of STC. Suppliers and subcontractors who understand and align with the need for discretion become valuable partners in the construction process. This collaborative approach enhances communication, coordination, and adaptability to challenges inherent in discreet construction projects. By building solid relationships with suppliers and subcontractors committed to upholding the confidentiality of the project, CSCM ensures a cohesive and well-aligned supply chain, contributing significantly to the overall success of the STC project.

In addition, tailored CSCM in STC contributes to risk mitigation, as Luthra et al. (2022) pointed out that sustainable practices in the supply chain aid smooth operations. The unique challenges of secrecy, uncertainty, and security are strategically addressed, rendering the supply chain resilient to potential disruptions. Proactive risk management strategies, adaptive supply chain practices, and the integration of advanced technologies contribute to a robust and secure supply chain that can navigate the intricacies of STC. This risk mitigation not only safeguards the project's confidentiality but also ensures the successful completion of the building within the specified parameters. The tailored approach enhances the overall resilience of the supply chain, allowing it to adapt to unforeseen circumstances and continue operating effectively.

Moving further, CSCM for STC contributes to cost savings over the project lifecycle. While initially requiring strategic planning and potentially customised logistics solutions, the long-term benefits include optimised resource utilisation and minimised waste (Tatlici-Kupeli & Sertyesilisik, 2024). The efficient flow of materials and the strategic selection of suppliers contribute to reducing unnecessary costs associated with delays, disruptions, and inefficient processes. By minimising the ecological footprint and optimising operational efficiency, the tailored CSCM approach aligns with sustainability goals while also enhancing the economic viability of the STC project.

Ultimately, the tailored approach to CSCM fosters innovation and adaptability. Construction projects that demand discretion often require creative solutions and adaptive strategies (Le et al., 2023). By tailoring the supply chain to meet the unique challenges of STC, the project team is encouraged to think innovatively and adapt to the project's evolving needs. This innovative mindset extends to integrating advanced technologies, such as BIM and real-time tracking systems, which further enhance the efficiency and effectiveness of the tailored supply chain. The result is a construction project that meets the discreet requirements and pushes the boundaries of industry norms through innovative and adaptive CSCM practices.

Integrating CSCM for STC

When integrating supply chain management into the construction process to realise STC, various aspects of the construction project must be considered. These encompass the development of the building's cross-section, the management of visibility, the control of energy transmission, and the implementation of countermeasures. The subsequent sections provide detailed insights into each of these crucial components.

Building Cross-Section Development

Integrating CSCM is pivotal in shaping STC's discreet and strategic objectives. A vital aspect of this integration involves meticulously examining building cross-section development. Building cross-section development entails an in-depth analysis of the construction shape, internal components, material usage, and other critical factors (Golewski, 2023). Through the lens of CSCM, this process becomes more than just a design consideration; it evolves into a strategic alignment of supply chain dynamics to ensure the covert integration of the structure into its environment.

CSCM facilitates seamless coordination between architects, designers, and suppliers, ensuring that the envisioned cross-sectional characteristics align with the discreet objectives of STC. As stated by Zarei et al. (2023), the supply chain becomes a conduit for sourcing materials that not only meet structural requirements but also contribute to the overall covert nature of the project. Strategic collaboration within the supply chain allows for the procurement of specialised materials and components tailored to achieve specific cross-sectional attributes, enhancing both the structural integrity and the inconspicuous integration of the construction. Furthermore, Vignali et al. (2021) illustrated that integrating advanced technologies within CSCM, such as BIM, offers a digital platform for visualising and coordinating building cross-section development. This digital representation enables stakeholders across the supply chain to engage in the planning collaboratively and optimise the cross-sectional aspects. Through real-time collaboration, potential challenges or discrepancies in the building cross-section can be identified and addressed early in construction. This ensures a more seamless and efficient integration that aligns with the discreet objectives of STC.

The Shape of the Building

Integrating CSCM is instrumental in shaping the discreet and strategic aspects of building cross-section development in STC. One crucial element within this integration is the consideration of the shape of the building. The shape is not merely an architectural choice but a strategic decision significantly impacting the project's unobtrusive integration into its environment (Zolfaghari et al., 2024). CSCM plays a crucial role in aligning the supply chain with the vision of the construction team, ensuring that the selected shape contributes to the overall objectives of STC.

As a component of building cross-section development, the shape of the building influences the structure's visual impact on its surroundings. As Le and Nguyen (2022) pointed out, through CSCM, construction teams can collaborate with architects and designers to source materials and components that facilitate the realisation of the desired shape. This collaborative effort ensures that the supply chain is attuned to the specific requirements of the chosen shape, contributing to the overall success of the STC project. By strategically managing the supply chain, materials that meet structural needs and enhance the building's discreet profile can be efficiently sourced and incorporated.

Advanced technologies, integrated into CSCM practices, further enhance the achievement of building cross-section development through the shape of the building. Sepasgozar et al. (2020) expressed that BIM, for instance, allows stakeholders across the supply chain to visualise and optimise the building's shape digitally. This digital representation facilitates a more accurate and collaborative approach to building design, ensuring that the shape aligns seamlessly with the overall goals of STC. CSCM, in conjunction with digital tools, transforms the shape of the building from a conceptual choice into a strategically managed aspect that contributes to the discreet integration of the structure.

Achieving cross-section building development in STC through the shape of the building is linked to the strategic application of CSCM. The supply chain becomes a dynamic force in realising the envisioned shape, ensuring that materials are sourced and integrated efficiently. CSCM's collaborative nature, coupled with advanced technologies, transforms the shape from a design element into a strategic component that aligns with the discreet objectives of STC. When managed strategically through CSCM, the shape of the building becomes a critical element of the seamless integration of the structure into its surroundings.

Internal Construction

Within STC, incorporating CSCM is pivotal in achieving building cross-section development, particularly concerning internal construction. Bibri (2022) added that the internal construction aspects, encompassing the design and arrangement of structural elements within the building, contribute significantly to the discreet integration of the structure into its environment. CSCM facilitates a strategic approach to internal construction by aligning the supply chain with the project's vision, ensuring that the chosen internal layout enhances functionality and the construction's inconspicuous nature.

As Liu et al. (2020) illustrated, supply chain management enables a collaborative effort between architects, designers, and suppliers to optimise the internal construction elements. This collaboration ensures that materials and components selected within the supply chain are tailored to meet the specific requirements of the internal layout. CSCM facilitates the strategic sourcing of materials contributing to internal construction's structural integrity and functionality. The supply chain becomes a dynamic channel for acquiring specialised materials that meet the project's structural needs and enhance internal construction's discreet and strategic objectives.

In achieving building cross-section development in STC through internal construction, the supply chain becomes a conduit for sourcing materials and components that enhance functionality and discreet integration. Through collaborative efforts and digital tools, CSCM transforms internal construction from a design consideration into a strategically managed process, aligning the supply chain with the overall goals of STC. When managed strategically through CSCM, the internal construction becomes a critical element of the seamless integration of the structure into its surroundings.

Material Usage

In the prospect of STC, integrating CSCM is essential in shaping the building cross-section development, specifically through the strategic consideration of material usage. As Rahim et al. (2020) mentioned, material selection and utilisation are pivotal to achieving a discreet and strategically integrated structure. Also, it facilitates a comprehensive approach to material usage by aligning the supply chain with the project's vision, ensuring that the chosen materials contribute to structural requirements and the overall objectives of STC.

Furthermore, it allows for a collaborative effort between architects, designers, and suppliers to optimise material usage (Meier et al., 2023). This collaborative approach ensures that the supply chain is attuned to the specific requirements of the chosen materials, contributing to both the structural integrity and the discreet profile of the construction. The strategic selection and sourcing of materials within the supply chain become instrumental in achieving the desired building cross-section. Through efficient material management, construction teams can source specialised materials that enhance the structural aspects and contribute to the overall inconspicuous integration of the structure. Furthermore, building cross-section development in STC through material usage is linked to the strategic application of CSCM. The supply chain becomes a dynamic force in realising the envisioned material choices, ensuring materials are sourced and integrated efficiently (Sutherland et al., 2020). This collaborative nature, coupled with advanced technologies, transforms material usage from a construction element into a strategic component that aligns with the discreet objectives of STC. The strategic selection and efficient utilisation of materials, when managed through CSCM, become integral to the seamless integration of the structure into its surroundings.

Construction Visibility

In STC, achieving visibility objectives is crucial in ensuring that structures seamlessly integrate into their surroundings while maintaining a discreet and strategic profile. CSCM emerges as a critical enabler in this endeavour, transforming traditional construction practices into a dynamic and strategic process. Visibility in STC extends beyond mere aesthetic considerations, encompassing elements such as colour, texture, and overall visual impact. By leveraging CSCM, construction teams can strategically navigate the complexities of material selection, sourcing, and project coordination to achieve visibility objectives that align with the discreet goals of STC. When integrated into the CSCM framework, visibility considerations become integral to the construction supply chain dynamics. The collaborative nature of CSCM allows stakeholders across the supply chain, including architects, designers, and suppliers, to shape the visual aspects of the construction collectively.

Colour and Texture of the Construction

As discussed previously, CSCM in STC emerges as a strategic driver in achieving visibility objectives, mainly through carefully considering the colour and texture of the construction in Mendivil's (2022) study. The integration of CSCM transforms these aesthetic elements into crucial components of the overall project, contributing to the discreet and strategic integration of the structure into its environment.

The colour and texture of the construction are pivotal aspects that influence the structure's visual impact. Through CSCM, construction teams collaborate with designers and suppliers to strategically select materials and finishes that align with the discreet objectives of STC. The supply chain becomes a conduit for sourcing materials that meet functional requirements and contribute to the overall aesthetic, ensuring a harmonious blend with the surrounding environment (Grant et al., 2017; Milind & Arti, 2024).

As illustrated by Stodt et al. (2024), the collaborative nature of CSCM allows stakeholders across the supply chain to align on the desired colour and texture specifications. This cooperative effort ensures that the chosen elements are visually appealing and strategically contribute to the overall stealth profile of the construction. By integrating these considerations into the supply chain dynamics, CSCM ensures that the colour and texture become integral components of the construction process rather than mere design choices.

Advanced technologies, such as BIM, integrated into CSCM practices further enhance the achievement of visibility objectives through the colour and texture of the construction. The digital representation provided by BIM allows for real-time visualisation and coordination, enabling stakeholders to assess the visual impact before the actual construction phase. CSCM, in conjunction with digital tools, transforms the colour and texture considerations into a dynamic and strategic process that aligns with the discreet goals of STC.

Visibility in STC through the supply chain becomes a dynamic force in realising the envisioned colour and texture specifications, ensuring that materials and finishes are sourced and integrated efficiently. CSCM's collaborative nature, coupled with advanced technologies, transforms these aesthetic elements from design choices into strategic components that align with the discreet objectives of STC. The colour and texture considerations, when managed through CSCM, become integral to the seamless integration of the structure into its surroundings.

Location

In STC, CSCM plays a pivotal role in orchestrating the strategic achievement of visibility objectives, mainly through carefully considering the construction's location. The integration of CSCM transforms the selection and coordination of construction sites into a dynamic process that aligns with the discreet and strategic integration of structures into their surroundings (Ghosh et al., 2022). Furthermore, the location of a construction project is a critical factor in determining its visibility impact on the environment. Aarikka-Stenroos et al. (2022) illustrated that through CSCM, construction teams collaborate with urban planners, environmental experts, and suppliers to strategically choose locations that meet logistical and functional requirements and contribute to the overall inconspicuous profile of the structure. The supply chain becomes a conduit for sourcing locations that align with the discreet objectives of STC, considering factors such as existing infrastructure, natural surroundings, and potential visibility points.

The collaborative nature of CSCM facilitates real-time coordination among stakeholders, ensuring that the chosen location enhances the discreet nature of the construction (Lu et al., 2021). By integrating location considerations into the supply chain dynamics, CSCM ensures that visibility objectives become a managed and deliberate process. This strategic approach to location selection, guided by CSCM principles, contributes to the overall success of STC projects, allowing structures to blend into their surroundings while strategically minimising visibility points seamlessly.

Interaction with the Environment

CSCM is a strategic catalyst in shaping the achievement of visibility objectives, mainly through the subtle consideration of the construction's interaction with the environment. The integration of CSCM transforms the relationship between a construction project and its surroundings into a dynamic and strategic process that aligns with the discreet integration of structures.

As Umoh et al. (2024) stated, the interaction of a construction project with its environment extends beyond physical placement, encompassing elements such as landscaping, sustainability features, and overall environmental impact. Through CSCM, construction teams collaborate with environmental experts, architects, and suppliers to strategically shape this interaction. The supply chain becomes a conduit for sourcing materials and technologies that not only meet functional requirements but also contribute to the overall discreet profile of the structure in its interaction with the environment. Furthermore, the collaborative nature of CSCM allows stakeholders across the supply chain to optimise the construction project's interaction with its environment. By integrating these considerations into the supply chain dynamics, CSCM ensures that visibility objectives become integral to the construction process. This strategic approach, guided by CSCM principles, contributes to the overall success of STC projects, allowing structures to seamlessly blend into their environment while strategically managing their interaction to minimise visibility and enhance discreet integration.

Energy Transmission

In discussing STC, achieving energy transmission objectives is crucial beyond traditional construction considerations. CSCM emerges as a strategic anchor in orchestrating the efficient and discreet energy flow throughout construction projects. This multifaceted approach integrates renewable energy sources, climate action planning, and energy-efficient technologies into the broader construction supply chain, aiming to align the energy transmission goals with the discreet and strategic integration of structures into their environments, as Sahabuddin et al. (2023) expressed.

Soltani et al. (2021) mentioned that efficient energy transmission is an environmental consideration and a critical component in maintaining the covert nature of construction projects. Through CSCM, construction teams collaborate with energy experts, technology providers, and suppliers to strategically plan and implement energy transmission systems. The supply chain becomes a dynamic channel for sourcing sustainable materials, energy-efficient technologies, and components that contribute not only to the functional requirements of the construction but also to the overall energy profile aligned with the discreet goals of STC.

The collaborative nature of CSCM facilitates real-time coordination among stakeholders, ensuring that the energy transmission objectives are seamlessly integrated into the construction process. By integrating energy considerations into the supply chain dynamics, CSCM ensures that energy transmission becomes an integral and deliberate aspect of the construction. This strategic approach, guided by CSCM principles, contributes to the overall success of STC projects, allowing structures to seamlessly operate with minimal environmental impact while strategically managing their energy transmission to align with discreet integration goals.

Renewable Energy

In STC, CSCM is essential in orchestrating the strategic achievement of energy transmission objectives, mainly through integrating renewable energy sources. The dynamic and collaborative nature of CSCM transforms the procurement and utilisation of renewable energy technologies into a process that aligns seamlessly with the discreet and strategic integration of structures into their environments (Le & Nguyen, 2024). Additionally, renewable energy sources, such as solar and wind power, play a central role in the energy transmission goals of STC projects. Through CSCM, construction teams collaborate with renewable energy providers, technology experts, and suppliers to strategically plan and implement sustainable energy solutions. The supply chain becomes a conduit for sourcing renewable energy technologies and components that meet functional requirements and contribute to the overall energy profile aligned with the discreet goals of STC.

Climate Action Planning

Within STC, CSCM is crucial in orchestrating the strategic integration of renewable energy sources to achieve efficient energy transmission. Abdelfattah and El-Shamy (2024) illustrated that renewable energy, encompassing solar and wind power sources, is a cornerstone in pursuing sustainability and discreet energy management within construction projects. Through CSCM, construction teams collaborate with specialised suppliers and technology experts, ensuring the seamless integration of renewable energy solutions into the construction supply chain.

Furthermore, supply chain management transforms the procurement and utilisation of renewable energy technologies into a streamlined process that aligns with the discreet and strategic integration of structures. The supply chain becomes a dynamic channel for sourcing high-quality renewable energy components, such as solar panels, wind turbines, hydroelectric equipment distributors, geothermal system providers, biomass energy component retailers and so on, ensuring they meet functional requirements and contribute to STC's overall sustainability and energy efficiency goals (Hoang & Nguyen, 2021). This collaborative effort enables construction teams to strategically manage the energy transmission aspect strategically, fostering a seamless blend of renewable energy solutions within the construction process.

Energy Management

Energy management in STC involves optimising energy sources, consumption patterns, and technological solutions. CSCM becomes the strategic enabler, transforming the procurement and utilisation of energy-efficient technologies into a dynamic process that aligns with structures' discreet and strategic integration into their environments (Iqbal et al., 2022). It facilitates collaboration among construction teams, energy experts, and technology providers to strategically plan and implement energy management systems. The supply chain becomes the conduit for sourcing cutting-edge technologies, such as smart grids, advanced building automation, and energy-efficient components, contributing to functional requirements and the overall energy profile aligned with the discreet goals of STC. By strategically managing the supply chain dynamics, CSCM ensures that energy management becomes an integral and deliberate aspect of the construction project.

The collaborative nature of CSCM ensures real-time coordination among stakeholders, allowing for the seamless integration of energy management systems into the construction process. Through this approach, construction teams can optimise energy consumption patterns, enhance the efficiency of building systems, and strategically plan for energy storage solutions. Its role in energy management extends beyond the construction phase, ensuring ongoing efficiency throughout the structure's life cycle. This strategic approach, guided by CSCM principles, contributes to the overall success of STC projects, allowing structures to operate with minimal environmental impact while strategically managing their energy transmission to align with discreet integration goals.

Integrating energy management into STC through CSCM involves strategic collaboration, advanced technology adoption, and ongoing efficiency considerations. By leveraging CSCM, construction projects can optimise energy transmission, enhance sustainability, and operate with minimal visibility, contributing to the overall success of STC endeavours.

Countermeasures

In the complex field of STC, the seamless implementation of robust countermeasures is paramount to safeguard structures and ensure their discreet integration into their respective environments. CSCM is a pivotal framework that provides a strategic approach to achieving countermeasure objectives. Countermeasures, encompassing a spectrum of security and protective measures, are vital components designed to mitigate potential risks and fortify the covert nature of construction projects (Alqahtani & Kumar, 2024). Within this context, CSCM becomes instrumental in transforming the procurement and deployment of countermeasure technologies and materials into a dynamic process that aligns seamlessly with the discreet goals of STC.

CSCM facilitates collaborative efforts among diverse stakeholders to strategically plan and implement robust countermeasure systems, including construction teams, security experts, technology providers, and suppliers. Hassija et al. (2020) and Kandarkar and Ravi (2024) asserted that the supply chain emerges as a critical channel, enabling the sourcing of cutting-edge security technologies, surveillance systems, and materials. These elements meet the stringent security requirements and contribute to the overall countermeasure profile aligned with the discreet goals of STC. By strategically managing the supply chain dynamics, CSCM ensures that countermeasures become integral and deliberate aspects of the construction project, enhancing the overall security and resilience of the structure.

Integrating countermeasures into STC through CSCM is a multifaceted and collaborative process, underlining the importance of strategic planning, technology adoption, and ongoing security considerations. CSCM emerges as the keystone, facilitating the implementation of countermeasures that safeguard construction projects and enhance their discreet and strategic integration. This strategic approach, guided by the principles of CSCM, is integral to the success of STC endeavours, providing a comprehensive framework for ensuring the security and resilience of structures throughout their lifecycle.

Functional Construction System

CSCM emerges as a crucial framework in orchestrating the seamless integration of these systems to enhance the security and discreet nature of construction projects. Functional construction systems, encompassing security features, surveillance infrastructure, and access control mechanisms, play a pivotal role in mitigating risks and fortifying the covert profile of structures (Olaniyi et al., 2023). Within this context, CSCM becomes instrumental in transforming the procurement and integration of these systems into a dynamic and collaborative process aligned with the discreet goals of STC.

Shojaei et al. (2023) illustrated that management facilitates collaboration among construction teams, security experts, technology providers, and suppliers to strategically plan and implement functional construction systems. The supply chain becomes a dynamic channel for sourcing cutting-edge security technologies, surveillance systems, and access control components. These components meet stringent security requirements and contribute to the overall countermeasure profile aligned with the discreet goals of STC. By strategically managing the supply chain dynamics, CSCM ensures that functional construction systems become integral and deliberate components of the construction project, enhancing the overall security and resilience of the structure.

Pro-active Practices

In STC, implementing effective countermeasures demands a proactive approach that anticipates and mitigates potential risks before they materialise. As Shadkam and Irannezhad (2024) illustrated, CSCM is a pivotal framework in orchestrating the seamless integration of proactive practices, ensuring construction projects' discreet and strategic fortification. Proactive countermeasure practices encompass a range of strategies, including thorough risk assessments, continuous monitoring protocols, and pre-emptive security measures, all strategically woven into the fabric of the construction supply chain. Within this context, CSCM becomes instrumental in transforming the procurement and deployment of these proactive countermeasure practices into a dynamic and collaborative process aligned with the discreet goals of STC.

It further facilitates collaboration among construction teams, security experts, technology providers, and suppliers to plan and implement proactive countermeasure practices strategically. The supply chain becomes a dynamic channel for sourcing cutting-edge security technologies, advanced surveillance systems, and materials specifically designed for proactive security measures (Hassija et al., 2020). These components contribute to reactive responses and the overall proactive countermeasure profile aligned with the discreet goals of STC. Strategically managing the supply chain dynamics ensures that proactive practices become integral and deliberate aspects of the construction project, enhancing the overall security and resilience of the structure.

CSCM Towards STC

CSCM plays a transformative role in STC projects by strategically coordinating stakeholders and aligning supply chain practices with discreet project objectives. It encompasses strategic procurement and deployment of materials, technologies, and services across pre-construction, construction, and post-construction stages, ensuring seamless integration and security throughout the construction lifecycle, as expressed by Cherian and Arun (2021). Pre-construction involves meticulous planning for security and sustainability and sourcing materials that contribute to secrecy and integration. During construction, it orchestrates the flow of resources to minimise visibility and enhance security, fostering efficient construction processes through real-time coordination among stakeholders. In post-construction stages, it encompasses ongoing maintenance, sustainability practices, and seamless integration into surroundings, ensuring long-term resilience and sustainability of stealth structures. While CSCM offers manifold benefits, such as enhanced efficiency and sustainability, challenges like dependencies on advanced technologies and dynamic environmental factors require a subtle and adaptive approach.

Environmental Protection

STC relies heavily on robust environmental protection measures to maintain its covert nature and ensure long-term sustainability. CSCM emerges as a pivotal framework in orchestrating strategic environmental practices throughout the construction process. By fostering collaboration among construction teams, suppliers, and environmental experts, CSCM seamlessly integrates sustainable practices into the supply chain, from material sourcing to waste management and energy efficiency (Tatlici & Sertyesilisik, 2021). Strategic sourcing of eco-friendly materials is a crucial component, guiding the selection process based on criteria such as recyclability and reduced carbon footprint, thereby contributing significantly to the project's overall sustainability.

Moreover, CSCM optimises transportation logistics to minimise environmental impact. By strategically planning material transportation, including selecting nearby suppliers and employing fuel-efficient transportation modes, CSCM reduces carbon emissions and energy consumption. Its collaborative nature extends beyond material sourcing to waste reduction and recycling initiatives, ensuring responsible disposal and reuse of materials (Milind & Arti, 2024; Rahim et al., 2020). Additionally, it facilitates the adoption of energy-efficient technologies and sustainable building practices, positioning construction projects as contributors to broader sustainability goals. Ultimately, implementing environmental protection measures through CSCM ensures that sustainability becomes an integral part of the construction process, enabling projects to achieve their covert objectives while advancing environmental goals.

Construction Safety

Yap and Lee (2020) and Xu et al. (2023) illustrated that safety is a paramount concern in construction, and in the specialised domain of STC, where precision and secrecy are paramount, ensuring robust safety measures becomes even more imperative. CSCM emerges as a strategic framework to meticulously plan, execute, and monitor safety practices throughout construction, safeguarding personnel and the project's integrity. Strategically selecting suppliers and contractors with vital safety track records fosters a safety-conscious culture within the project and promotes stakeholder collaboration to uphold stringent safety measures.

Moreover, it contributes to safety by optimising material and equipment transportation logistics, minimising the potential for accidents during transit to and within the construction site. Real-time coordination within the supply chain enables swift responses to safety concerns, enhancing the overall safety profile of the project. Additionally, it facilitates procuring state-of-the-art safety equipment and technologies, ensuring compliance with industry standards, and keeping safety measures aligned with technological advancements. From advanced personal protective equipment (PPE) to sophisticated monitoring systems, safety remains at the forefront of construction practices through CSCM.

Furthermore, CSCM fosters a collaborative approach to safety training and awareness programs, ensuring that all personnel involved in the project are well-versed in safety protocols and emergency response procedures (Ahmed et al., 2020).

Integrating safety considerations into the design and planning phases of the project ensures that safety features are seamlessly incorporated into the initial project plans, enhancing overall security and well-being. Also, implementing safety measures in STC through CSCM involves a multifaceted, collaborative, and proactive process, integrating safety practices into every aspect of the construction supply chain to create a secure environment for personnel and the successful execution of projects.

Construction Speed

STC, where precision, speed, and discretion are paramount, achieving optimal construction speed emerges as a strategic imperative. CSCM, a transformative and indispensable framework, orchestrates the planning, coordination, and execution of construction processes with a singular focus on speed and efficiency (Shao et al., 2021). Within this framework, the collaborative nature of the supply chain becomes instrumental in streamlining construction speed while upholding the covert nature of the project.

As mentioned in their study, Ailus (2023) stated that CSCM optimises construction speed through strategic supplier and contractor selection. By empowering construction teams to choose entities that offer quality materials, operate efficiently, and adhere strictly to project timelines, the supply chain becomes instrumental in expediting materials sourcing, reducing delays, and streamlining the construction process. Additionally, it facilitates real-time coordination among stakeholders, leveraging advanced technologies to ensure seamless information flow and agile decision-making, thus minimising bottlenecks and enhancing overall construction speed.

Furthermore, CSCM optimises transportation logistics to accelerate construction speed (Le et al., 2020). Through strategically planning transportation routes and implementing just-in-time delivery strategies, materials and equipment are promptly delivered to the construction site, reducing travel time and on-site storage needs. Moreover, it enables procuring cutting-edge construction technologies that expedite the building process while maintaining the discreet nature of STC. Collaboration among stakeholders, proactive risk management, and a cohesive approach to construction activities further contribute to the efficient and timely completion of projects, making CSCM an integral aspect of enhancing construction speed in STC projects.

Economy

Gawusu et al. (2022) mentioned that through strategic integration within the supply chain, CSCM helps achieve financial objectives while maintaining the covert nature of the project. It empowers construction teams to strategically source materials and resources, identifying cost-effective suppliers and materials without compromising quality. Collaborations within the supply chain facilitate negotiations, bulk purchasing, and the exploration of alternatives, fostering cost savings and ensuring the project stays within budgetary constraints, thus establishing an economically sound foundation for STC.

Moreover, thinking towards real-time cost monitoring and control throughout construction, leveraging advanced technologies and data analytics, and changing construction supply can provide insights into cost fluctuations (Abideen et al., 2021; Geng et al., 2023). This proactive approach to cost management prevents budget overruns, ensuring economic efficiency without compromising the discreet nature of STC. The seamless integration of economic considerations into the construction process through CSCM establishes a robust framework for managing financial aspects efficiently, contributing to the project's overall success.

Furthermore, optimising transportation and logistics within the supply chain is another critical aspect of achieving economic efficiency through CSCM. Strategic planning of transportation routes and just-in-time delivery strategies coordinated through the supply chain reduce transit times and minimise associated costs (Yang et al., 2021). This logistical optimisation enhances cost-effectiveness and streamlines the construction process, aligning economic considerations with operational efficiency. CSCM also facilitates the procurement of cost-effective technologies and construction methodologies, ensuring that the project benefits from technological advancements while maintaining economic viability. Also, the collaboration among stakeholders, including cost-sharing initiatives and proactive risk management, further enhances economic efficiency, establishing a cooperative environment conducive to the success of STC projects.

Aesthetics

In the novel STC, where precision and discretion are paramount, achieving aesthetic excellence poses a unique challenge. The CSCM framework extends beyond logistical and economic considerations to orchestrate the aesthetic aspects of a project (Saccani et al., 2023). Through strategic integration within the supply chain, CSCM facilitates a harmonious blend of form and function while preserving the covert nature of the project. This ensures that aesthetic goals are seamlessly integrated into the construction process without compromising the project's discreet objectives.

CSCM facilitates real coordination and communication within the supply chain, ensuring that design intent is translated into construction reality. Leveraging advanced technologies and communication systems, CSCM fosters collaboration between designers, architects, and construction teams, allowing for adjustments and refinements that guarantee the aesthetic outcome reflects the envisioned appeal. Additionally, its application in transportation and logistics within the supply chain preserves the integrity of delicate or aesthetically sensitive materials during transit, enhancing aesthetic outcomes and safeguarding against potential disruptions that could compromise the project's visual appeal.

Practical Applications of CSCM in STC

The summarised applications of CSCM in STC are explained below.

Strategic Material Sourcing and Procurement

One practical application of CSCM in STC is the strategic sourcing and procurement of materials. CSCM allows construction teams to identify suppliers that align with the discreet nature of the project while ensuring the timely and cost-effective delivery of materials. By fostering collaborations within the supply chain, construction teams can negotiate favourable terms and bulk purchase materials and identify alternative suppliers, contributing to cost savings and efficiency. This strategic approach ensures that the materials used in construction meet quality standards and align with STC's aesthetic and functional requirements.

Real-time Collaboration and Communication

CSCM facilitates real-time collaboration and communication among stakeholders involved in STC projects. This application ensures that design intent, logistical considerations, and aesthetic requirements are seamlessly communicated and coordinated. Leveraging advanced technologies and communication systems, construction teams, architects, and suppliers can collaborate in real-time, making informed decisions promptly. This dynamic collaboration enhances the efficiency of the construction process, allowing for adjustments, refinements, and risk mitigation strategies, which are crucial in maintaining the discreet nature of STC.

Optimisation of Transportation and Logistics

The optimisation of transportation and logistics within the supply chain is another practical application of CSCM in STC. The framework enables the strategic planning of transportation routes to minimise transit times and reduce the risk of damage to sensitive materials during transit. Coordinated through the supply chain, just-in-time delivery strategies contribute to logistical optimisation. This not only enhances the efficiency of the construction process but also safeguards against potential disruptions that could compromise the discreet nature of the project. The careful planning and execution of transportation logistics through CSCM contribute to the overall success of STC projects.

Case Study Application

Case Study 1: Advanced material sourcing and procurement

Objective: Enhancing STC through strategic material sourcing and procurement.

In this case study, a large-scale government facility project can aim to incorporate innovative construction practices to achieve efficiency and discretion. The construction team can optimise the sourcing and procurement of materials by implementing advanced sensors and data analytics within the CSCM framework. After this, real-time tracking of inventory levels and demand forecasts can allow for proactive decision-making. Also, the system can identify suppliers that adhere to security protocols and offer cost-effective solutions without compromising quality. This application of smart construction within the CSCM framework not only streamlined the procurement process but also ensured that materials aligned with the discreet nature of the project.

Case Study 2: Real-time collaboration and communication hub

Objective: Enhancing communication and collaboration for seamless STC

In a high-profile corporate headquarters project, the construction team can implement a centralised communication and collaboration hub as part of their smart construction strategy. Also, leveraging BIM and IoT technologies, the hub can facilitate real-time collaboration between architects, construction teams, and suppliers. This can facilitate change in design plans, logistical considerations, and aesthetic requirements, allowing quick adjustments when communicated instantly. This smart construction approach will not only improve the overall efficiency of the construction process but also ensure that every stakeholder is aligned with the discreet goals of the project.

Case Study 3: Construction industry sustainable supply chain management

Objective: Optimising sustainable construction supply chain integration

A construction firm can aim to integrate sustainable practices into its supply chain to minimise environmental impact and enhance project sustainability. The project can involve collaborating with suppliers to source eco-friendly materials, adopt renewable energy solutions for construction operations, and implement waste reduction strategies. The firm can improve its reputation, attract environmentally conscious clients, and mitigate regulatory risks by promoting sustainable procurement practices and reducing carbon emissions through an integrated supply chain system.

Case study 4: Highway construction and maintenance project

Objective: Enhancing transportation connectivity and safety by building new highways or renovating existing road infrastructure

Supply chain management (SCM) is fundamental for managing construction materials such as asphalt, concrete, steel, and signage required for highway development. It involves coordinating transportation logistics, inventory management, and procurement processes to meet project schedules and specifications. Integration of SCM also includes managing subcontractors, implementing quality assurance protocols, and optimising maintenance operations to ensure the long-term durability and performance of highway infrastructure.

Managing the Supply Chain for STC During the Construction Phase

In STC, the construction stage emerges as a critical phase where precision and efficiency are paramount. CSCM takes centre stage during this phase, orchestrating the seamless flow of materials, technologies, and labour. The strategic procurement and timely delivery of materials become focal points. To begin with, it employs advanced technologies such as Radio Frequency Identification (RFID) tracking, global positioning system (GPS) systems, and data analytics to ensure a streamlined and efficient process. Furthermore, the discreet nature of STC necessitates carefully sourcing materials that align with security protocols, aesthetic requirements, and functional specifications, which can be navigated in the supply chain stages. In addition, its role in these stages extends beyond mere logistics; it becomes an enabler of operational excellence, ensuring that every element of the construction process adheres to the project's covert objectives.

Additionally, the justification for applying CSCM in the construction stage lies in its ability to address STC's unique demands and intricacies effectively. During construction, the emphasis on precision and efficiency is at its peak, making the strategic procurement of materials and logistics optimisation crucial. CSCM, with its focus on material sourcing, logistics streamlining, and real-time collaboration, seamlessly aligns with the dynamic requirements of the construction stage. Implementing advanced technologies within the CSCM framework ensures that construction teams can adhere to tight schedules and make informed decisions promptly.

In contrast, the pre-construction stage revolves around strategic planning, design considerations, and initial material selections. While CSCM plays a role in material planning during this phase, its full potential is realised during the construction stage, where the execution of meticulously laid plans demands a synchronised and efficient supply chain. Post construction stages, focusing on maintenance and sustainability, shift the emphasis away from real-time collaboration and logistics optimisation. Therefore, the construction stage stands out as the optimal arena for applying CSCM, where its capabilities significantly contribute to achieving the discreet goals of STC. The framework ensures a smooth and efficient building process, where every component operates harmoniously to bring the project to fruition while maintaining the necessary level of secrecy.

Conclusion

In the comprehensive exploration of CSCM within the context of STC, this discussion has highlighted the pivotal role that CSCM plays in orchestrating the dance of materials, logistics, and collaboration during the construction stage. Its practical applications in strategic material sourcing, real-time collaboration, and logistics optimisation have been demonstrated through case studies, showcasing how advanced technologies within this framework contribute to the discreet and efficient execution of construction projects. The justification for its emphasis on the construction stage, as opposed to pre or post-construction stages, lies in its unique ability to address the dynamic requirements of precision and efficiency while ensuring the covert nature of the project is maintained. In summary, by seamlessly integrating advanced technologies, CSCM becomes not only a logistical tool but an enabler of operational excellence, contributing significantly to the success of STC endeavours. As construction practices continue to evolve, their role stands out as a backbone in achieving the discreet goals of STC projects, ensuring a harmonious and efficient construction process while upholding the necessary level of secrecy.

References

Aarikka-Stenroos, L., Chiaroni, D., Kaipainen, J., & Urbinati, A. (2022). Companies' circular business models enabled by supply chain collaborations: An empirical-based framework, synthesis, and research agenda. *Industrial Marketing Management, 105*, 322–339. https://doi.org/10.1016/j.indmarman.2022.06.015

Abdelfattah, I., & El-Shamy, A. M. (2024). Review on the escalating imperative of zero liquid discharge (ZLD) technology for sustainable water management and environmental resilience. *Journal of Environmental Management, 351*, 1–24. https://doi.org/10.1016/j.jenvman.2023.119614

Abideen, A. Z., Pyeman, J., Sundram, V. P. K., Tseng, M. L., & Sorooshian, S. (2021). Leveraging capabilities of technology into a circular supply chain to build circular business models: A state-of-the-art systematic review. *Sustainability, 13*(16), 1–26. https://doi.org/10.3390/su13168997

Ahmed, M., Thaheem, M. J., & Maqsoom, A. (2020). Barriers and opportunities to greening the construction supply chain management: Cause-driven implementation strategies for developing countries. *Benchmarking: An International Journal, 27*(3), 1211–1237.

Ailus, J. (2023). *The effect of a successful supplier selection and purchasing strategy*. https://lutpub.lut.fi/handle/10024/165287

Alqahtani, H., & Kumar, G. (2024). Machine learning for enhancing transportation security: A comprehensive analysis of electric and flying vehicle systems. *Engineering Applications of Artificial Intelligence, 129*, 1–24. https://doi.org/10.1016/j.engappai.2023.107667

Arowosegbe, O. B., Olutimehin, D. O., Odunaiya, O. G., & Soyombo, O. T. (2024). Risk management in global supply chains: Addressing vulnerabilities in shipping and logistics. *International Journal of Management & Entrepreneurship Research, 6*(3), 910–922.

Bacchetta, A. V. B., Krümpel, V., & Cullen, E. (2021). Transparency with blockchain and physical tracking technologies: Enabling traceability in raw material supply chains. *Materials Proceedings, 5*(1), 1–13. https://doi.org/10.3390/materproc2021005001

Bibri, S. E. (2022). Eco-districts and data-driven smart eco-cities: Emerging approaches to strategic planning by design and spatial scaling and evaluation by technology. *Land Use Policy, 113*, 1–16. https://doi.org/10.1016/j.landusepol.2021.105830

Biersteker, E., Koppenjan, J., & van Marrewijk, A. (2021). Translating the invisible: Governing underground utilities in the Amsterdam airport Schiphol terminal project. *International Journal of Project Management, 39*(6), 581–593.

Blanchard, D. (2021). *Supply chain management best practices*. John Wiley & Sons.

Bontempi, E., Sorrentino, G. P., Zanoletti, A., Alessandri, I., Depero, L. E., & Caneschi, A. (2021). Sustainable materials and their contribution to the sustainable development goals (SDGs): A critical review based on an Italian example. *Molecules, 26*(5), 1–26. https://doi.org/10.3390/molecules26051407

Cherian, T. M., & Arun, C. J. S. (2021). Digital transformation in supply chain management: A conceptual framework for construction industry. *Indian Journal of Economics and Business, 20*(3), 1167–1187.

Cigolini, R., Gosling, J., Iyer, A., & Senicheva, O. (2022). Supply chain management in construction and engineer-to-order industries. *Production Planning & Control, 33*(9–10), 803–810.

Dutta, G., Kumar, R., Sindhwani, R., & Singh, R. K. (2020). Digital transformation priorities of India's discrete manufacturing SMEs – A conceptual study in perspective of Industry 4.0. *Competitiveness Review: An International Business Journal, 30*(3), 289–314.

Einizinab, S., Khoshelham, K., Winter, S., Christopher, P., Fang, Y., Windholz, E., Radanovic, M., & Hu, S. (2023). Enabling technologies for remote and virtual inspection of building work. *Automation in Construction, 156*, 1–20. https://doi.org/10.1016/j.autcon.2023.105096

Fatorachian, H., & Kazemi, H. (2021). Impact of Industry 4.0 on supply chain performance. *Production Planning & Control*, *32*(1), 63–81.

Gawusu, S., Zhang, X., Jamatutu, S. A., Ahmed, A., Amadu, A. A., & Djam Miensah, E. (2022). The dynamics of green supply chain management within the framework of renewable energy. *International Journal of Energy Research*, *46*(2), 684–711.

Geng, S., Luo, Q., Liu, K., Li, Y., Hou, Y., & Long, W. (2023). Research status and prospect of machine learning in construction 3D printing. *Case Studies in Construction Materials*, *18*, 1–19. https://doi.org/10.1016/j.cscm.2023.e01952

George, A. S., & George, A. H. (2024). Riding the wave: An exploration of emerging technologies reshaping modern industry. *Partners Universal International Innovation Journal*, *2*(1), 15–38.

Ghosh, A., Bhola, P., & Sivarajah, U. (2022). Emerging associates of the circular economy: Analysing interactions and trends by a mixed methods systematic review. *Sustainability*, *14*(16), 1–41. https://doi.org/10.3390/su14169998

Golewski, G. L. (2023). The phenomenon of cracking in cement concretes and reinforced concrete structures: The mechanism of cracks formation, causes of their initiation, types and places of occurrence, and methods of detection – A review. *Buildings*, *13*(3), 1–34. https://doi.org/10.3390/buildings13030765

Golpîra, H. (2020). Optimal integration of the facility location problem into the multi-project multi-supplier multi-resource construction supply chain network design under the vendor managed inventory strategy. *Expert Systems with Applications*, *139*, 1–12. https://doi.org/10.1016/j.eswa.2019.112841

Grant, D. B., Wong, C. Y., & Trautrims, A. (2017). *Sustainable logistics and supply chain management: Principles and practices for sustainable operations and management*. Kogan Page Publishers.

Harris, F., McCaffer, R., Baldwin, A., & Edum-Fotwe, F. (2021). *Modern construction management*. John Wiley & Sons.

Hassija, V., Chamola, V., Gupta, V., Jain, S., & Guizani, N. (2020). A survey on supply chain security: Application areas, security threats, and solution architectures. *IEEE Internet of Things Journal*, *8*(8), 6222–6246.

Hoang, A. T., & Nguyen, X. P. (2021). Integrating renewable sources into energy system for smart city as a sagacious strategy towards clean and sustainable process. *Journal of Cleaner Production*, *305*, 1–33. https://doi.org/10.1016/j.jclepro.2021.127161

Hossain, M. U., Ng, S. T., Antwi-Afari, P., & Amor, B. (2020). Circular economy and the construction industry: Existing trends, challenges, and prospective framework for sustainable construction. *Renewable and Sustainable Energy Reviews*, *130*, 1–15. https://doi.org/10.1016/j.rser.2020.109948

Hossain, K. N. (2023). *Supply Chain Collaboration as a facilitator of circular supply chain with logistics management in cement companies in Bangladesh*. https://www.theseus.fi/handle/10024/805498

Huq, F. A., & Stevenson, M. (2020). Implementing socially sustainable practices in challenging institutional contexts: Building theory from seven developing country supplier cases. *Journal of Business Ethics*, *161*(2), 415–442.

Hussein, M., Eltoukhy, A. E., Karam, A., Shaban, I. A., & Zayed, T. (2021). Modelling in off-site construction supply chain management: A review and future directions for sustainable modular integrated construction. *Journal of Cleaner Production*, *310*, 1–32. https://doi.org/10.1016/j.jclepro.2021.127503

Iqbal, M., Ma, J., Ahmad, N., Hussain, K., Waqas, M., & Liang, Y. (2022). Sustainable construction through energy management practices: An integrated hierarchal framework of drivers in the construction sector. *Environmental Science and Pollution Research*, *29*(60), 90108–90127.

Kandarkar, P.C. & Ravi, V. (2024). Investigating the impact of smart manufacturing and interconnected emerging technologies in building smarter supply chains. *Journal of*

Manufacturing Technology Management, *2*(1), 1–26. https://doi.org/10.1108/JMTM-11-2023-0498

Kedir, F., & Hall, D. M. (2021). Resource efficiency in industrialized housing construction – A systematic review of current performance and future opportunities. *Journal of Cleaner Production*, *286*, 1–15. https://doi.org/10.1016/j.jclepro.2020.125443

Kim, S. Y., & Nguyen, V. T. (2022). Supply chain management in construction: Critical study of barriers to implementation. *International Journal of Construction Management*, *22*(16), 3148–3157.

Koctas-Cotur, O., Ozen, Y. D. O., & Ozturkoglu, Y. (2024). A new perspective for construction supply chains in digital era. *International Journal of Sustainable Construction Engineering and Technology*, *15*(2), 136–152.

Lăzăroiu, G., Ionescu, L., Uţă, C., Hurloiu, I., Andronie, M., & Dijmărescu, I. (2020). Environmentally responsible behavior and sustainability policy adoption in green public procurement. *Sustainability*, *12*(5), 1–12. https://doi.org/10.3390/su12052110

Le, T. T., Behl, A., & Graham, G. (2023). The role of entrepreneurship in successfully achieving circular supply chain management. *Global Journal of Flexible Systems Management*, *24*(4), 537–561.

Le, P. L., Elmughrabi, W., Dao, T. M., & Chaabane, A. (2020). Present focuses and future directions of decision-making in construction supply chain management: A systematic review. *International Journal of Construction Management*, *20*(5), 490–509.

Le, P. L., Jarroudi, I., Dao, T. M., & Chaabane, A. (2021). Integrated construction supply chain: An optimal decision-making model with third-party logistics partnership. *Construction management and economics*, *39*(2), 133–155.

Le, P. L., & Nguyen, N. T. D. (2022). Prospect of lean practices towards construction supply chain management trends. *International Journal of Lean Six Sigma*, *13*(3), 557–593.

Le, P. L., & Nguyen, D. T. (2024). Exploring lean practices' importance in sustainable supply chain management trends: An empirical study in Canadian Construction Industry. *Engineering Management Journal*, *36*(1), 66–91.

Liu, Y., Dong, J., & Shen, L. (2020). A conceptual development framework for prefabricated construction supply chain management: An integrated overview. *Sustainability*, *12*(5), 1–29. https://doi.org/10.3390/su12051878

Lu, W., Li, X., Xue, F., Zhao, R., Wu, L., & Yeh, A. G. (2021). Exploring smart construction objects as blockchain oracles in construction supply chain management. *Automation in Construction*, *129*, 1–14. https://doi.org/10.1016/j.autcon.2021.103816

Luthra, S., Sharma, M., Kumar, A., Joshi, S., Collins, E., & Mangla, S. (2022). Overcoming barriers to cross-sector collaboration in circular supply chain management: A multi-method approach. *Transportation Research Part E: Logistics and Transportation Review*, *157*, 1–25. https://doi.org/10.1016/j.tre.2021.102582

Meier, O., Gruchmann, T., & Ivanov, D. (2023). Circular supply chain management with blockchain technology: A dynamic capabilities view. *Transportation Research Part E: Logistics and Transportation Review*, *176*, 1–14. https://doi.org/10.1016/j.tre.2023.103177

Melenbrink, N., Werfel, J., & Menges, A. (2020). On-site autonomous construction robots: Towards unsupervised building. *Automation in Construction*, *119*, 1–21. https://doi.org/10.1016/j.autcon.2020.103312

Mendivil, A. R. (2022). *Towards the circular economy – A qualitative study of the barriers and drivers of reusing materials in the construction industry* [Master's thesis, Høgskulen på Vestlandet].

Milind, S., & Arti, S. (2024). Importance of material selection to achieve sustainable construction. In N. Haddad, W. A. Hammad, K. Figueiredo (Eds.), *Materials selection for sustainability in the built environment* (pp. 43–70). Woodhead Publishing.

Mishra, M., Lourenço, P. B., & Ramana, G. V. (2022). Structural health monitoring of civil engineering structures by using the Internet of Things: A review. *Journal of Building Engineering*, *48*, 1–20. https://doi.org/10.1016/j.jobe.2021.103954

Modak, N. M., Sinha, S., Raj, A., Panda, S., Merigó, J. M., & de Sousa Jabbour, A. B. L. (2020). Corporate social responsibility and supply chain management: Framing and pushing forward the debate. *Journal of Cleaner Production*, *273*, 1–20. https://doi.org/10.1016/j.jclepro.2020.122981

Montecchi, M., Plangger, K., & West, D. C. (2021). Supply chain transparency: A bibliometric review and research agenda. *International Journal of Production Economics*, *238*, 1–15. https://doi.org/10.1016/j.ijpe.2021.108152

Munaro, M. R., Tavares, S. F., & Bragança, L. (2020). Towards circular and more sustainable buildings: A systematic literature review on the circular economy in the built environment. *Journal of Cleaner Production*, *260*, 1–25. https://doi.org/10.1016/j.jclepro.2020.121134

Olaniyi, O. O., Okunleye, O. J., Olabanji, S. O., & Asonze, C. U. (2023). IoT security in the era of ubiquitous computing: A multidisciplinary approach to addressing vulnerabilities and promoting resilience. *Asian Journal of Research in Computer Science*, *16*(4), 1–18.

Olaniyi, O., Shah, N. H., Abalaka, A., & Olaniyi, F. G. (2023). Harnessing predictive analytics for strategic foresight: A comprehensive review of techniques and applications in transforming raw data to actionable insights. *Asian Journal of Economics, Business, and Accounting*, *23*(22), 441–459.

Omer, M. A., & Noguchi, T. (2020). A conceptual framework for understanding the contribution of building materials in the achievement of Sustainable Development Goals (SDGs). *Sustainable Cities and Society*, *52*, 1–14. https://doi.org/10.1016/j.scs.2019.101869

Oyewole, A. T., Okoye, C. C., Ofodile, O. C., & Ejairu, E. (2024). Reviewing predictive analytics in supply chain management: Applications and benefits. *World Journal of Advanced Research and Reviews*, *21*(3), 568–574.

Pham, T., & Pham, H. (2021). Improving green performance of construction projects through supply chain integration: The role of environmental knowledge. *Sustainable Production and Consumption*, *26*, 933–942. https://doi.org/10.1016/j.spc.2021.01.004

Powers, K. E., Clore, D. M., Davidson, G. M., & Harris, R. C. (2022). A bird's-eye view: Novel use of drone images to quantify differences in altitudinal reflections in bird-window collision studies. *The American Midland Naturalist*, *187*(1), 51–61.

Rahim, A. A., Musa, S. N., Ramesh, S., & Lim, M. K. (2020). A systematic review on material selection methods. *Proceedings of the Institution of Mechanical Engineers, Part L: Journal of Materials: Design and Applications*, *234*(7), 1032–1059.

Saccani, N., Bressanelli, G., & Visintin, F. (2023). Circular supply chain orchestration to overcome circular economy challenges: An empirical investigation in the textile and fashion industries. *Sustainable Production and Consumption*, *35*, 469–482. https://doi.org/10.1016/j.spc.2022.11.020

Sahabuddin, M., Alam, M. S., Khokhar, M., Hossain, M. A., Alam, M. F., & Khan, W. (2023). Circular value creation: Business models and supply chain strategies. *Environmental Science and Pollution Research*, *30*(47), 103898–103909.

Sepasgozar, S. M., Hui, F. K. P., Shirowzhan, S., Foroozanfar, M., Yang, L., & Aye, L. (2020). Lean practices using Building Information Modeling (BIM) and digital twinning for sustainable construction. *Sustainability*, *13*(1), 1–22. https://doi.org/10.3390/su13010161

Shadkam, E., & Irannezhad, E. A. (2024). *Comprehensive review of simulation optimization methods in agricultural supply chains and transition towards an agent-based intelligent digital framework for Agriculture 4.0*. https://ssrn.com/abstract=4791793

Shao, X. F., Liu, W., Li, Y., Chaudhry, H. R., & Yue, X. G. (2021). Multistage implementation framework for smart supply chain management under industry 4.0. *Technological Forecasting and Social Change*, *162*, 1–11. https://doi.org/10.1016/j.techfore.2020.120354

Shojaei, R. S., Oti-Sarpong, K., & Burgess, G. (2023). Enablers for the adoption and use of BIM in main contractor companies in the UK. *Engineering, Construction and Architectural Management*, *30*(4), 1726–1745.

Shooshtarian, S., Maqsood, T., Caldera, S., & Ryley, T. (2022). Transformation towards a circular economy in the Australian construction and demolition waste management system. *Sustainable Production and Consumption*, *30*, 89–106. https://doi.org/10.1016/j.spc.2021.11.032

Soltani, M., Kashkooli, F. M., Souri, M., Rafiei, B., Jabarifar, M., Gharali, K., & Nathwani, J. S. (2021). Environmental, economic, and social impacts of geothermal energy systems. *Renewable and Sustainable Energy Reviews*, *140*, 1–25. https://doi.org/10.1016/j.rser.2021.110750

Stodt, F., Maisch, N., Ruf, P., Lechler, A., Riedel, O., & Reich, C. (2024, February 21). *Collaborative smart production supply chains with blockchain based digital product passports*. Preprints.org. https://doi.org/10.20944/preprints202402.1194.v1

Sutherland, J. W., Skerlos, S. J., Haapala, K. R., Cooper, D., Zhao, F., & Huang, A. (2020). Industrial sustainability: Reviewing the past and envisioning the future. *Journal of Manufacturing Science and Engineering*, *142*(11), 1–16.

Tatlici, G., & Sertyesilisik, B. (2021). Integrating performance measurement systems into the global lean and sustainable construction supply chain management: Enhancing sustainability performance of the construction industry. In G. Tatlici, B. Sertyesilisik (Eds.), *Research anthology on environmental and societal well-being considerations in buildings and architecture* (pp. 160–177). IGI Global.

Tatlici-Kupeli, G., & Sertyesilisik, B. (2024). Integrated thinking of the construction supply chain and project management: Performance domains and delivery principles. In J. Sarkis (Ed.), *The Palgrave handbook of supply chain management* (pp. 183–200). Springer International Publishing.

Ullah, Z., Raza, B., Shah, H., Khan, S., & Waheed, A. (2022). Towards blockchain-based secure storage and trusted data sharing scheme for IoT environment. *IEEE Access*, *10*, 36978–36994.

Umoh, A. A., Adefemi, A., Ibewe, K. I., Etukudoh, E. A., Ilojianya, V. I., & Nwokediegwu, Z. Q. S. (2024). Green architecture and energy efficiency: A review of innovative design and construction techniques. *Engineering Science & Technology Journal*, *5*(1), 185–200.

Vignali, V., Acerra, E. M., Lantieri, C., Di Vincenzo, F., Piacentini, G., & Pancaldi, S. (2021). Building information Modelling (BIM) application for an existing road infrastructure. *Automation in Construction*, *128*, 1–10. https://doi.org/10.1016/j.autcon.2021.103752

Wuni, I. Y., & Shen, G. Q. (2023). Exploring the critical success determinants for supply chain management in modular integrated construction projects. *Smart and Sustainable Built Environment*, *12*(2), 258–276.

Xu, J., Cheung, C., Manu, P., Ejohwomu, O., & Too, J. (2023). Implementing safety leading indicators in construction: Toward a proactive approach to safety management. *Safety Science*, *157*, 1–29. https://doi.org/10.1016/j.ssci.2022.105929

Yang, J., Xie, H., Yu, G., & Liu, M. (2021). Achieving a just–in–time supply chain: The role of supply chain intelligence. *International Journal of Production Economics*, *231*, 1–34. https://doi.org/10.1016/j.ijpe.2020.107878

Yap, J. B. H., & Lee, W. K. (2020). Analysing the underlying factors affecting safety performance in building construction. *Production Planning & Control*, *31*(13), 1061–1076.

Zarei, H., Rasti-Barzoki, M., Altmann, J., & Egger, B. (2023). Cooperation, coordination, or collaboration? A structured review of buyers' partnerships to support sustainable sourcing in supply chains. *Environmental Science and Pollution Research*, *30*(31), 76491–76514.

Zhou, X., Jiang, J., Zhou, C., Li, X., & Yin, M. (2024). Circular supply chain management: Antecedent effect of social capital and big data analysis capability and their impact on sustainable performance. *Sustainable Development Early View*, 1–21. https://onlinelibrary.wiley.com/doi/abs/10.1002/sd.2963

Zimon, D., Tyan, J., & Sroufe, R. (2020). Drivers of sustainable supply chain management: Practices to alignment with un sustainable development goals. *International Journal for Quality Research, 14*(1), 219–236.

Zolfaghari, S., Kristoffersson, A., Folke, M., Lindén, M., & Riboni, D. (2024). Unobtrusive cognitive assessment in smart homes: Leveraging visual encoding and synthetic movement traces data mining. *Sensors, 24*(5), 1–23. https://doi.org/10.3390/s24051381

Chapter Six

Lean Practices for Stealth Construction

Abstract

The chapter provided a comprehensive overview of lean construction as a transformative paradigm within the building industry. It delved into the core principles, tools, and techniques of lean construction, emphasising its advantages and the challenges associated with its implementation. Furthermore, it highlighted the pivotal role of lean construction principles in streamlining building excellence during the construction stage. The chapter also explored the concept of lean construction for stealth construction, presenting practical applications and a case study to illustrate its efficacy. Overall, it offered a synthesised understanding of lean construction's significance, potential, and challenges, concluding with a general summary of its implications for the building industry.

Keywords: Lean construction; building industry; streamlining excellence; stealth construction; sustainability; resilient construction

Introduction

According to Hamzeh et al. (2021), lean construction represents a profound transformation in how construction projects are conceived, managed, and executed. Emerging as an extension of the broader lean methodology that originated in manufacturing, lean construction is founded on a commitment to efficiency, value, and continuous improvement. Its fundamental principles challenge traditional construction practices by prioritising waste elimination, resource optimisation, and value creation for clients and stakeholders (Francis & Thomas, 2020). In this introduction, we explore the core tenets of lean construction, its evolution, and the impact it has had on the construction industry.

Stealth Construction: Integrating Practices for Resilience and Sustainability, 129–161

Published under exclusive licence by Emerald Publishing Limited
doi:10.1108/978-1-83608-182-120251006

Core Principles of Lean Construction

Oakland and Marosszeky (2022) and Lohne et al. (2022) stated that lean construction is underpinned by five core principles that guide its philosophy and methodology. To begin with, it is centred around value – understanding what clients truly value in a construction project and ensuring that those aspects are delivered precisely. Furthermore, lean construction vehemently pursues waste elimination, aiming to eradicate inefficiencies, redundancies, and non-value-adding activities from the construction process. In addition, it thrives on continuous improvement, cultivating a culture of innovation and enhancement. Respect for people, another core principle, underscores the importance of empowering and respecting all stakeholders, from workers on-site to clients. Finally, lean construction employs pull planning, which ensures work is only carried out when required, thus avoiding stockpiling of materials and minimising project lead times.

Value-centred Focus

According to Balaraman (2022), the first and most critical lean construction principle is an unwavering value-focused focus. Lean construction begins by defining value from the client's perspective, emphasising the importance of understanding what the client truly values in a construction project. This goes beyond the mere physical attributes of a building and extends to factors like functionality, quality, and how well the project aligns with the client's objectives and expectations. By clearly understanding value, lean construction ensures that all project efforts deliver precisely what the client values most efficiently.

Elimination of Waste

Rosli et al. (2023) asserted that the second principle revolves around the relentless pursuit of waste elimination. Lean construction identifies waste as anything that does not add value to the project. This encompasses non-value-adding activities, materials, processes, and resources that consume time, money, and resources without benefiting the project or the client. Common types of waste targeted in lean construction include overproduction, excess inventory, unnecessary movements, waiting times, overprocessing, underutilised human potential, and defects. By identifying and eliminating these wasteful elements, lean construction optimises project efficiency, reduces delays, and minimises costs.

Continuous Improvement

Continuous improvement is the third core principle of lean construction. It fosters a culture of ongoing enhancement by consistently reflecting on work processes, searching for better ways to accomplish tasks, and learning from experiences, as defined by Albalkhy and Sweis (2022). This principle embraces the Plan-Do-Check-Act (PDCA) cycle and encourages project teams to set aside time for regular reviews and improvements (Esfahani, 2021). The continuous improvement mindset seeks to identify opportunities for streamlining processes, reducing waste, and delivering higher value to the client. It recognises that there is always room for progress and that embracing change is vital for achieving excellence.

Respect for People

According to Solaimani and Sedighi (2020), respect for people is the fourth core principle of lean construction and emphasises the importance of valuing and empowering individuals involved in the construction process. Lean construction recognises that all project stakeholders' collective knowledge, experience, and creativity, from labourers to designers, are invaluable resources. This principle encourages open communication, collaboration, and shared decision-making among all team members. It fosters an atmosphere where everyone's contributions and insights are respected and leveraged for the project's benefit.

Pull Planning

The fifth and final core principle is pull planning, which seeks to coordinate work based on the actual demand and interdependencies within the project (Saad et al., 2021). Pull planning focuses on executing tasks only when needed and in the correct sequence, avoiding stockpiles of materials or overproduction of work. This approach optimises project schedules, reduces lead times, and minimises inefficiencies. Pull planning ensures that work is performed in response to a specific need rather than pushing work forward without regard for actual project requirements.

The five core principles of lean construction collectively constitute a transformative approach that seeks to optimise the construction process by eliminating waste, enhancing efficiency, delivering value, fostering a culture of continuous improvement, and respecting the contributions of all project stakeholders. These principles provide a strong foundation for lean construction's methodology, tools, and techniques, ultimately leading to more cost-effective, efficient, and client-focused construction projects.

Lean Construction Tools and Techniques

Lean construction employs a range of practical tools and techniques to facilitate the implementation of its core principles, ultimately leading to more efficient project management and the optimisation of construction processes, as summarised in Fig. 3. These tools and methods are indispensable in creating a lean construction environment that targets waste reduction, value maximisation, and enhanced productivity. They are expressed below.

Last Planner System (LPS)

According to Warid and Hamani (2023) and Govindasamy and Bekker (2024), the LPS is a pivotal tool in lean construction, representing a collaborative scheduling and project management system that empowers project teams to work cohesively. It facilitates the coordination of work, enables teams to establish realistic and achievable goals, and allows for the adaptation of plans as circumstances evolve.

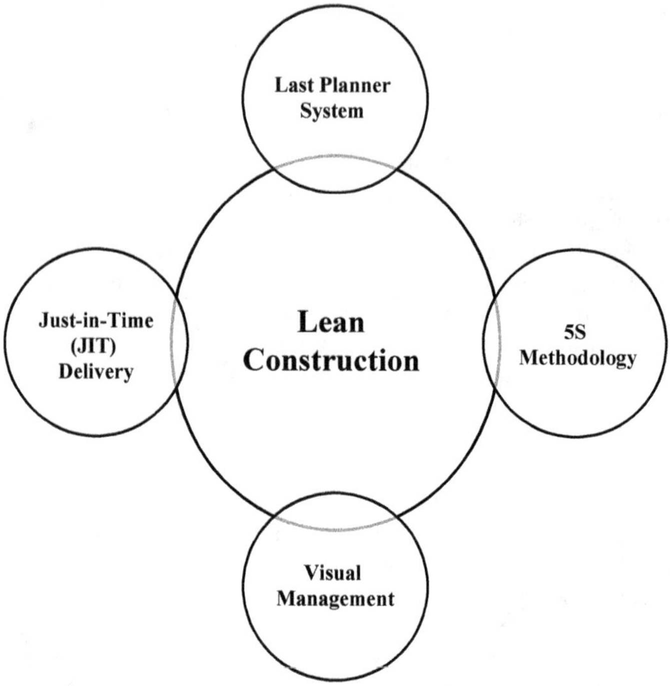

Fig. 3. Lean Construction Tools and Techniques.

A distinctive feature of the LPS is its emphasis on decentralising decision-making, granting workers on-site a say in how tasks are executed and when they are completed, as reiterated by Włodarkiewicz-Klimek (2021). This inclusivity and autonomy encourage a sense of ownership among the workforce, fostering a culture of collaboration and shared responsibility.

5S Methodology

The 5S methodology is another fundamental tool used in lean construction to optimise organisation, cleanliness, and safety on the construction site (Singh & Kumar, 2021). The five steps, Sort, Set in order, Shine, Standardise, and Sustain, provide a structured approach to creating an efficient and safe work environment. Noorzai (2023) added that by decluttering and organising work areas, designating specific places for tools and materials, maintaining cleanliness, standardising processes, and sustaining these practices, lean construction ensures that the workspace is conducive to productivity, reducing the risk of accidents, and promoting an organised and efficient workflow.

Visual Management

Singh and Kumar (2021) stated that visual management tools are vital in ensuring project teams have real-time insights into project progress and potential bottlenecks. Tools such as Kanban boards, which provide a visual representation of tasks, and Gantt charts, which offer a timeline view of project activities, allow teams to monitor work progress and identify any issues that may arise (Alaidaros et al., 2021). Visual management tools facilitate rapid problem-solving and decision-making by making it easy to identify areas that require attention or where delays might occur. This transparency enhances communication, aligns teams, and ensures project activities stay on track.

Just-in-Time (JIT) Delivery

In their study, Hussein and Zayed (2021) asserted that lean construction promotes the concept of JIT delivery, ensuring that materials and resources are supplied precisely when needed for a particular stage of the construction process. JIT delivery minimises the need for extensive inventories, reducing costs associated with storing surplus materials. By synchronising material delivery with the construction schedule, lean construction mitigates the risk of excess inventory and minimises waste (Le & Nguyen, 2022). It streamlines project logistics and contributes to cost-effectiveness and waste reduction, aligning with lean construction principles.

Advantages of Lean Construction

Lean construction, with its unwavering focus on waste reduction, resource optimisation, and value delivery, promises various benefits to the construction industry. These advantages span various facets, touching on project management, economics, collaboration, quality, sustainability, and client satisfaction, offering a multifaceted approach to enhancing the construction process and its outcomes.

Improved Efficiency

Lean construction aims to enhance project efficiency through waste elimination and process optimisation, as Nwaki and Eze (2020) stated. By identifying and eliminating non-value-adding activities, redundancies, and inefficiencies, lean construction ensures that construction projects run smoothly and encounter fewer disruptions. This efficiency directly impacts the timely completion of projects, reducing delays and cost overruns and leading to more predictable and reliable outcomes. The result is an industry that meets deadlines consistently and enhances its reputation for efficient project execution.

Cost Reduction

Furthermore, Gómez-Cabrera et al. (2020) illustrated that the cornerstone of lean construction is its commitment to cost reduction. By targeting waste and optimising resource allocation, lean construction delivers cost-effective projects. This cost-efficiency is a win-win, benefiting both construction companies and clients. For construction firms, lean practices improve profit margins, making projects more financially attractive. For clients, it means more efficient budget management, ensuring that every dollar spent is directed towards value-adding activities. In essence, lean construction aims to make every penny count and delivers projects that meet financial objectives while maintaining high standards of quality and efficiency.

Enhanced Collaboration

According to McHugh et al. (2022) and Bayhan et al. (2023), lean construction fosters a collaborative environment where all project stakeholders, from contractors to designers and clients, work together seamlessly. Open communication and shared goals reduce misunderstandings and conflicts, leading to smoother project execution. This enhanced collaboration is not merely a goal but a transformative approach, creating a sense of shared responsibility and mutual trust among team members. Projects progress with less friction when everyone works cohesively, and the shared sense of achievement enhances project outcomes. The result is more successful projects and more substantial unity and camaraderie among all parties involved.

Higher Quality

Lean construction's unwavering focus on value delivery and continuous improvement directly impacts project quality (Esfahani, 2021). Lean practices ensure that every aspect of a project aligns with client needs and expectations. This results in a higher-quality product that fulfils the client's vision to the highest standards. Quality is not sacrificed in the pursuit of efficiency; somewhat, it is enhanced by focusing on value. The outcome is not just a construction project completed on time and within budget but a project that exceeds expectations, leading to higher client satisfaction.

Shorter Project Timelines

Lean construction's ability to reduce waste and optimise workflows leads to shorter project timelines (Rosli et al., 2023). Projects can be completed swiftly with fewer bottlenecks, reduced downtime, and more efficient scheduling. This offers significant cost savings, as less time equates to lower costs, and ensures clients see returns on their investments sooner. It can be a pivotal advantage, particularly in industries where time-to-market or the rapid realisation of returns is crucial.

Sustainability

In an era where sustainability is increasingly vital, Mellado and Lou (2020) stated that lean construction practices align with the broader goals of environmental responsibility. Lean construction minimises waste and resource consumption, and its focus on efficiency and value aligns with the ethos of eco-friendliness. This results in more sustainable and environmentally responsible construction projects, making them attractive to clients who prioritise green practices and contribute to the overall sustainability efforts of the construction industry.

Increased Client Satisfaction

Lean construction aims to deliver projects that meet and exceed client expectations (Nwaki et al., 2021). By focusing on value delivery, quality, and efficient project execution, lean construction enhances client satisfaction. Satisfied clients are more likely to provide repeat business and positive referrals, leading to the long-term success of construction firms and projects. The emphasis on client satisfaction underscores the client-centric nature of lean construction, where the client's needs, desires, and vision are at the forefront of every project.

Challenges in Implementing Lean Construction

While promising significant benefits, adopting lean construction practices has its fair share of challenges in the construction industry. These challenges often stem from the industry's traditional practices, entrenched mindsets, and the complexity of construction projects. To successfully implement lean construction, stakeholders must proactively address these challenges.

Resistance to Change

Mano et al. (2020) state that one of the most formidable challenges in implementing lean construction is the resistance to change within the industry. The construction sector often adheres to traditional practices and resists adopting new methodologies. Construction professionals, including workers, contractors, and project managers, may be apprehensive about altering established routines and processes (Hatoum et al., 2021). This resistance can slow the adoption of lean construction practices, as individuals must buy into the new approach and be willing to embrace change. Overcoming this resistance requires effective change management strategies, clear communication, and practical demonstration of the benefits of lean construction.

Lack of Education and Training

Another challenge is the scarcity of education and training on lean construction principles and methodologies (Al-Balkhy et al., 2021). Many construction professionals may not be adequately informed or trained in lean practices, leading to a knowledge gap. Lean construction involves specific tools and techniques that necessitate training and skill development. Construction firms must invest in education and training programs to equip their workforce with the necessary knowledge and skills to implement lean practices effectively. This investment in education may initially require time and resources but is essential for the long-term success of lean construction adoption.

Project Complexity

Construction projects are inherently complex, often involving various stakeholders, intricate designs, and numerous variables (Xing et al., 2021). The dynamic nature of construction makes it challenging to implement lean practices consistently across all project phases. Adaptability and flexibility in lean construction can be complex in such an environment. Project complexities may include changing client requirements, unforeseen site conditions, and needing specialised materials or equipment. Navigating these complexities requires a robust understanding of lean construction principles and the ability to apply them to each unique project.

Integrating Technology

The construction industry has historically been slow to embrace technology, and integrating technology in lean construction can present challenges. Lean construction relies on digital tools for real-time project monitoring, data analysis, and communication among project teams (Evans et al., 2023). Adopting new software and technologies can be met with resistance, particularly in smaller construction firms. Overcoming this challenge requires investment in technology infrastructure, training for personnel to use these tools effectively, and a commitment to digital transformation.

Lean Practices for Stealth Construction

The concept of 'stealth construction', where the construction industry meets invincibility, calls for an innovative and efficient approach to building that aligns closely with lean construction principles. Stealth construction, often associated with ensuring environmental protection, safety, speed, economy, and aesthetic appeal, requires an approach that not only conceals the construction process but also optimises it for superior outcomes. Integrating lean construction methodologies into the framework of stealth construction will enhance the industry's ability to navigate the unique challenges and opportunities it presents.

According to Rosli et al. (2023) and Allouzi and Al Jaafreh (2023), lean construction's unwavering focus on waste elimination, resource optimisation, and value delivery can be promising for stealth construction. Whether it is concealing the environmental impact of construction, enhancing safety protocols to minimise visibility, or optimising the efficiency of the building process, lean principles are directly aligned with the goals of stealth construction. For example, lean construction's commitment to efficient resource allocation can minimise the environmental footprint by reducing waste. In contrast, its emphasis on lean supply chains can facilitate discreet and timely material deliveries. Moreover, lean's dedication to quality control and value delivery can ensure that the final product remains aesthetically pleasing and environmentally sound, fulfilling the vision of stealth construction.

In stealth construction, the need for a swift and covert process is of paramount importance. Lean construction, committed to reducing delays, enhancing workflow efficiency, and accelerating project timelines, is a natural fit for achieving these goals. By identifying and eliminating bottlenecks and non-value-adding activities, lean practices ensure that construction projects proceed smoothly and swiftly, reducing the time the construction process may be exposed to public view.

Lean construction's JIT delivery system, which ensures materials and resources are supplied precisely when needed, further contributes to a rapid and discreet construction process, aligning with the requirements of stealth construction (Hussein & Zayed, 2021).

Furthermore, stealth construction often involves stringent environmental protection requirements, including minimising the ecological impact and conserving resources. Lean construction principles, focusing on resource optimisation and waste reduction, align seamlessly with these goals. By efficiently allocating resources, minimising overproduction, and carefully managing the construction process, lean practices ensure that the environmental footprint of construction remains discreet and eco-friendly. Whether through reduced waste, efficient resource allocation, or sustainable construction practices, lean principles can help the construction industry meet the challenges of stealth construction while delivering efficient, economical, and aesthetically pleasing outcomes (Hamzeh et al., 2021; Oakland & Marosszeky, 2022; Bayhan et al., 2023).

Several construction parts must be considered when incorporating lean principles into achieving stealth construction. These include building cross-section development, visibility, energy emission, and countermeasures. These variables cover construction in shapes, aesthetics, energy, and security. These are expatiated below.

Building Cross-Section Development

Building cross-section development is fundamental to architectural and structural design, offering a comprehensive view of a structure's internal composition and dimensions, as Maksoud et al. (2024) stated. It involves the creation of detailed representations that showcase the structure's intricate layers, materials, and elements, both vertically and horizontally. Architects, engineers, and designers use these cross-sections to understand better how the building components interact, from the foundation to the roof. These cross-sections are crucial for evaluating structural integrity, ensuring compliance with building codes, and making informed decisions during the design and construction phases. Ultimately, building cross-section development provides a visual roadmap that realises safe, functional, and aesthetically pleasing structures. For this section, the shape and size of the building, internal construction, and material usage are used as building cross-section development determinants as the construction industry moves towards a stealth one.

Shape and Size of the Building

Stealth construction is a unique and specialised field within the construction industry, often driven by the need to maintain low visibility or minimise the environmental impact of a building project. Xu et al. (2023) illustrated that the shape and size of a structure play pivotal roles in achieving the goals of stealth construction. This comprehensive approach encompasses architectural, engineering, and construction considerations to create structures that blend seamlessly into their surroundings, whether they aim to harmonise with the environment, enhance safety, or maintain a discreet presence.

Internal Construction

Internal construction, although typically hidden from view, plays a crucial role in stealth construction projects, where the primary objective is to maintain low visibility or minimise the environmental impact of a building (Eapen & Finkenstadt, 2024). Stealth construction focuses on ensuring that the internal elements of a structure not only fulfil their intended functions but also align with the project's overarching goals of invisibility, safety, security, or environmental protection. From architectural design to material selection and interior layout, internal construction becomes a concealed foundation upon which the success of stealth construction projects is built.

Material Usage

Material selection in stealth construction projects is critical beyond mere structural considerations. These projects demand careful thought and strategic planning to ensure that materials serve functional needs and align with the overarching goals of maintaining low visibility, enhancing safety and security, minimising environmental impact, or harmonising with the architectural context (Xu et al., 2023). The choice of materials and their integration into the construction process plays a pivotal role in achieving the objectives of stealth construction.

Material usage in stealth construction is a dynamic, multifaceted process that balances form and function. Material selection is not just a matter of structural integrity but a strategic consideration aimed at achieving the goals of low visibility, safety, security, environmental responsibility, and architectural harmony. The integration of materials, guided by these objectives, exemplifies the intricate and innovative nature of stealth construction, where even the choice of materials plays a vital role in achieving buildings that are not only functional but also purposefully concealed, robust, and sustainable.

Visibility

In stealth construction, where the primary goal is to maintain low visibility or minimise the environmental impact of a building project, colour and texture must be considered. These aesthetic elements are not merely decorative considerations but integral components in achieving the overarching objectives of stealth construction (Rayworth, 2022). Whether blending with the natural surroundings, harmonising with the urban context, or enhancing the visual appeal while remaining discreet, the choice of colour and texture plays a pivotal role in the success of construction projects. Furthermore, visibility in stealth construction entails colour and texture, the location of the building, interaction with the environment, and using materials that reduce reflections or absorb light. Additionally, strategically placing vegetation and natural elements can further mask the structure, ensuring it remains inconspicuous. By carefully selecting and integrating these factors, architects and engineers can create buildings seamlessly blending into their surroundings, achieving the delicate balance between functionality, sustainability, and stealth.

Colour and Texture

The appearance of a building can dictate a lot about the project. From colour opted for due to the climatic temperate of the region to the texture (smooth, rough, sliding, etc.) designed with respect to the aesthetic or compliance with environmental design (mimicry), these are very essential in designing projects, especially in the growing concern for ecological protection. Colour and texture are integral elements in stealth construction, serving aesthetic and functional purposes. These elements go beyond traditional notions of design and decor to become strategic considerations supporting project objectives achievement. Whether it is blending with the environment, enhancing safety and security, harmonising with an urban context, or aligning with sustainability goals, the choice of colour and texture exemplifies the intricate and innovative nature of stealth construction, where even the aesthetics are purposefully designed to be discreet, elegant, and aligned with the project's goals.

Location of the Building

The location of a building is a fundamental consideration in stealth construction, where the primary goal is to maintain low visibility or minimise the environmental impact of a construction project. The choice of location is a strategic decision that goes beyond mere geography. It directly influences a project's success in achieving aesthetics, safety and security objectives, environmental protection, urban integration, or defence. The concept of location in stealth construction is dynamic and multifaceted, encompassing various aspects, including site selection, orientation, context, and access.

The location of a building in stealth construction is not a static choice but a dynamic, strategic, and multifaceted decision that impacts the project's ability to meet its objectives. It comprehensively evaluates site selection, orientation, context, and access. Whether the goal is environmental protection, safety and security, urban integration, or defence, the location serves as a foundational element upon which the success of stealth construction projects is built. This integration of geographical and contextual considerations underscores the intricate and innovative nature of stealth construction, where the building's position is as purposefully designed as any other element to ensure low visibility, safety, and effectiveness.

Interaction with the Environment

In the construction world, interaction with the environment holds special significance in stealth construction projects. These endeavours are characterised by their primary goal of maintaining low visibility or minimising the environmental impact of a building. Interacting with the environment, in this context, is a multifaceted process that involves careful consideration of the surrounding ecosystems, climate, natural features, and more. Balancing the objectives of aesthetics, safety and security, environmental protection, urban integration, or defence with the requirements of discreet presence is a dynamic and innovative challenge in stealth construction.

Interacting with the environment is a dynamic and multifaceted process that underscores stealth construction's intricate and innovative nature. It is a balance between fulfilling project objectives related to aesthetics, safety, security, environmental protection, urban integration, or defence while minimising ecological impact. Whether focusing on preserving ecosystems, enhancing security, blending with urban surroundings, or minimising radar detection, the interaction with the environment exemplifies stealth construction's comprehensive and strategic approach. This dynamic and evolving field continues to push the boundaries of how buildings can interact with their surroundings to achieve discreet and effective outcomes.

Energy Emission

Energy efficiency is a crucial aspect of stealth construction, where the primary objective is to maintain low visibility or minimise the environmental impact of a building project (Varjovi & Babaie, 2020). In this context, energy efficiency is multifaceted, enhancing the structure's functionality while aligning with the overarching goals of aesthetics, safety and security, environmental protection, urban integration, or defence. The pursuit of energy efficiency in stealth construction is a dynamic and innovative challenge encompassing various elements, from architectural design to systems integration. This is represented below in energy efficiency and communication.

Energy Efficiency

Energy efficiency is a cornerstone of sustainable building practices in stealth construction projects focusing on environmental protection (Liu & Ren, 2020). The architectural design is oriented to take advantage of natural resources, such as sunlight and wind, to reduce energy consumption. Sustainable materials and construction methods that enhance insulation and reduce heat gain are employed to minimise the building's carbon footprint. Integrating renewable energy sources, such as solar panels and wind turbines, supports the project's environmental goals. Overall, energy efficiency is a commitment to environmental responsibility, ensuring the building operates with minimal environmental impact. Energy efficiency is a dynamic element in stealth construction, serving functional and environmental purposes. It goes beyond mere power consumption to encompass a strategic approach aimed at achieving the goals of aesthetics, safety, security, environmental responsibility, and urban integration. The discreet pursuit of energy efficiency exemplifies the intricate and innovative nature of stealth construction, where even the way buildings consume energy is purposefully designed to be efficient, discreet, and aligned with the project's objectives.

Communication

Communication is fundamental to any construction project but takes on a unique and specialised role in stealth construction (Lensjø, 2024). The primary goal of these projects is to maintain low visibility or minimise the environmental impact of a building while achieving other objectives related to aesthetics, safety and security, environmental protection, urban integration, or defence (Lucchi & Buda, 2022).

Effective communication becomes the linchpin, ensuring all project stakeholders work harmoniously to achieve these multifaceted goals while maintaining discretion. Communication in stealth construction projects is a multifaceted process that goes beyond conveying information as it fosters collaboration. It ensures all stakeholders work together to achieve the project's complex goals. Whether the focus is on environmental protection, safety and security, urban integration, or defence, effective communication is the backbone that maintains the project's discretion and success. It exemplifies the intricate and innovative nature of stealth construction, where stakeholders' communication is purposefully designed to be efficient, discreet, and aligned with the project's multifaceted objectives.

Countermeasures

Countermeasures play a critical role in stealth construction, where the primary goal is to maintain low visibility or minimise the environmental impact of a building project while meeting other objectives related to aesthetics, safety and security, environmental protection, urban integration, or defence (Eapen & Finkenstadt, 2024). Countermeasures encompass various strategies and technologies designed to protect the project's objectives while maintaining discretion and security. These countermeasures can be categorised into multiple facets, each tailored to address specific challenges and goals in stealth construction. They are environmental protection, functional construction systems, tactics (methods of construction), and strategies and management (before and after construction).

Environmental Protection

In stealth construction projects emphasising environmental protection, countermeasures are essential to safeguard the site's ecological integrity. These measures may include erosion control strategies, water runoff management, and pollution prevention to mitigate the environmental impact during construction. Countermeasures also involve the careful selection of construction materials that are environmentally friendly and sustainable. Monitoring and inspections throughout construction ensure compliance with environmental regulations and the project's sustainability goals.

Countermeasures in stealth construction address challenges and safeguarding project objectives. Whether focusing on environmental protection, safety and security, urban integration, or defence, these countermeasures are pivotal in ensuring the project remains discreet and secure while meeting its multifaceted goals. Countermeasures exemplify the intricate and innovative nature of stealth construction, where the approach to protection and security is purposefully designed to maintain unobtrusiveness and effectiveness.

Functional Construction Systems

Functional construction systems are at the heart of any building project but take on a distinctive role in stealth construction. Functional construction systems become the backbone that ensures the discreet execution of these multifaceted, sustainable, and resilient project delivery. Functional construction systems are an integral and dynamic aspect of stealth construction. Whether the focus is on environmental protection, safety and security, urban integration, or defence, these systems are designed to meet the project's multifaceted goals. They exemplify the intricate and innovative nature of stealth construction, where every aspect of a building is purposefully designed to be discreet, efficient, and aligned with the project's specific objectives. Functional construction systems represent the essence of elegance with purpose, striking a delicate balance between achieving the building's functional needs and its multifaceted goals in stealth construction.

Tactics (Methods of Construction)

Tactics in stealth construction are a set of strategic methods and approaches designed to ensure that the primary objective of maintaining low visibility or minimising the environmental impact of a building is achieved while addressing other goals related to aesthetics, safety and security, environmental protection, urban integration, or defence. These tactics are multifaceted, spanning various construction stages and aspects of the project. Let us explore how tactics are tailored to the specific needs of stealth construction.

- ***Environmental protection***

  Sobkowiak (2023) and Cardou and Vellend (2023) stated that in stealth construction projects prioritising environmental protection, tactics are aimed at reducing the ecological impact of the construction process. Site selection tactics involve choosing areas with minimal environmental impact, avoiding sensitive habitats, and respecting natural features. Construction tactics may include erosion control measures, pollutant containment, and stormwater management to protect nearby ecosystems. Material selection tactics prioritise sustainable and eco-friendly materials, and construction processes may employ environmentally responsible practices like low-waste construction and minimal site disturbance.

- ***Safety and security***

  In safety and security-focused stealth construction, tactics are critical to safeguarding the project from potential threats while maintaining low visibility. Access control tactics include restricted access points, security screenings, and biometric authentication (Singh & Singh, 2023). Surveillance tactics involve discreetly integrating cameras, alarms, and motion sensors to monitor the site. The choice of construction materials and methods is strategic, focusing on blast-resistant materials, ballistic-resistant structures, and secure entry and exit points. Fire safety tactics may involve using advanced fire suppression systems that do not raise suspicion.

- ***Urban stealth and aesthetics***

  In projects that aim to blend into an urban context while maintaining an aesthetically pleasing presence, tactics revolve around harmonising with the surroundings. Design tactics include architectural styles that complement neighbouring buildings, colour schemes that blend with the local palette, and facades that mirror the area's architectural character (Hong et al., 2022). Construction tactics employ techniques that minimise disruption to the urban environment and ensure that the building's design aligns with the local context. Noise control tactics may involve using soundproofing materials and construction schedules that reduce noise pollution.

- ***Innovative technologies***

Advances in technology have introduced innovative tactics in stealth construction. These include using advanced surveillance and security technologies like thermal imaging, biometrics, and intrusion detection systems, as Prajapati et al. (2021) stated. As further stated by Zhuang et al. (2020), sustainable construction tactics may involve integrating intelligent building systems, which provide real-time monitoring and control of energy use, lighting, and security. Tactical use of transparent solar panels and photovoltaic materials allows for harnessing solar energy without compromising aesthetics or the project's objectives.

Strategies and Management (Before and After Construction)

Strategies and management in stealth construction are essential for ensuring that complex project objectives are met before and after construction. The primary goal in these projects is to maintain low visibility or minimise the environmental impact of a building while addressing other key objectives. Practical strategies and management encompass various activities that must be meticulously planned and executed. In exploring how these aspects are tailored to the specific needs of stealth construction, they are discussed below.

1. Before Construction

a. Project planning and design strategies

Strategic planning is essential to determine the project's goals and objectives in the initial stages of stealth construction (Oppong, 2020). Detailed architectural and engineering designs are crafted to align with these objectives. The choice of materials, construction methods, and technologies is carefully considered to ensure they serve the project's discreet and multifaceted goals. Design strategies include developing a detailed project brief, conducting feasibility studies, and creating a clear roadmap for the project.

b. Environmental protection management

According to Nugroho (2021), environmental management strategies are critical in projects prioritising environmental protection to safeguard local ecosystems. Environmental impact assessments are conducted to identify potential risks and develop mitigation strategies. Management activities include erosion control measures, wetland protection, and establishing protective barriers to prevent pollution during construction.

c. Safety and security management

In safety and security-focused stealth construction, safety and security management strategies are vital to protect the project from potential threats. Security risk assessments are carried out to identify vulnerabilities and security management plans are developed to mitigate risks (Ganin et al., 2020). Access control strategies are put in place to regulate personnel and visitor entry. Security protocols include visitor screening, background checks, and biometric authentication to ensure that only authorised personnel can access the site.

d. Urban stealth and aesthetics management

In projects focused on urban stealth and aesthetics, management strategies revolve around creating a building that seamlessly integrates with its surroundings. Malekpour et al. (2021) illustrated that urban planners and city officials collaborate to develop management plans that adhere to local regulations and design guidelines. These plans include zoning and permitting activities and coordination with neighbouring property owners. Construction activities are managed to minimise disruption to the urban environment and ensure that the building design aligns with the local architectural character.

e. Military and defence management

Management strategies focus on maintaining security and radar invisibility in military and defence applications of stealth construction. Risk management strategies are utilised to identify potential threats and develop mitigation plans (Bakhtiari et al., 2024). Communication and coordination with defence agencies ensure the project aligns with defence standards. Management activities include strategically using radar-absorbing materials and technologies to achieve radar invisibility.

f. Technology and innovation management

Stealth construction often incorporates innovative technologies, and their management is integral to project success. Technology management includes selecting and integrating intelligent building systems for real-time monitoring and control of energy use, security systems, and building performance, as stated by Eini et al. (2021). Innovation management involves identifying cutting-edge materials and construction techniques that align with the project's goals.

2. After Construction

a. Environmental monitoring and maintenance

After construction, environmental monitoring is vital to minimise the building's ecological impact. Management activities include routine inspections, landscaping maintenance, and water runoff management (Mali et al., 2020). Ongoing monitoring ensures that the building continues to align with sustainability goals.

b. Safety and security post-construction

Post-construction management involves regular security assessments, alarm system maintenance, and personnel training in safety and security-focused projects. Security management plans are updated to address evolving threats and maintain the project's security objectives.

c. Urban stealth and aesthetics post-construction

Projects focused on urban stealth and aesthetics require post-construction management to preserve the building's integration with the urban context. Management activities involve architectural maintenance, facade cleaning, and adherence to local design guidelines. Noise and light pollution are addressed as part of post-construction management.

d. Military and defence post-construction

In military and defence applications, post-construction management involves radar cross-section testing and maintenance of radar-absorbing materials. Security systems and defence technology are continuously monitored and updated to ensure the building meets defence standards.

e. Technology and innovation post-construction

Post-construction management of technology and innovation in stealth construction projects focuses on sustaining excellence and ensuring the building meets its performance and efficiency goals (Salliou et al., 2023; Khoza & Haupt, 2021). This phase involves regular maintenance and inspections of hardware and software components, with upgrades and updates implemented to adapt to technological advancements. Furthermore, energy efficiency optimisation is a priority, with systems fine-tuned to reduce operational costs and environmental impact. Security and surveillance systems are continually evaluated and upgraded to counter evolving threats, and innovative building materials, such as those with radar-absorbing properties, are monitored for their effectiveness. Also, ongoing training and knowledge transfer are emphasised to ensure that building personnel can operate and maintain advanced systems effectively.

This meticulous post-construction management approach underscores the commitment to excellence and long-term sustainability in stealth construction projects. It ensures that the building remains at the forefront of technological advancements while adhering to its multifaceted goals, which may encompass environmental protection, safety and security, urban integration, or defence, depending on the project's specific objectives.

Lean Construction Towards Stealth Construction

This is designed towards environmental protection, Construction Safety, Construction Speed (duration), Construction Economy, and Aesthetic, as discussed below.

Environmental Protection

Lean construction principles, primarily associated with efficiency, waste reduction, and value enhancement, can drive environmental protection efforts within stealth construction. These principles can propel projects towards more sustainable, eco-friendly, and environmentally responsible outcomes when strategically applied. This section explores how lean construction can catalyse environmental protection in stealth construction.

Waste minimisation and material efficiency (Rosli et al., 2023): Lean construction strongly emphasises waste minimisation, which aligns seamlessly with environmental protection goals. In stealth construction, the optimisation of materials and resources is crucial. Lean principles encourage the efficient use of materials, reducing waste and minimising the project's environmental footprint. For instance, value stream mapping techniques can identify areas where material waste can be reduced without compromising quality or security. This approach reduces costs and ensures that resources are used responsibly and sustainably.

Streamlined supply chain (Le & Nguyen, 2022): A lean supply chain management approach can significantly reduce the environmental impact of stealth construction projects. Efficient supply chain practices mean that materials are procured and delivered just in time, reducing the need for extensive on-site storage. This minimises the risk of exposure to specialised materials and lowers the environmental impact associated with storage and transportation. By optimising the supply chain, lean construction helps mitigate the carbon footprint and energy consumption associated with material procurement.

Eco-friendly construction methods (Cardou & Vellend, 2023): Lean construction principles encourage using environmentally responsible construction methods. For instance, selecting sustainable building materials, energy-efficient construction techniques, and environmentally friendly construction processes can align with environmental protection objectives. Incorporating renewable energy sources, such as solar panels or wind turbines, can be part of a lean construction approach. Energy-efficient building systems and innovative technologies can be integrated to minimise resource consumption and reduce the ecological impact of the building.

Smart building systems and monitoring (Kumar et al., 2021): Lean construction promotes using smart building systems for real-time monitoring and control of energy use, lighting, and security. These systems not only enhance the efficiency of the building but also contribute to environmental protection. By optimising energy consumption and managing resources more effectively, smart building systems reduce the building's ecological impact. Lean construction ensures these systems are incorporated into the project and align with sustainability goals.

Urban stealth and aesthetics (Sidorova, 2022; Xu et al., 2023): In stealth construction projects in urban environments, lean principles drive architectural and design choices that seamlessly integrate with the local context. This integration is not only aesthetic but also environmentally responsible. It involves designing buildings that harmonise with their surroundings and adhere to local architectural character, reducing the visual and environmental impact of the project on the urban landscape.

Construction Safety

Safety is paramount in stealth construction projects, often involving sensitive materials, intricate security systems, and discreet operations. Lean construction, focusing on efficiency and waste reduction, can enhance safety measures and protocols within stealth construction. By adopting lean principles, projects can proactively address safety concerns, reduce risks, and create a secure environment that aligns with the project's objectives.

Efficient operations for reduced safety risks (Khoza & Haupt, 2021; Staves et al., 2024): Lean construction eliminates waste and optimises operations. By streamlining processes and minimising unnecessary activities, safety risks are inherently reduced. For example, lean principles encourage value stream mapping to identify non-value-adding activities. In stealth construction, this can lead to the elimination of redundant security protocols that may increase the likelihood of confusion or accidents. By reducing such redundancies, the safety of the construction process is improved.

JIT construction for safety (Hussein & Zayed, 2021): JIT construction, a core component of lean principles, is vital for maintaining safety in stealth construction. JIT scheduling ensures that construction tasks are executed precisely when needed, reducing the risk of accidents or exposure to sensitive information. For instance, JIT scheduling can be applied to the installation of security systems, ensuring that they are only put in place when required. This approach minimises the time-sensitive systems' activity, lowering the potential for security breaches.

Proactive risk identification and mitigation (Ganin et al., 2020): Lean construction encourages a proactive risk identification and mitigation approach. Risk management is a critical component of safety in stealth construction projects. By systematically assessing potential risks and vulnerabilities, project teams can develop strategies to mitigate these risks. For instance, in projects with stringent security requirements, lean principles facilitate the early identification of security vulnerabilities and the development of protocols to address them. This proactive approach ensures that safety remains a top priority throughout the project's lifecycle.

Ongoing training and knowledge transfer (Chapman & Schott, 2020): Safety in stealth construction is closely linked to the competence and preparedness of personnel. Lean construction encourages ongoing training and knowledge transfer, ensuring that all project stakeholders, from construction workers to security personnel, are well-prepared and informed. Training in security protocols, emergency response, and operating advanced security systems is crucial to maintaining a secure and safe environment.

Efficient security and surveillance systems (Kapoor, 2023): Lean principles promote the efficient integration of security and surveillance systems into construction. These systems are crucial for safety and security in stealth construction. Lean construction ensures these systems are seamlessly incorporated, regularly inspected, and efficiently maintained. These systems' real-time monitoring and control are integral to maintaining a safe and secure environment.

Construction Speed (Duration)

Stealth construction projects often operate on tight timelines, requiring precision and discretion. Lean construction principles, focusing on efficiency, waste reduction, and streamlined processes, can significantly contribute to accelerating the construction speed in these specialised projects. Here, the section explores how lean construction can propel construction speed in the context of stealth construction.

Eliminating non-value-adding activities (Kumar et al., 2023): Lean construction emphasises eliminating non-value-adding activities. This principle is particularly relevant in stealth construction, where every action must contribute to the project's objectives. Construction teams can reduce delays and bottlenecks by identifying and removing redundant tasks and processes. This lean practice ensures that construction operations are concentrated on essential activities, expediting project timelines.

JIT construction scheduling (Abbasi et al., 2020): JIT construction is a core component of lean principles. It involves scheduling construction tasks to align precisely with project timelines, reducing idle time and downtime. In stealth construction, where discretion and precision are essential, JIT ensures that construction activities are coordinated efficiently. For example, it can be applied to installing advanced security systems, ensuring they are deployed precisely when needed and minimising the time when sensitive information or materials are exposed.

Efficient supply chain management (Le & Nguyen, 2022): Lean principles advocate for a streamlined and efficient supply chain. In stealth construction, where procurement and delivery of specialised materials are often critical, an optimised supply chain ensures that materials are delivered just in time. This minimises the need for extensive on-site storage and reduces the risk of exposure. It also leads to a reduction in project lead times, accelerating construction speed.

Value stream mapping for process optimisation (Zahraee et al., 2020): Lean construction encourages value stream mapping to analyse and optimise construction processes. In stealth construction, this practice can help identify areas where process efficiency can be improved without compromising security or discretion. By eliminating unnecessary steps and improving workflows, projects can proceed more rapidly, meeting tight deadlines while maintaining low visibility.

Efficient integration of technology and innovation (Li et al., 2022): Lean construction promotes the efficient integration of technology and innovation, which is particularly relevant in stealth construction. Advanced construction technologies and innovative building materials can accelerate construction speed, ensuring security and precision. Technology for real-time monitoring and control of various building functions, such as lighting, heating, ventilation, and air conditioning (HVAC), and security, can enhance efficiency and reduce delays.

Risk mitigation and proactive problem-solving (Meng, 2020): Lean construction encourages a proactive risk identification and mitigation approach. By addressing potential challenges early in the project, teams can minimise disruptions and delays. In stealth construction, where security breaches or exposure of sensitive materials can result in significant setbacks, lean principles ensure that risks are identified and mitigated promptly, ensuring smoother project execution.

Construction Economy

Economic considerations are fundamental in stealth construction, as these projects often involve intricate security measures, specialised materials, and tight budgets. Lean construction's efficiency and waste reduction principles can contribute to achieving a stealth construction economy. This explores how lean construction can drive cost-efficiency without compromising the project's multifaceted objectives.

Waste reduction for cost savings (Hannan et al., 2020): One of the primary tenets of lean construction is eliminating waste. In stealth construction, where precision and discretion are paramount, minimising waste is essential for achieving a construction economy. Lean principles encourage identifying and eliminating non-value-adding activities, reducing redundancies and inefficiencies. This, in turn, leads to cost savings by streamlining construction processes and minimising material waste, thereby enhancing the sustainable development goals (SDGs) vision of 2030.

Efficient supply chain management (Aamer et al., 2020): An optimised supply chain is integral to the construction economy in stealth projects. Lean construction practices enable JIT procurement and delivery of materials. This reduces the risk of sensitive materials being exposed or compromised and minimises the costs associated with on-site storage. A well-managed supply chain ensures that materials are available precisely when needed, contributing to cost efficiency.

Value stream mapping for process efficiency (Jamil et al., 2020): Value stream mapping, a lean construction technique, focuses on identifying and eliminating non-value-adding activities. In stealth construction, this practice can help optimise construction processes. By reducing unnecessary steps and improving workflows, projects can save costs while maintaining security and discretion. This streamlining of processes ensures that resources are used more efficiently, reducing project costs.

Proactive risk mitigation for budget control (Ahmadi-Javid et al., 2020): Lean construction promotes a proactive risk identification and mitigation approach. This is vital in stealth construction, where unforeseen issues can result in costly setbacks. By addressing potential challenges early in the project, lean principles ensure that risks are identified and mitigated promptly. Proactive problem-solving contributes to budget control and cost savings by preventing costly disruptions and delays.

Efficient integration of technology and innovation (You & Feng, 2020): Lean construction encourages the efficient integration of technology and innovation. In stealth construction, advanced technologies and innovative building materials can enhance cost-efficiency without compromising the project's objectives. The use of technology for real-time monitoring and control of building functions, such as energy use, lighting, and security, can lead to operational cost savings.

Eco-friendly construction methods (Jose & Sia, 2022): Lean principles also align with environmentally responsible practices, which can contribute to long-term cost savings. Sustainability measures, such as using eco-friendly building materials and energy-efficient construction techniques, can reduce operational costs and resource consumption, enhancing the construction economy while aligning with broader sustainability goals.

Aesthetic

Although often overshadowed by security and discretion in stealth construction, aesthetics ensure that projects blend seamlessly with their surroundings and maintain a low profile. Lean construction can enhance the aesthetic aspects of stealth construction without compromising the project's objectives.

Urban integration and contextual aesthetics (Nielsen et al., 2021): In stealth construction projects in urban environments, lean principles encourage architectural and design choices that seamlessly integrate with the local context. Lean construction emphasises the importance of adapting the project to its surroundings, ensuring it adheres to local architectural character, scales, and aesthetics. This approach helps projects blend in and avoid drawing unnecessary attention.

Efficient use of materials for aesthetic appeal (Umoh et al., 2024): Lean principles promote the efficient use of materials, which can also contribute to the project's aesthetics. In stealth construction, specialised building materials and finishes are often used to meet security requirements. Lean construction ensures these materials are used effectively, reducing waste and maximising their aesthetic potential. This results in projects that meet security standards and maintain an aesthetically pleasing appearance.

Value stream mapping for streamlined aesthetics (Azzat et al., 2024): Value stream mapping, a lean construction technique, focuses on identifying and eliminating non-value-adding activities. In aesthetics, this practice can help streamline the design and construction process. By reducing unnecessary design elements and construction steps, and projects can achieve a cleaner and more cohesive aesthetic, ensuring that security features are seamlessly integrated into the design.

Smart building systems for aesthetic control (Cannavale et al., 2020): Lean construction encourages using intelligent systems for real-time monitoring and control of various building functions, including lighting, HVAC, and security. These systems can be leveraged not only for efficiency but also for aesthetic control. Lighting can be managed to create the desired aesthetic ambience, ensuring that the project's visual appeal aligns with its security objectives.

Advanced technology for hidden aesthetics (Lehtinen, 2020): In some stealth construction projects, aesthetics are hidden beneath security features. Lean principles promote the efficient integration of advanced technology and innovation. This can include integrating concealed surveillance systems, hidden doors, or architectural features that maintain aesthetic appeal while serving security functions.

Seamless project integration (Onungwa et al., 2021): Lean construction encourages streamlined processes, efficient scheduling, and close collaboration among project stakeholders. This cohesion ensures that aesthetics is seamlessly integrated into the project from design to construction. By eliminating communication gaps and promoting collaborative decision-making, lean construction ensures that the aesthetic aspects align with security measures and other project requirements.

Practical Application of Lean Construction Practices Towards Stealth Construction (Case Study)

From the sections above, this chapter explains the integration of lean construction practices for improved and resilient construction. Using variables such as building cross-section development, countermeasures, strategies and management (before and after construction), and tactics (construction methods), the chapter was designed to help achieve environmental protection, construction safety, construction speed, economy, and construction aesthetics in stealth construction. Some case studies were analysed below to add more practicality to the discussion, as stated in the article by The Constructor (2023).

CASE STUDY ONE: The Edge, Amsterdam (PLP Architecture, 2024)

Located centrally in Amsterdam, The Edge is a beacon of sustainability and forward-thinking design. Crafted by PLP Architecture, renowned for its commitment to eco-friendly workspaces, The Edge has earned global acclaim for its green credentials. Its construction prioritises energy efficiency and conservation, utilising a blend of materials like wood and concrete to capture and produce heat while enabling natural temperature regulation. A pioneering sun shading system enhances both comfort and energy efficiency. Beyond its innovative design, The Edge integrates various eco-friendly features, including solar panels, rainwater harvesting, and electric vehicle charging stations, all aimed at minimising its environmental impact. Moreover, its green roof provides insulation and fosters green spaces, exemplifying a holistic approach to sustainable architecture.

The Edge, Amsterdam, Netherlands, combines lean construction practices with stealth construction methodologies across various critical variables.

Environmental Protection: The Edge in Amsterdam demonstrates a strong commitment to environmental protection through its construction and design choices. The utilisation of sustainable materials such as wood and concrete, coupled with energy-efficient systems like solar panels and rainwater harvesting, significantly reduces its environmental footprint. Lean construction practices, such as modular construction and efficient resource management, enhance its eco-friendly attributes by minimising waste and optimising energy usage. The integration of green roofs provides insulation and fosters green spaces, exemplifying a holistic approach to sustainable architecture that aligns with environmental protection goals.

Construction Safety: The construction of The Edge prioritised safety measures to ensure the well-being of workers and minimise accidents. The project maintained a safe working environment through proactive safety training, stringent protocols, and regular hazard assessments. Lean construction principles facilitated efficient workflow planning and organisation, enabling early identification and mitigation of safety hazards. This proactive approach to safety management contributed to a successful construction process with minimal incidents, showcasing the integration of safety considerations into lean construction practices.

Construction Speed: Lean construction practices played a crucial role in expediting the construction timeline of The Edge. The project achieved timely completion by optimising processes and reducing inefficiencies while minimising disruptions. Modular construction techniques allowed simultaneous work on different components, accelerating overall progress. Additionally, efficient communication and collaboration among project teams, facilitated by lean principles, ensured smooth workflow coordination, further enhancing construction speed without compromising quality.

Economy: The Edge exemplifies how lean construction practices can contribute to economic efficiency. The project achieved cost savings while maintaining high-quality standards by minimising waste, optimising resource allocation, and adopting cost-effective techniques such as value engineering. The efficient use of materials and streamlined construction processes helped control expenses and maximise the value of investments. Furthermore, the project's focus on sustainability and energy efficiency aligns with long-term cost-saving objectives, demonstrating the economic benefits of integrating lean practices into construction projects.

Construction Aesthetic: The aesthetic appeal of The Edge reflects a thoughtful integration of design and construction techniques, showcasing the potential of lean practices in achieving construction aesthetic goals. Sustainable design elements such as green roofs and energy-efficient facades contribute to its visual appeal while promoting environmental sustainability. Modular construction techniques offer flexibility in design, allowing for customisation and creating aesthetically pleasing structures that harmonise with the surrounding environment. The seamless blend of form and function exemplifies how lean construction practices can enhance aesthetics while maximising efficiency and sustainability.

Towards a Stealth Construction

Integrating the variables above into stealth construction requires a cohesive approach prioritising efficiency, sustainability, and seamless integration with the surroundings. Environmental protection can be achieved by selecting sustainable materials and energy-efficient systems, as demonstrated by The Edge, while lean construction practices further enhance eco-friendliness by minimising waste and optimising resource utilisation. Concurrently, construction safety remains paramount, necessitating proactive safety measures, rigorous training, and hazard assessments to ensure the well-being of workers. By integrating safety considerations into lean construction principles, projects can maintain a safe working environment while maximising efficiency.

Construction speed is a critical aspect of stealth construction to minimise disruptions and maintain a low profile. Lean practices, such as modular construction and efficient workflow planning, facilitate timely project completion without compromising quality. Additionally, prioritising economy through cost-effective techniques like value engineering and optimised resource allocation contributes to the project's overall efficiency. Meanwhile, construction aesthetic considerations are seamlessly integrated through sustainable design elements and flexible construction techniques, ensuring the project blends harmoniously with its surroundings. By integrating these variables cohesively, stealth construction projects can achieve their objectives while minimising environmental impact, ensuring safety, optimising construction speed, maximising economic efficiency, and enhancing aesthetic appeal.

CASE STUDY TWO: Bosco Verticale, Milan (Zappa, 2024)

The Bosco Verticale, also known as the Vertical Forest, situated in Milan, Italy, stands as an emblem of innovative sustainable architecture, comprising two residential towers of 110 and 76 metres in height adorned with 9,000 trees, 13,000 shrubs, and 5,000 plants. This groundbreaking project, a collaboration between Stefano Boeri Architetti and the City of Milan, aims to foster a sustainable urban habitat, generating oxygen, mitigating pollution, and nurturing local wildlife. Constructed with natural materials like cement, aluminium, and glass, the towers boast insulation properties, minimising energy consumption. Surrounding green spaces spanning 15,000 square metres further augment the building's biodiversity, epitomising its commitment to environmental stewardship and urban livability.

The Bosco Verticale in Milan, Italy, combines lean construction practices with stealth construction methodologies across various critical variables.

Environmental Protection: The Bosco Verticale project in Milan exemplifies a harmonious integration of lean construction practices with environmental protection objectives. By incorporating sustainable design elements such as extensive greenery comprising thousands of trees, shrubs, and plants, the project serves as a green oasis within the urban landscape, mitigating pollution and promoting biodiversity. Lean construction principles are evident in the efficient use of materials and resources, minimising waste and optimising energy efficiency. The project's focus on natural materials for construction, such as cement, aluminium, and glass, further underscores its commitment to environmental stewardship by reducing the ecological footprint of the building. The Bosco Verticale project showcases how lean construction practices can be leveraged to achieve significant environmental benefits while creating sustainable urban habitats.

Construction Safety: Safety considerations are paramount in constructing vertical structures like the Bosco Verticale towers, and lean construction practices are instrumental in ensuring a safe working environment. Proactive safety measures, rigorous training programs, and adherence to strict safety protocols are implemented to minimise the risk of accidents and injuries during construction. By integrating safety considerations into lean construction principles, such as efficient workflow planning and organisation, hazards can be identified and mitigated early in the construction process. Additionally, the project emphasises collaboration among project teams to promote effective communication and problem-solving, further enhancing construction safety outcomes. Thus, the Bosco Verticale project demonstrates how lean construction practices contribute to maintaining a safe working environment while achieving construction objectives.

Construction Speed: Lean construction practices facilitate efficient project delivery without compromising quality, as evidenced by the construction speed of the Bosco Verticale project. Modular construction techniques and streamlined workflows enable simultaneous work on different components of the towers, accelerating overall progress. Additionally, effective communication and collaboration among project teams, facilitated by lean principles, ensure smooth coordination and minimise delays. The project achieves timely completion by optimising processes and reducing inefficiencies, minimising disruptions and maximising efficiency. Thus, the Bosco Verticale project illustrates how lean construction practices can expedite construction timelines while maintaining high-quality standards.

Economy: The Bosco Verticale project underscores the economic benefits of lean construction practices through cost-effective techniques and efficient resource utilisation. The project controls expenses and maximises the value of investments by minimising waste and optimising resource allocation. Lean construction principles, such as value engineering and JIT delivery of materials, contribute to cost savings while ensuring the project's economic viability. Moreover, using natural materials for construction, which offer insulation properties and reduce energy costs, further enhances the project's economic efficiency. The Bosco Verticale project demonstrates how lean construction practices can achieve economic objectives while promoting sustainability.

Construction Aesthetic: The aesthetic appeal of the Bosco Verticale towers reflects a thoughtful integration of design and construction techniques, showcasing the potential of lean practices in achieving construction aesthetic goals. The extensive greenery adorning the towers creates a visually striking and unique appearance, blending seamlessly with the surrounding urban landscape. Lean construction principles, such as modular construction and efficient workflow planning, allow for customisation and flexibility in design, contributing to the project's aesthetic appeal. The project enhances its visual impact by prioritising sustainability and incorporating green spaces while promoting environmental stewardship. Thus, the Bosco Verticale project exemplifies how lean construction practices can achieve construction aesthetic objectives while maximising efficiency and sustainability.

Towards a Stealth Construction

Integrating lean construction practices with the objectives of stealth construction, the Bosco Verticale project in Milan offers a compelling example of how environmental, safety, speed, economy, and aesthetic considerations can converge seamlessly. By prioritising environmental protection, the project strategically utilises sustainable materials and extensive greenery to mitigate pollution, promote biodiversity, and enhance the urban environment, aligning with the principles of stealth construction. Lean construction techniques, such as efficient resource management and modular construction, expedite project timelines without compromising quality, minimising disruptions and maintaining a low profile during construction. Concurrently, stringent safety measures and proactive hazard assessments ensure a safe working environment for personnel, reinforcing the project's commitment to safety and efficiency.

Moreover, the economic efficiency of the Bosco Verticale project underscores the compatibility of lean construction practices with stealth construction objectives. Cost-effective techniques, including value engineering and optimised resource allocation, contribute to cost savings while maximising the project's value and sustainability. The aesthetic appeal of the towers, characterised by their unique greenery and harmonious integration with the urban landscape, exemplifies how lean practices can enhance construction aesthetic goals while promoting environmental stewardship. By integrating these variables cohesively, the Bosco Verticale project demonstrates how lean construction practices can achieve the objectives of stealth construction, minimising environmental impact, ensuring safety, optimising construction speed, maximising economic efficiency, and enhancing aesthetic appeal.

Lean Construction Practice at the Construction Stage

Implementing lean construction practices is better suited at the construction stage than before or after due to its direct impact on project execution and resource utilisation. During construction, Lean principles such as JIT delivery and minimising waste streamline operations on-site. When material flow and scheduling tasks are efficient, construction teams can significantly reduce delays and avoid bottlenecks, ultimately speeding up the project timeline. For instance, lean techniques like pull planning enable real-time adjustments to construction schedules based on actual progress and emerging challenges, fostering adaptability and responsiveness.

Moreover, focusing on lean construction during construction allows for immediate identification and resolution of issues, leading to enhanced quality and cost savings. With the emphasis on continuous improvement and empowering frontline workers to identify and address inefficiencies promptly, construction teams can prevent rework and minimise defects, resulting in higher-quality outcomes. Additionally, the emphasis on collaboration and communication inherent in lean construction fosters more vital teamwork and coordination among project stakeholders, leading to smoother operations and reduced conflicts during construction. Therefore, by integrating lean principles into construction practices, projects can achieve higher efficiency, improved quality, and greater satisfaction for all involved parties.

General Summary

In this comprehensive discussion, the chapter explored the multifaceted realm of stealth construction, a field where precision, security, and discretion are paramount. It is a domain where traditional construction practices merge with advanced technologies and methodologies to achieve the delicate balance between invisibility and functionality. It began with an intriguing topic that delved into the fusion of construction and invincibility, examining how Stealth Technology (ST) in aviation and military contexts shares commonalities with the challenges and goals of stealth construction. This set the stage for exploring construction practices, wherein the chapter uncovered how lean construction, smart construction, sustainable construction, and other principles come together to create an innovative approach known as stealth construction.

Furthermore, the chapter delved into various facets of stealth construction, including its relationship with environmental protection, safety, speed, economy, aesthetics, and integration with advanced construction practices. Lean construction emerged as a potent catalyst for achieving multiple objectives within stealth projects, from enhancing safety and accelerating construction speed to optimising the construction economy and elevating aesthetic appeal. Throughout these discussions, the role of lean principles in waste reduction, efficient supply chain management, value stream mapping, and proactive risk mitigation became evident. The chapter also explored how lean practices synergise with advanced technologies, hidden aesthetics, and eco-friendly construction methods to maintain the low profile essential to stealth projects.

In summary, stealth construction represents the intersection of innovation, precision, and security. It is a field that recognises the need for projects to remain discreet, efficient, and aesthetically pleasing while meeting stringent security requirements. Lean construction, with its core principles of efficiency, waste reduction, and proactive problem-solving, emerges as a driving force in propelling stealth projects towards these multifaceted goals. By efficiently integrating advanced technology, streamlining construction processes, and minimising waste, lean practices offer an adaptable and holistic approach that aligns with the complex objectives of stealth construction.

Conclusion

The journey through the world of stealth construction has unveiled a dynamic and complex domain where construction meets invincibility. From the initial exploration of Stealth Technology (ST) and its relevance to aviation and construction practices, it transitioned into the concept of stealth construction. This multifaceted approach skilfully combines traditional construction principles with advanced methodologies to meet the unique demands of projects that demand invisibility and functionality. Lean construction stood out as a pivotal factor in achieving the multifaceted goals of stealth projects, spanning safety, construction speed, economy, aesthetics, and environmental responsibility. Also, lean construction's core principles of efficiency, waste reduction, and proactive problem-solving provide a versatile framework that caters to stealth construction's intricate and diverse requirements. As the chapter ventured further into the multifaceted aspects of stealth construction, it was found that it offers much more than mere invisibility. It encompasses environmental protection by minimising waste and optimising materials, safety through efficient operations and risk management, construction speed by eliminating non-value-adding activities and ensuring JIT scheduling, construction economy through cost-effective resource management, aesthetics by integrating with the surroundings, and environmental responsibility by promoting sustainable practices. Furthermore, the adaptable nature of lean construction ensures that it aligns seamlessly with the multifaceted goals of stealth construction, offering a holistic approach that caters to these projects' unique and challenging requirements. In essence, stealth construction represents the fusion of ingenuity, security, and construction prowess. Lean construction is a guiding force to harmonise these intricate elements and propel projects towards excellence in function and invisibility.

References

Aamer, A., Eka Yani, L., & Alan Priyatna, I. (2020). Data analytics in the supply chain management: Review of machine learning applications in demand forecasting. *Operations and Supply Chain Management: An International Journal*, *14*(1), 1–13.

Abbasi, S., Taghizade, K., & Noorzai, E. (2020). BIM-based combination of takt time and discrete event simulation for implementing just in time in construction scheduling under constraints. *Journal of Construction Engineering and Management*, *146*(12), 1–15.

Ahmadi-Javid, A., Fateminia, S. H., & Gemünden, H. G. (2020). A method for risk response planning in project portfolio management. *Project Management Journal*, *51*(1), 77–95.

Alaidaros, H., Omar, M., & Romli, R. (2021). The state of the art of agile kanban method: Challenges and opportunities. *Independent Journal of Management & Production*, *12*(8), 2535–2550.

Albalkhy, W., & Sweis, R. (2022). Assessing lean construction conformance amongst the second-grade Jordanian construction contractors. *International Journal of Construction Management*, *22*(5), 900–912.

Al-Balkhy, W., Sweis, R., & Lafhaj, Z. (2021). Barriers to adopting lean construction in the construction industry – The case of Jordan. *Buildings*, *11*(6), 222–245. https://doi.org/10.3390/buildings11060222

Allouzi, M. A., & Al Jaafreh, M. B. (2023). Enhancing project efficiency: A comprehensive analysis of lean construction management implementation in Saudi megaprojects. *Elm və İnnovativ Texnologiyalar Jurnalı Nömrə*, *27*, 39–77. https://doi.org/10.30546/2616-4418.27.2023.39

Azzat, N. N., Kusrini, E., Arifin, M., & Romadhon, M. S. (2024, January). Productivity improvement using lean manufacturing. In 2024 *ASU International Conference in Emerging Technologies for Sustainability and Intelligent Systems (ICETSIS)* (pp. 1144–1148). IEEE.

Bakhtiari, V., Piadeh, F., Chen, A. S., & Behzadian, K. (2024). Stakeholder analysis in the application of cutting-edge digital visualisation technologies for urban flood risk management: A critical review. *Expert Systems with Applications*, *236*, 1–15. https://doi.org/10.1016/j.eswa.2023.121426

Balaraman, V. (2022). *Framework to integrate Industry 4.0 and lean methodologies: Operational excellence in the automotive industry* [Doctoral dissertation, Wayne State University].

Bayhan, H. G., Demirkesen, S., Zhang, C., & Tezel, A. (2023). A lean construction and BIM interaction model for the construction industry. *Production Planning & Control*, *34*(15), 1447–1474.

Cannavale, A., Ayr, U., Fiorito, F., & Martellotta, F. (2020). Smart electrochromic windows to enhance building energy efficiency and visual comfort. *Energies*, *13*(6), 1–17. https://doi.org/10.3390/en13061449
Cardou, F., & Vellend, M. (2023). Stealth advocacy in ecology and conservation biology. *Biological Conservation*, *280*, 1–8. https://doi.org/10.1016/j.biocon.2023.109968
Chapman, J. M., & Schott, S. (2020). Knowledge coevolution: Generating new understanding through bridging and strengthening distinct knowledge systems and empowering local knowledge holders. *Sustainability Science*, *15*(3), 931–943.
Eapen, T., & Finkenstadt, D. J. (2024). *Strategic prominence management*. https://papers.ssrn.com/sol3/papers.cfm?abstract_id=4754403
Eini, R., Linkous, L., Zohrabi, N., & Abdelwahed, S. (2021). Smart building management system: Performance specifications and design requirements. *Journal of Building Engineering*, *39*, 1–15. https://doi.org/10.1016/j.jobe.2021.102222
Esfahani, A. K. (2021). *A comprehensive review on lean project management, transformation & implementation methods, and its utilization in the construction industry (Lean Construction)* [Master's thesis, uis].
Evans, M., Farrell, P., Elbeltagi, E., & Dion, H. (2023). Barriers to integrating lean construction and integrated project delivery (IPD) on construction megaprojects towards the global integrated delivery (GID) in multinational organisations: Lean IPD&GID transformative initiatives. *Journal of Engineering, Design and Technology*, *21*(3), 778–818.
Francis, A., & Thomas, A. (2020). Exploring the relationship between lean construction and environmental sustainability: A review of existing literature to decipher broader dimensions. *Journal of Cleaner Production*, *252*, 1–14. https://doi.org/10.1016/j.jclepro.2019.119913
Ganin, A. A., Quach, P., Panwar, M., Collier, Z. A., Keisler, J. M., Marchese, D., & Linkov, I. (2020). Multicriteria decision framework for cybersecurity risk assessment and management. *Risk Analysis*, *40*(1), 183–199.
Gómez-Cabrera, A., Salazar, L. A., Ponz-Tienda, J. L., & Alarcón, L. F. (2020, July). Lean tools proposal to mitigate delays and cost overruns in construction projects. In *Proceedings of the 28th Annual Conference of the International Group for Lean Construction (IGLC28)*, Berkeley, CA, USA (pp. 6–10).
Govindasamy, T., & Bekker, M. C. (2024). Lean construction: Implementing the Last Planner System (LPM) on mining projects. *Journal of the Southern African Institute of Mining and Metallurgy*, *124*(1), 15–24.
Hamzeh, F., González, V. A., Alarcon, L. F., & Khalife, S. (2021). Lean Construction 4.0: Exploring the challenges of development in the AEC Industry. In L. F. Alarcon & V. A. González (Eds.), *Proc. 29th Annual conference of the international group for lean construction (IGLC29)*, Lima, Peru (pp. 207–216). https://doi.org/10.24928/2021/0181
Hannan, M. A., Begum, R. A., Al-Shetwi, A. Q., Ker, P. J., Al Mamun, M. A., Hussain, A., Basri, H., & Mahlia, T. M. I. (2020). Waste collection route optimisation model for linking cost saving and emission reduction to achieve sustainable development goals. *Sustainable Cities and Society*, *62*, 1–11. https://doi.org/10.1016/j.scs.2020.102393
Hatoum, M. B., Nassereddine, H., & Badurdeen, F. (2021, July). Reengineering construction processes in the era of construction 4.0: A lean-based framework. In *Proc. 29th Annual Conference of the International Group for Lean Construction (IGLC)* (pp. 403–412).
Hong, X., Ji, X., & Wu, Z. (2022). Architectural colour planning strategy and planning implementation evaluation of historical and cultural cities based on different urban zones in Xuzhou (China). *Colour Research & Application*, *47*(2), 424–453.
Hussein, M., & Zayed, T. (2021). Critical factors for successful implementation of just-in- time concept in modular integrated construction: A systematic review and meta- analysis. *Journal of Cleaner Production*, *284*, 1–25. https://doi.org/10.1016/j.jclepro.2020.124716

Jamil, N., Gholami, H., Mat Saman, M. Z., Streimikiene, D., Sharif, S., & Zakuan, N. (2020). DMAIC-based approach to sustainable value stream mapping: Towards a sustainable manufacturing system. *Economic research-Ekonomska istraživanja*, *33*(1), 331–360.

Jose K, A., & Sia, S. K. (2022). Theory of planned behavior in predicting the construction of eco-friendly houses. *Management of Environmental Quality: An International Journal*, *33*(4), 938–954.

Kapoor, M. K. (2023). *Security by design: Protecting buildings and public places against crime and terror*. Taylor & Francis.

Khoza, J. D., & Haupt, T. C. (2021, February). Measuring health and safety performance of construction projects in South Africa. In *IOP Conference Series: Earth and environmental science* (Vol. 654, No. 1, p. 012031). IOP Publishing. https://doi.org/10.1088/1755-1315/654/1/012031

Kumar, A., Sharma, S., Goyal, N., Singh, A., Cheng, X., & Singh, P. (2021). Secure and energy-efficient smart building architecture with emerging technology IoT. *Computer Communications*, *176*, 207–217.

Kumar, D., Devadasan, S. R., Elangovan, D., & Santhosh Ranganathan, V. (2023). Laying foundation for successful implementation of lean manufacturing through the elimination of non-value adding activities in conventional engineering products manufacturing companies. *Proceedings of the Institution of Mechanical Engineers, Part E: Journal of Process Mechanical Engineering*, *237*(5), 2060–2073.

Le, P. L., & Nguyen, N. T. D. (2022). Prospect of lean practices towards construction supply chain management trends. *International Journal of Lean Six Sigma*, *13*(3), 557–593.

Lehtinen, S. (2020). Living with urban everyday technologies. *ESPES*, *9*(2), 81–89.

Lensjø, M. (2024). Vocational teachers' craft knowledge and working-life experiences in building and construction: A narrative study of embodied and tacit learning. *Vocations and Learning*, *1*(1), 1–18.

Li, J., Chen, L., Chen, Y., & He, J. (2022). Digital economy, technological innovation, and green economic efficiency – Empirical evidence from 277 cities in China. *Managerial and Decision Economics*, *43*(3), 616–629.

Liu, Q., & Ren, J. (2020). Research on the building energy efficiency design strategy of Chinese universities based on green performance analysis. *Energy and Buildings*, *224*, 1–13. https://doi.org/10.1016/j.enbuild.2020.110242

Lohne, J., Torp, O., Andersen, B., Aslesen, S., Bygballe, L., Bolviken, T., Drevland, F., Engebø, A., Fosse, R., Holm, T., H., Hunn, L., K., Kalsaas, B., T., Klakegg, O., J., Knotten, V., Kristensen, K., H., Olsson, N., O., Rolstadås, A., Skaar, J., Svalestuen, F., … Laedre, O. (2022). The emergence of lean construction in the Norwegian AEC industry. *Construction Management and Economics*, *40*(7–8), 585–597.

Lucchi, E., & Buda, A. (2022). Urban green rating systems: Insights for balancing sustainable principles and heritage conservation for neighbourhood and cities renovation planning. *Renewable and Sustainable Energy Reviews*, *161*, 1–19. https://doi.org/10.1016/j.rser.2022.112324

Maksoud, A., Alawneh, S. I. A. R., Hussien, A., Abdeen, A., & Abdalla, S. B. (2024). Computational design for multi-optimized geometry of sustainable flood-resilient urban design habitats in Indonesia. *Sustainability*, *16*(7), 1–41.

Malekpour, S., Tawfik, S., & Chesterfield, C. (2021). Designing collaborative governance for nature-based solutions. *Urban Forestry & Urban Greening*, *62*, 1–13. https://doi.org/10.1016/j.ufug.2021.127177

Mali, R., Kirloskar, S. G., Thakur, C., Kokare, Y., & Patil, N. (2020). Management of storm water using Bioretention filter technique. *International Journal of Engineering Research & Technology (IJERT)*, 9(3), 160–165.

Mano, A. P., Gouvea da Costa, S. E., & Pinheiro de Lima, E. (2020). Criticality assessment of the barriers to lean construction. *International Journal of Productivity and Performance Management*, *70*(1), 65–86.

McHugh, K., Dave, B., & Koskela, L. (2022). On the role of lean in digital construction. *Industry 4.0 for the Built Environment: Methodologies, Technologies and Skills*, *1*(1), 207–226.

Mellado, F., & Lou, E. C. (2020). Building information modelling, lean and sustainability: An integration framework to promote performance improvements in the construction industry. *Sustainable Cities and Society*, *61*, 1–13. https://doi.org/10.1016/j.scs.2020.102355

Meng, X. (2020). Proactive management in the context of construction supply chains. *Production Planning & Control*, *31*(7), 527–539.

Nielsen, M., Sumich, J., & Bertelsen, B. E. (2021). Enclaving: Spatial detachment as an aesthetics of imagination in an urban sub-Saharan African context. *Urban Studies*, *58*(5), 881–902.

Noorzai, E. (2023). Evaluating lean techniques to improve success factors in the construction phase. *Construction Innovation*, *23*(3), 622–639.

Nugroho, W. (2021). Relationship between environmental management policy and the local wisdom of indigenous peoples in the handling of COVID-19 in Indonesia. *Onati Socio-Legal Series*, *11*(3), 860–882.

Nwaki, W. N., & Eze, C. E. (2020). Lean construction as a panacea for poor construction projects performance. *ITEGAM-JETIA*, *6*(26), 61–72.

Nwaki, W., Eze, E., & Awodele, I. (2021). Major barriers assessment of lean construction application in construction projects delivery. *CSID Journal of Infrastructure Development*, *4*(1), 63–82.

Oakland, J., & Marosszeky, M. (2022). Lean quality in construction project delivery – A new model and principles. In D. Dalcher (Ed.), *Rethinking project management for a dynamic and digital world* (pp. 33–40). Routledge.

Onungwa, I., Olugu-Uduma, N., & Shelden, D. R. (2021). Cloud BIM technology as a means of collaboration and project integration in smart cities. *Sage Open*, *11*(3), 1–9.

Oppong, D. G. (2020). *External stakeholder management at the planning stage of construction projects in Ghana: Consultants' perspective*. https://theses.lib.polyu.edu.hk/handle/200/10457

PLP Architecture. (2024). *The Edge, Amsterdam, The Netherlands – PLP Architecture*. https://plparchitecture.com/the-edge/

Prajapati, P., Bhatt, B., Zalavadiya, G., Ajwalia, M., and Shah, P. (2021). A review on recent intrusion detection systems and intrusion prevention systems in IoT. *2021 11th International conference on cloud computing, data science & engineering (Confluence)*, Noida, India, 2021 (pp. 588–593). https://doi.org/10.1109/Confluence51648.2021.9377202

Rayworth, A. (2022). *Investigating the phenomenology of design objects within a specific environment* [Doctoral dissertation, Staffordshire University].

Rosli, M. F., Muhammad Tamyez, P. F., & Zahari, A. R. (2023). The effects of suitability and acceptability of lean principles in the flow of waste management on construction project performance. *International Journal of Construction Management*, *23*(1), 114–125.

Saad, D. A., Masoud, M., & Osman, H. (2021). Multi-objective optimization of lean-based repetitive scheduling using batch and pull production. *Automation in Construction*, *127*, 1–12. https://doi.org/10.1016/j.autcon.2021.103696

Salliou, N., Arborino, T., Nassauer, J. I., Salmeron, D., Urech, P., Vollmer, D., & Grêt-Regamey, A. (2023). Science-design loop for the design of resilient urban landscapes. *Socio-Environmental Systems Modelling*, *5*, 1–21. https://doi.org/10.18174/sesmo.18543

Sidorova, L. (2022). *Enhancing aesthetics in Industrial Architecture: Integrating 3D-printed facade structures in the design of an incineration plant in Tromso*. https://www.politesi.polimi.it/handle/10589/214551

Singh, I., & Singh, B. (2023). Access management of IoT devices using access control mechanism and decentralized authentication: A review. *Measurement: Sensors, 25*, 1–8. https://doi.org/10.1016/j.measen.2022.100591

Singh, S., & Kumar, K. (2021). A study of lean construction and visual management tools through cluster analysis. *Ain Shams Engineering Journal, 12*(1), 1153–1162.

Sobkowiak, M. (2023). The making of imperfect indicators for biodiversity: A case study of UK biodiversity performance measurement. *Business Strategy and the Environment, 32*(1), 336–352.

Solaimani, S., & Sedighi, M. (2020). Toward a holistic view on lean sustainable construction: A literature review. *Journal of Cleaner Production, 248*, 1–14. https://doi.org/10.1016/j.jclepro.2019.119213

Staves, A., Gouglidis, A., Maesschalck, S., & Hutchison, D. (2024). Risk-based safety scoping of adversary-centric security testing on operational technology. *Safety Science, 174*, 1–21. https://doi.org/10.1016/j.ssci.2024.106481

The Constructor. (2023). *The top 10 most sustainable buildings around the world.* https://theconstructor.org/architecture/the-top-10-most-sustainable-buildings-around-the-world

Umoh, A. A., Adefemi, A., Ibewe, K. I., Etukudoh, E. A., Ilojianya, V. I., & Nwokediegwu, Z. Q. S. (2024). Green architecture and energy efficiency: A review of innovative design and construction techniques. *Engineering Science & Technology Journal, 5*(1), 185–200.

Varjovi, A. E., & Babaie, S. (2020). Green Internet of Things (GIoT). Vision, applications and research challenges. *Sustainable Computing: Informatics and Systems, 28*, 1–9. https://doi.org/10.1016/j.suscom.2020.100448

Warid, O., & Hamani, K. (2023). Lean Construction in the UAE: Implementation of Last Planner System (LPS). *Lean Construction Journal, 1*(1), 1–20.

Włodarkiewicz-Klimek, H. (2021). New models of work organization in an industry 4.0 enterprise-evolution of the form of work. *European Research Studies, 24*(3), 1095–1105.

Xing, W., Hao, J. L., Qian, L., Tam, V. W., & Sikora, K. S. (2021). Implementing lean construction techniques and management methods in Chinese projects: A case study in Suzhou, China. *Journal of Cleaner Production, 286*, 1–13. https://doi.org/10.1016/j.jclepro.2020.124944

Xu, Z., Li, J., Li, J., Du, J., Li, T., Zeng, W., Qiu, J., & Meng, F. (2023). Bionic structures for optimizing the design of stealth materials. *Physical Chemistry Chemical Physics, 25*(8), 5913–5925.

You, Z., & Feng, L. (2020). Integration of industry 4.0 related technologies in construction industry: A framework of cyber-physical system. *IEEE Access, 8*, 122908–122922.

Zahraee, S. M., Tolooie, A., Abrishami, S. J., Shiwakoti, N., & Stasinopoulos, P. (2020). Lean manufacturing analysis of a Heater industry based on value stream mapping and computer simulation. *Procedia Manufacturing, 51*, 1379–1386.

Zappa, G. (2024). *The Vertical Forest: A concrete utopia for the world, from Milan.* https://www.infrajournal.com/en/w/the-vertical-forest-a-concrete-utopia-for-the-world

Zhuang, H., Zhang, J., CB, S., & Muthu, B. A. (2020). Sustainable smart city building construction methods. *Sustainability, 12*(12), 1–17. https://doi.org/10.3390/su12124947

Chapter Seven

Smart Infrastructure and Development in Stealth Construction

Abstract

The chapter explored integrating smart construction techniques in achieving stealth construction objectives, emphasising the development of building cross-sections, visibility management, energy transmission optimisation, and countermeasure implementation. It delved into the multifaceted aspects of smart construction towards achieving stealth construction goals, including environmental protection, enhanced construction safety, accelerated construction duration, cost-effectiveness, and aesthetic considerations. Furthermore, the chapter underscores the importance of leveraging innovative approaches and advanced technologies to meet the evolving demands of stealth construction projects and pave the way for sustainable, safe, and aesthetically pleasing built environments.

Keywords: Smart construction; stealth construction; construction 4.0; construction innovation; resilient building

Introduction

Sepasgozar (2021) and Rahimian et al. (2021) stated that smart, intelligent, or digital construction represents a paradigm shift in how buildings and infrastructure are designed, built, and operated. Kor et al. (2023) illustrated that one of the critical elements of smart construction is using digital twins, virtual representations of physical structures, enabling real-time monitoring, predictive analysis, and holistic lifecycle management. Through technologies such as Building Information Modelling (BIM), collaboration among stakeholders is enhanced, and potential conflicts are detected early in the design phase, reducing rework and optimising cost estimates. Furthermore, integrating Internet of Things (IoT)

Stealth Construction: Integrating Practices for Resilience and Sustainability, 163–193

Published under exclusive licence by Emerald Publishing Limited
doi:10.1108/978-1-83608-182-120251007

devices brings a new level of connectivity to construction sites, with sensors and wearables providing real-time data on various parameters, as Dian et al. (2020) discussed. Robotics and automation contribute to increased efficiency as machines take on labour-intensive tasks, ensuring precision and safety. Augmented Reality (AR) and Virtual Reality (VR) technologies facilitate design visualisation and serve as valuable tools for training and remote collaboration.

Smart Infrastructure and Development for Stealth Construction

Stealth construction represents a pioneering approach in the construction industry, where traditional building practices seamlessly merge into a unified and advanced process. At its core, stealth construction is the intersection where construction meets invincibility, ushering in a new era of efficiency, precision, and sustainability. This innovative concept strategically integrates construction practices across distinct construction, resulting in a comprehensive and interconnected construction lifecycle. The goal is to create an invisible shield against inefficiencies, delays, and challenges, ultimately enhancing the resilience and effectiveness of the entire construction process, as depicted in Fig. 4.

Smart construction within the paradigm of stealth construction signifies a transformative leap into a new era of advanced building processes, where the integration of cutting-edge technologies seamlessly aligns with construction stages using construction practices. In the pre-construction phase, the deployment of innovative tools such as BIM emerges as a foundational element. By generating intricate digital representations of the project, stakeholders gain a collaborative platform enriched with predictive capabilities through artificial intelligence (AI) and data analytics. This pre-emptive insight empowers decision-makers to anticipate challenges, optimise resource allocation, and forge a robust foundation for the impending construction stages, embodying a shield of invincibility against potential setbacks.

The construction phase under the stealth construction paradigm unfolds with a strategic blend of robotics and automation, exemplifying the principles of advanced efficiency. Robotics assumes responsibility for labour-intensive tasks,

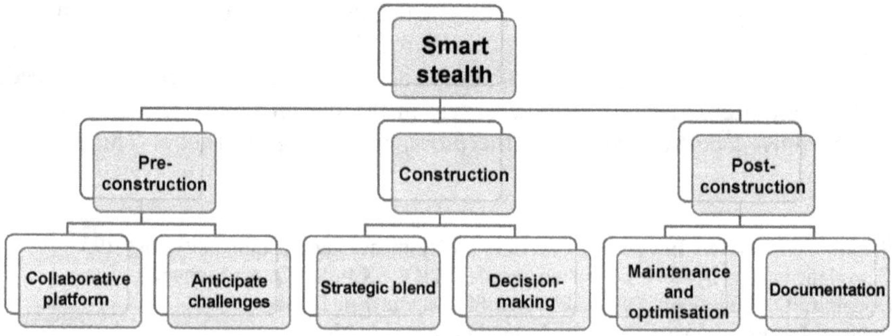

Fig. 4. Smart Stealth Construction Concept.

freeing human workers to focus on complex aspects of construction. AR and VR technologies elevate decision-making on-site, offering real-time information in an immersive environment. The IoT, with its network of embedded sensors, orchestrates real-time monitoring, ensuring synchronisation across the construction site. This interconnected relationship creates an invisible shield, guarding against inefficiencies and disruptions, thus solidifying the construction process as invincible.

Post-construction, the smart construction approach within stealth construction extends its influence into the maintenance and optimisation phase, perpetuating the theme of invincibility. Digital twins, acting as virtual mirrors of the physical structure, sustain real-time visibility into the building's performance and condition. Predictive maintenance, a product of IoT sensors and data analytics, anticipates potential issues, contributing to the structure's longevity and minimising long-term maintenance costs. Blockchain technology, renowned for its security and transparency, takes centre stage by documenting maintenance activities, warranties, and certifications. This robust digital ledger adds invincibility to the post-construction phase, ensuring the ongoing integrity of the constructed asset.

Nevertheless, the path to realising stealth construction through smart practices has its share of challenges, with some already discussed above. The initial investment in advanced technologies and the requisite training for the workforce pose significant hurdles. Establishing standardised protocols and ensuring interoperability across diverse smart systems are imperative for a seamless integration that embodies stealth construction principles. Overcoming these challenges necessitates industry-wide collaboration, a commitment to continual innovation, and a collective vision of constructing invincible structures that transcend conventional boundaries. In addition, the potential benefits are profound in the broader context of smart construction for stealth construction. Efficiency, precision, and sustainability become the cornerstones of a construction process that meets immediate needs and proactively anticipates and adapts to future challenges. As the construction industry wholeheartedly embraces these innovative approaches, the realisation of an invincible construction process evolves from a visionary concept into a tangible reality, ushering in an era where construction practices are not just advanced but invincibly so.

Several construction parts must be considered when incorporating smart construction principles into achieving stealth construction. This includes building cross-section development, visibility, energy emission, and countermeasures. These are expatiated below.

Building Cross-Section Development

Building cross-section development stands as an anchor in pursuing stealth construction through the lens of smart construction methodologies. This refined and comprehensive approach involves a cautious examination and optimisation of various facets, encompassing the shape of the building, internal construction elements, material utilisation, and several other critical factors, all strategically interwoven to foster a more advanced and invincible construction process. The building cross section is conceptualised under the shape of the building, internal construction, material usage, and other relevant concepts.

Smart construction, a paradigm that integrates advanced technologies and innovative methodologies, places building cross-section development at the forefront of its strategic approach, envisioning the realisation of stealth construction, Musarat et al. (2024) illustrated. This comprehensive and meticulous strategy involves an in-depth examination and optimisation of various facets, including the shape of the building, internal construction components, material usage, and numerous other critical factors. By seamlessly integrating these elements, smart construction aims to cultivate a process that exemplifies efficiency and embodies inherent resilience, akin to the stealth characteristics that render it less susceptible to disruptions and challenges.

The Shape of the Building

Jafari and Alipour (2021) and Hamidavi et al. (2020) stated that the shape of the building assumes paramount importance in cross-section development. Within smart construction, it transforms into a strategic element for infusing stealth characteristics. Leveraging advanced design tools such as BIM and parametric design, architects and engineers can intricately mould and analyse the building's shape. This optimisation process considers many considerations, including energy efficiency, structural integrity, and aesthetic appeal. The result is a meticulously designed structure that meets functional requirements and seamlessly integrates with its environment, embodying the principles of stealth construction by efficiently using space and resources.

Internal Construction

Internally, cross-section development delves into the intricacies of the building's internal construction components (Vagtholm et al., 2023). Smart construction capitalises on digital twins, 3D modelling, and virtual simulations to scrutinise the behaviour of these components. This advanced analysis identifies potential clashes or inefficiencies before the physical construction commences. When internal construction is optimised during cross-section development, smart construction ensures that the building's internal framework is robust and strategically designed for seamless integration. This attribute significantly contributes to the stealthiness of the construction process.

Material Usage

As depicted by Santos (2020) and Khan et al. (2024), material usage is a critical facet of cross-section development and is anchored in sustainability within the smart construction paradigm. Materials are meticulously selected based on lifecycle impacts, embodied energy, and recyclability. Pursuing eco-friendly and sustainable materials is paramount to minimising the environmental footprint of construction projects. This approach extends the building's lifecycle and contributes to the stealthiness of the construction process by mitigating its ecological impact. Furthermore, integrating sensors and smart technologies in the cross-section allows real-time monitoring of material performance, contributing to predictive maintenance and ensuring the longevity of the construction.

Visibility

Visibility as a way of achieving stealth construction is divided into the colour and texture of the construction, interaction with the environment, and other related physical attributes of the projected construction.

Colour and Texture of the Construction

In smart construction, achieving visibility in stealth construction is a cautious process, and one essential facet is the consideration of the colour and texture of the construction. As Sidorova (2022) and Mun (2023) added, this element goes beyond mere aesthetics, becoming a strategic tool to seamlessly integrate the structure into its surroundings while embodying the principles of stealth construction. Through smart approaches, the colour and texture choices become essential in creating structures that harmonise with their environment, minimising visual impact and fostering an inconspicuous yet sophisticated presence.

Furthermore, Zhang and Kim (2023) illustrated that the colour palette chosen for a construction project within smart construction principles is a deliberate decision influenced by the natural surroundings. Instead of opting for attention-grabbing hues, smart construction prioritises colours that either blend with or complement the local landscape. This strategic selection considers the immediate surroundings and factors in seasonal changes, ensuring the construction maintains its stealthiness throughout the year. For instance, muted earth tones or greens may mimic the natural environment's colours, providing a visual continuity that helps the construction seamlessly merge with its context.

In addition, Guesmi et al. (2023) added that texture plays a significant role in the stealth construction approach within smart construction as an integral component of visibility management. The surface finish and detailing of the building are carefully considered to enhance its blending capabilities. Textures that mimic natural elements, such as stone or wood finishes, are often employed to connect visually with the environment. This not only contributes to the aesthetic appeal but also aids in breaking up the visual mass of the structure, making it less conspicuous. In addition, the texture can be designed to interact with light in a way that reduces reflections and shadows, further enhancing the subtlety of the construction.

Interaction with the Environment

In complex smart construction, the pursuit of visibility in stealth construction is intricately linked to the deliberate and strategic interaction of the construction with its environment. This interaction encompasses a multifaceted approach, integrating sustainability principles, ecological harmony, and a seamless fusion with the natural surroundings. Smart construction views the construction's relationship with the environment as a pivotal element in minimising visual disruption and creating a structure that epitomises the principles of stealth construction.

D'Auria et al. (2018) illustrated that sustainability forms a fundamental pillar of the interaction strategy within smart construction, which relates to stealth construction. The construction process is imbued with eco-friendly practices, from selecting sustainable materials to implementing energy-efficient design. By adhering to green building standards and incorporating renewable energy sources, smart construction ensures that the building's environmental impact is minimised and aligned with ethical considerations. This commitment to sustainability not only supports environmental stewardship but also contributes to the stealthiness of the structure by reducing its overall ecological footprint.

Furthermore, Catalano et al. (2021) added that ecological integration emerges as a central aspect of how smart construction achieves visibility in stealth construction. The building is purposefully designed to seamlessly meld with the natural surroundings, incorporating landscaping and ecological features that enhance its integration into the environment. Green roofs, vertical gardens, and porous surfaces are common elements employed by smart construction to contribute to environmental health and visually unite the structure with its surroundings. This integration fosters an inconspicuous presence, making the construction appear organic and harmonious within the broader landscape.

In addition, smart construction's deliberate use of native vegetation in landscaping is a strategic approach towards achieving stealth construction objectives. An environmentally sustainable landscaping approach is fostered by opting for indigenous flora that is well-suited to the local climate and requires minimal water and upkeep (Desha et al., 2024). This seamless integration of native plants extends the natural environment and minimises visual disparities between the constructed and natural landscapes. Water management practices are also paramount in the interaction strategy of smart construction for stealth construction purposes. Incorporating sustainable water strategies like rainwater harvesting and porous surfaces aids in reducing water runoff and conserving local water resources. Moreover, the facade design of the construction is meticulously curated to harmonise with its environment, utilising materials, textures, and design elements that complement rather than clash with the natural surroundings. Incorporating sun shading devices and smart lighting systems driven by smart technologies further enhances the structure's adaptability and responsiveness to its surroundings. Integrating sensors and adaptive systems allows the construction to interact dynamically with its environment, minimising its visual impact and reinforcing the inconspicuous presence crucial for stealth construction goals.

Energy Transmission

In the contemporary landscape of smart construction, the imperative to blend innovation with sustainability has led to the evolution of construction practices prioritising efficiency and embracing eco-friendly solutions (Umoh et al., 2024). One of the pivotal aspects of this approach is the strategic transmission of energy within construction projects. When executed through smart construction methodologies, this detailed process goes beyond mere functionality, aiming to achieve stealth construction by integrating renewable energy sources seamlessly into the built environment. This aspect delves into how smart construction ensures efficient energy transmission by combining renewable energy, contributing to the functionality of structures and their sustainability and inconspicuous coexistence within their surroundings.

Renewable Energy

Smart construction significantly emphasises harnessing renewable energy sources to achieve sustainable and stealthy energy transmission within buildings. Photovoltaic (PV) solar panels stand out as a prominent feature in this paradigm, adorning rooftops or integrated into building facades to capture sunlight and convert it into electricity (Grover, 2021). The strategic placement of these solar panels enhances the building's energy efficiency and contributes to overall stealthiness by seamlessly integrating this technology into the architectural design.

In addition to solar energy, Bagherian and Mehranzamir (2020) illustrated that smart construction explores various other renewable sources, such as wind and geothermal, to diversify the energy mix. Small-scale wind turbines strategically positioned on building rooftops or integrated into the architectural design harness wind energy, contributing to the overall energy needs of the construction. Similarly, geothermal systems leverage the Earth's natural heat, providing an efficient and inconspicuous energy source for heating and cooling within buildings. When seamlessly integrated, these renewable energy solutions ensure that the energy transmission process is sustainable and aligns with the principles of stealth construction.

Energy storage technologies are pivotal in smart construction's stealthy energy transmission strategy, with advanced battery systems seamlessly integrated into building structures to store surplus energy from renewable sources. These stored reserves ensure uninterrupted power supply during low renewable energy production or high demand periods while avoiding bulky external installations. Smart construction employs sophisticated energy management systems driven by AI and machine learning to optimise renewable energy utilisation by analysing consumption patterns and building occupancy. Architects and engineers collaborate to embed sustainability into building structures, seamlessly integrating renewable energy solutions into both aesthetic and functional aspects. This approach enhances energy efficiency and visual unity within the environment, aligning with stealth construction principles. Moreover, smart construction emphasises community education through interactive displays and signage, fostering environmental awareness and portraying the structure as a conscientious community asset.

Climate Action Planning

As a proactive smart construction strategy for energy transmission, climate action planning involves an intricate understanding of the local climate, weather patterns, and environmental conditions (Thurman et al., 2020). This encompassing approach enables the construction to optimise energy consumption and transmission, resulting in structures that function efficiently and harmonise inconspicuously with their surroundings.

Ozarisoy's (2022) study showed that passive design strategies can take centre stage in the climate action planning of smart construction. Buildings are designed to minimise reliance on mechanical systems for temperature control through strategic orientation, natural shading elements, and optimised ventilation. This passive design not only enhances energy efficiency but also seamlessly integrates into the architectural aesthetics, contributing to the stealthiness of the construction. The advancement of climate-responsive building envelopes is another cornerstone of energy transmission in smart construction. The construction dynamically responds to environmental conditions by employing cutting-edge materials and technologies, such as thermochromic or photochromic coatings on windows. This adaptive building envelope ensures that the energy transmission process aligns inconspicuously with the changing climate, enhancing efficiency and stealth.

Furthermore, renewable energy integration is foundational to smart construction, guided by climate action planning (Vukovic et al., 2021). By meticulously assessing local climate and solar potential, renewable energy infrastructure like solar panels or wind turbines is strategically placed within the building design, optimising energy transmission and reducing environmental impact. Dynamic climate-responsive control systems, powered by real-time sensor data and AI, adjust energy usage based on occupancy and environmental conditions, ensuring seamless adaptation to climate variations. Also, water management strategies, including rainwater harvesting and permeable surfaces, are seamlessly integrated into smart construction's climate action planning, enhancing water conservation and energy efficiency. Community engagement and education are prioritised, with interactive displays and signage to raise awareness about sustainable practices and foster environmental responsibility, portraying the construction as a conscientious and harmoniously integrated element within the broader environmental landscape.

Energy Management

Energy management in the strategic framework of smart construction, is steering the course of dynamic and efficient energy transmission. Central to this approach is the deployment of advanced sensor networks strategically positioned throughout the building (Sofi et al., 2022). These sensors, ranging from occupancy detectors to environmental monitors, collect real-time data that forms the foundation for informed decision-making within the energy management system.

Integrating AI and machine learning elevates energy management to a sophisticated level in smart construction. Himeur et al. (2021) mentioned that AI algorithms process the wealth of data obtained from sensors, analysing patterns and trends to understand the unique energy consumption behaviour of the building. This adaptive intelligence enables the system to dynamically adjust and optimise energy usage, contributing to the inconspicuous and efficient energy transmission characterising stealth construction. Furthermore, intricately woven into energy management strategies, dynamic control systems utilise AI's insights to govern various building systems. These systems orchestrate the operation of lighting, heating, ventilation, and air conditioning (HVAC), responding in real-time to changing conditions. For instance, lighting levels are adjusted based on the availability of natural light, and HVAC systems adapt to occupancy patterns, ensuring energy is utilised precisely when and where needed. In addition, demand response strategies further enhance the agility of energy management in stealth construction. Korkas et al. (2022) illustrated that smart buildings curtail energy consumption during peak demand periods by actively participating in demand-side management programs, contributing to grid stability. This aligns with broader sustainability goals and ensures that the building operates inconspicuously, minimising its energy footprint during heightened demand.

Qayyum et al. (2023) and Nazari et al. (2020) mentioned that integrating smart buildings with advanced energy grids and utility-scale management systems significantly enhances energy management in stealth construction. Participation in grid-balancing initiatives and demand response programs optimises energy transmission within the building, contributing to the stability and efficiency of the broader energy grid. Furthermore, community engagement is crucial, as transparent communication about energy-saving initiatives fosters environmental responsibility, encouraging active participation in energy-saving practices. The strategic implementation of energy management systems within smart construction marks a transformative shift towards inconspicuous yet highly efficient energy transmission.

Energy Efficiency

Energy efficiency takes centre stage in smart construction's strategy for achieving inconspicuous and optimal energy transmission (Ali et al., 2022). At its core lies a meticulous blend of design considerations, advanced technologies, and a holistic approach to minimising waste and maximising the efficacy of energy utilisation. Passive design principles are foundational in achieving energy efficiency within stealth construction. The strategic orientation of structures, coupled with intelligent shading solutions and optimised natural lighting, reduces reliance on artificial lighting and HVAC systems. These passive design elements enhance energy efficiency and seamlessly integrate into architectural aesthetics, embodying the essence of stealth construction.

Advanced building materials and technologies play a pivotal role in the energy efficiency strategies of smart construction (Vagtholm et al., 2023). High-performance insulation, energy-efficient windows, and innovative facade designs contribute to minimised energy loss and optimised temperature control. These features not only enhance the overall energy performance of the building but also operate inconspicuously, aligning seamlessly with the principles of stealth construction. Also, infusing smart technologies augments energy efficiency by enabling real-time monitoring and control. Building Management Systems (BMS) and IoT devices offer granular insights into energy consumption patterns, allowing for real-time dynamic adjustments, as illustrated by Merabet et al. (2022). For instance, intelligent lighting systems adapt to occupancy, and heating and cooling systems optimise based on real-time environmental conditions, ensuring efficient energy transmission without compromising occupant comfort.

In addition, Ding et al. (2020) and Mohsan et al. (2022) stipulated that energy-efficient lighting solutions, including light-emitting diodes (LEDs) and advanced lighting controls, are integral to the quest for stealthy energy transmission. Smart construction embraces these technologies to reduce energy consumption and create ambient and inconspicuous lighting environments. Lighting schemes are designed to align with natural circadian rhythms, promoting occupant well-being while optimising energy use. Also, renewable energy sources, seamlessly integrated into the building's design, contribute to energy efficiency and sustainability. Vijayan et al. (2023) mentioned that PV solar panels positioned on rooftops or integrated into facades harness clean energy, providing a renewable source that operates inconspicuously. Smart construction strategically places these renewable energy systems to minimise visual impact while maximising efficiency.

Countermeasures

In the dynamic view of smart construction, the strategic implementation of countermeasures is essential, particularly within stealth construction. Countermeasures encompass a proactive set of strategies designed to address challenges and enhance the functionality and invisibility of buildings.

Functional Construction System

Integrating highly functional construction systems in smart construction becomes a cornerstone in deploying effective countermeasures. This explanation explains the multifaceted strategies employed in smart construction to achieve countermeasures through the seamless integration of functional construction systems that align with the principles of stealth construction. Integrating functional construction systems is achieved through the following practices.

Integrated Security Systems (Charles & Mishra, 2021; Hammi et al., 2022)

The implementation of countermeasures begins with the integration of advanced security systems. Functional construction systems include state-of-the-art access control mechanisms, surveillance cameras, and biometric identification. These elements bolster the safety of occupants and contribute to the building's overall security and discreet operation.

Smart Surveillance and Monitoring (Gao et al., 2021; Ibrahim et al., 2021)

Smart construction leverages the IoT and advanced technologies to enhance surveillance and monitoring capabilities. Strategic placement of smart cameras, sensors, and monitoring devices ensures real-time detection and response to potential security threats. This proactive approach reinforces the security and inconspicuous functioning of the building.

Adaptive Lighting Systems (Seyedolhosseini et al., 2020; Putrada et al., 2022)

Functional construction systems incorporate adaptive lighting solutions that respond intelligently to environmental conditions and occupancy patterns. This countermeasure not only optimises energy efficiency but also contributes to the inconspicuous operation of the building. Lighting systems are designed to adapt to natural light levels, reducing visibility from the exterior and maintaining a low-profile presence.

Noise Reduction Technologies (Liu et al., 2024; Taheri et al., 2022)

Smart construction integrates noise reduction technologies into its functional systems to counter potential disturbances and enhance occupant comfort. This includes soundproofing materials, acoustic design considerations, and smart HVAC systems for quiet operation. These countermeasures contribute to a serene and inconspicuous building environment.

Automated Maintenance and Repairs (Sajid et al., 2021)

Smart construction employs predictive maintenance and automated repair systems as countermeasures against potential disruptions. These systems use IoT sensors and data analytics to predict and address maintenance issues before they escalate. This proactive approach ensures the seamless and inconspicuous operation of building systems.

Innovative Building Materials (Song et al., 2021; Vagtholm et al., 2023)

Countermeasures against wear and tear, weathering, and environmental factors involve using innovative building materials within functional construction systems. Self-healing materials, for example, automatically repair minor damages, contributing to the longevity and inconspicuous maintenance of the building.

Dynamic Climate Control Systems (Zhang et al., 2022)

Smart construction integrates dynamic climate control systems as countermeasures against external climate variations. These systems adapt to changing weather conditions, ensuring optimal indoor comfort while minimising the visible impact of HVAC operations. Integrating these functional systems contributes to the overall inconspicuousness of the building.

Efficient Waste Management Systems (Kabirifar et al., 2020)

Counteracting environmental impact involves integrating efficient waste management systems into the functional construction framework. Smart waste disposal and recycling systems minimise the visual effects of waste, contributing to the overall cleanliness and inconspicuous operation of the building.

Invisible Technology Integration (Li et al., 2020; Loo & Wong, 2023)

To counteract the visual impact of technology, smart construction seamlessly integrates devices into the building's design. Concealed wiring, hidden sensors, and inconspicuous technology placement ensure that the technological infrastructure supports building functionality without compromising its aesthetic appeal.

Modular and Flexible Design (Andersen et al., 2023; Quach, 2022)

Functional construction systems are designed with modularity and flexibility, acting as countermeasures against changing needs and potential modifications. This adaptability allows for seamless adjustments and expansions, ensuring the building can evolve unobtrusively.

Achieving countermeasures through functional construction systems represents a strategic and proactive approach to smart construction. Integrating advanced security, surveillance, adaptive lighting, noise reduction, automated maintenance, innovative materials, climate control, waste management, invisible technology, and flexible design embodies the principles of stealth construction. These countermeasures address potential challenges and contribute to the overall functionality, security, and unnoticeable operation of buildings, defining the future of smart and seamlessly integrated construction practices.

Method of Construction (Tactics)

In smart construction, deploying countermeasures assumes a strategic significance, especially considering stealth construction's discrete and secure nature. Countermeasures are proactive strategies integrated into the construction process to mitigate challenges and elevate the overall functionality. The method of construction, treated as a tactical approach, emerges as a pivotal element in executing effective countermeasures. This section discusses strategies employed by smart construction to achieve countermeasures through specific tactics in the construction method, harmonising seamlessly with the principles of stealth construction.

Integrated Security Measures (Salor & Baeza, 2023; Shurrab et al., 2022)

Tactically incorporating security measures into the construction method ensures a cohesive integration. Access control systems, surveillance infrastructure, and biometric identification mechanisms are seamlessly embedded, fortifying the building's security from its inception and contributing to a discreet and secure environment.

Smart Surveillance Integration (Ahsan, 2024; Pan & Zhang, 2023)

Leveraging the method of construction as a tactical tool, smart construction integrates surveillance infrastructure seamlessly. Cameras, sensors, and monitoring devices are strategically placed during construction, endowing the building with real-time surveillance capabilities. This proactive approach addresses security concerns and aligns with the inconspicuous operation characteristic of stealth construction.

Adaptive Lighting Design (Alkhatib et al., 2021; Wang et al., 2020)

Construction tactics influence lighting design, incorporating adaptive solutions into the construction process. Lighting systems are strategically integrated to respond intelligently to environmental conditions and occupancy patterns. This tactical approach ensures that lighting is discreetly designed, adapting to natural light levels and minimising visibility from the exterior.

Predictive Maintenance Integration (Brintrup et al., 2020; Marocco & Garofolo, 2021)

Construction tactics involve the integration of predictive maintenance to counter potential disruptions. Smart construction predicts and addresses maintenance issues pre-emptively using sensors and data analytics during construction. This tactical strategy ensures the building's seamless and inconspicuous operation by preventing potential maintenance-related challenges.

Strategic Material Selection (Rane et al., 2023)

Construction methods strategically select materials to counter wear and tear and environmental factors. Integrating innovative, self-healing materials into the construction process ensures automatic repair of minor damages, contributing to the longevity and inconspicuous maintenance of the building.

Efficient Waste Management Design (Oluleye et al., 2022; Zoghi & Kim, 2020)

Construction tactics encompass the design of efficient waste management systems. Smart construction strategically plans waste disposal and recycling systems during construction, minimising the visual impact of waste and contributing to the overall cleanliness and inconspicuous operation of the building.

Invisible Technology Embedment (Jones et al., 2022; Mars & Kohlstedt, 2020)

Construction methods are tactically employed to embed technology seamlessly into the building's design. Concealed wiring, hidden sensors, and placement of technology elements ensure that the technological infrastructure supports building functionality without compromising its aesthetic appeal.

Achieving countermeasures through construction tactics in smart construction represents a strategic and proactive approach. The method of construction, viewed as a tactical element, influences the seamless integration of security measures, surveillance, adaptive lighting, noise reduction, predictive maintenance, innovative materials, climate control, waste management, invisible technology, and flexible design. These construction tactics address potential challenges and contribute to the overall functionality, security, and inconspicuous operation of buildings, defining the future of smart and seamlessly integrated construction practices.

Smart Infrastructure and Development Towards Stealth Construction

Smart construction represents a paradigm shift in how buildings are conceptualised, designed, and executed, ushering in an era where technological integration converges with sustainability and functionality. Within intelligent construction, the pursuit of stealth construction emerges as a critical objective, emphasising the seamless integration of structures with their surroundings and minimising their visual impact. This section delves into how smart construction, with its innovative technologies and strategic methodologies, contributes to realising stealth construction objectives and creating buildings that harmoniously blend into their environments.

Smart construction, at its core, seeks to optimise every phase of the building lifecycle, from design and construction to operation and maintenance (Pan & Zhang, 2023). This optimisation is crucial when aspiring towards stealth construction, a concept where the visibility and impact of built structures are intentionally reduced, aligning with aesthetic, environmental, and security considerations. Introducing smart construction methodologies into stealth construction represents a synergistic approach where technology enables inconspicuous and efficient building practices.

Ahsan (2024) mentioned that one of the fundamental principles of smart construction contributing to stealth objectives is the integration of advanced technologies, such as BIM, digital twins, and the IoT. These technologies provide a comprehensive understanding of the built environment, facilitating meticulous planning and design considering aesthetic and functional aspects. Introducing these technologies ensures that every building element is strategically placed, optimising visibility and seamlessly blending the structure into its surroundings. Moreover, smart construction emphasises the importance of sustainability, another crucial dimension of stealth construction. Introducing eco-friendly materials, energy-efficient systems, and green construction practices ensures that buildings operate inconspicuously and minimise their environmental footprint. This integration aligns with the principles of stealth construction, where buildings coexist harmoniously with nature, embodying a visual and ecological subtlety.

Furthermore, Akhras (2000), Anwar et al. (2017), and Zhang et al. (2019) illustrated that introducing smart construction methodologies also address security concerns, a paramount consideration in stealth construction. Advanced surveillance systems, access control mechanisms, and intelligent lighting solutions are seamlessly integrated during construction. These measures ensure that the building operates discreetly, enhancing security without compromising the hidden nature of the structure. Furthermore, introducing modular and flexible construction approaches ensures adaptability, allowing buildings to evolve unobtrusive over time. Smart construction incorporates methodologies that enable easy modifications and expansions, aligning with the environment's and inhabitants' changing needs. This adaptability is a crucial aspect of stealth construction, where structures seamlessly integrate with the dynamic nature of their surroundings.

From the functionalities of smart construction towards stealth construction, we have coined five (5) central sustainable practices in environmental protection, safety, duration, economy, and aesthetics. These practices are directed towards implementing stealth construction technologies into construction practices right from the onset of construction to completion and the whole life cycle of the project.

Environmental Protection

Integrating smart construction methodologies in the ever-evolving construction landscape heralds a new era where environmental protection is not merely a consideration but a fundamental essence. Stealth construction, emphasising harmonious integration with the environment, finds a natural ally in smart construction practices. This section explains how environmental protection is achieved in stealth construction through the strategic implementation of smart construction technologies and methodologies.

Sustainable Material Selection (Desha et al., 2024; Figueiredo et al., 2021)

Smart construction prioritises using sustainable materials, a cornerstone in achieving environmental protection within stealth construction. The selection of materials with low environmental impact, recycled content, and a reduced carbon footprint contributes to the overall sustainability of the construction process. By choosing materials that align with eco-friendly standards, smart construction ensures that the built environment seamlessly coexists with the natural surroundings.

Energy-Efficient Systems (Metallidou et al., 2020; Umoh et al., 2024)

Implementing energy-efficient systems is a pivotal aspect of smart construction, furthering the cause of environmental protection in stealth construction. By integrating technologies such as smart lighting, HVAC systems, buildings operate with optimal energy efficiency. This minimises the ecological footprint and enhances the inconspicuous operation of the structure, aligning seamlessly with the principles of stealth construction.

Renewable Energy Integration (Ali et al., 2022; Vijayan et al., 2023)

Smart construction strongly emphasises integrating renewable energy sources, an essential strategy for achieving environmental protection in stealth construction. Incorporating solar panels, wind turbines, and geothermal systems ensures that buildings harness clean and sustainable energy. Smart construction contributes to a greener and more inconspicuous built environment by reducing reliance on traditional energy sources.

Smart Water Management (Obaideen et al., 2022; Wang, 2022)

Efficient water management is another dimension where smart construction contributes to environmental protection within stealth construction. Implementing smart irrigation systems, rainwater harvesting, and water recycling technologies minimises water wastage and ensures a sustainable approach to water usage. These practices not only align with eco-friendly principles but also enhance the inconspicuous and sustainable operation of the building.

Green Roofs and Vertical Gardens (Ogut et al., 2022; Tseng et al., 2022)

Integrating green roofs and vertical gardens is a distinctive feature of smart construction that significantly contributes to environmental protection in stealth construction. These elements not only enhance the aesthetic appeal of the building but also serve as natural insulation, reducing energy consumption.

Moreover, they contribute to biodiversity, promoting a healthier and more natural integration of the structure with its environment.

Digital Twins for Resource Optimisation (Jiang et al., 2022; Kor et al., 2023)

Smart construction leverages digital twins, virtual replicas of physical structures, to optimise resource usage and enhance environmental protection in stealth construction. By simulating construction processes and analysing real-time data, digital twins enable precise resource management, reducing waste and minimising the ecological impact of construction activities. This technology ensures the construction process aligns with stealth construction's inconspicuous and eco-friendly goals.

Waste Reduction and Recycling (Hu et al., 2022; Nalon et al., 2022)

Smart construction embraces waste reduction and recycling practices as integral components of environmental protection in stealth construction. Through advanced waste management systems and recycling technologies, construction sites minimise waste generation and ensure responsible materials disposal. This commitment to sustainable waste practices aligns with stealth construction's inconspicuous and eco-conscious essence.

Construction Safety

Safety is a paramount concern in construction, and its importance is further accentuated in stealth construction, where inconspicuous operations demand meticulous planning and execution. Smart construction emerges as a transformative force in enhancing safety protocols within stealth construction projects. This delves into the extensive ways in which safety is achieved through the strategic implementation of smart construction technologies and methodologies, ensuring that the construction process aligns seamlessly with the principles of both safety and stealth.

Risk Mitigation Through Predictive Analytics (Gondia et al., 2020; Okpala et al., 2020)

Smart construction employs predictive analytics to anticipate potential risks and hazards, significantly enhancing safety in stealth construction. By leveraging historical data, real-time monitoring, and machine learning algorithms, smart construction platforms can predict and identify potential safety issues before they escalate. This proactive approach allows for the implementation of preventive measures, reducing the likelihood of accidents and ensuring the well-being of construction personnel.

IoT-Enabled Safety Monitoring (Elghaish et al., 2021; Gao et al., 2021)

The IoT is pivotal in augmenting safety measures in stealth construction projects. Through the deployment of IoT-enabled sensors and wearables, construction sites can continuously monitor the health and safety of workers. These devices can detect unsafe conditions, track the movement of personnel, and even alert authorities in the event of an emergency. Real-time data from IoT devices empowers construction teams to respond promptly to potential safety threats, creating a secure working environment.

AR for Safety Training (Getuli et al., 2022; Oke & Arowoiya, 2022)

Smart construction incorporates AR as an innovative tool for safety training. AR applications provide immersive and interactive training experiences, allowing workers to simulate and practice safety protocols in a virtual environment. This enhances their understanding of potential risks and proper safety procedures, contributing to a safer work environment in stealth construction where adherence to safety protocols is paramount.

BIM for Collaborative Safety Planning (Siddiqui et al., 2021; Rane, 2023)

BIM is a cornerstone in collaborative safety planning within smart construction. BIM platforms enable multidisciplinary teams to visualise the entire construction process in a digital environment. This collaborative approach allows stakeholders to identify and address potential safety issues during the planning phase, minimising risks and ensuring that safety considerations are seamlessly integrated into the construction process.

Drones for Site Surveillance (Gohari et al., 2022; Shanti et al., 2022)

Smart construction harnesses the power of drones for comprehensive site surveillance, enhancing safety in stealth construction. Drones equipped with cameras and sensors provide real-time aerial views of construction sites, enabling teams to monitor activities and identify potential safety hazards. Drones also facilitate the inspection of hard-to-reach areas, ensuring that safety protocols are consistently upheld throughout construction.

Intelligent Equipment and Robotics (Baduge et al., 2022; Liang et al., 2021)

Integrating intelligent equipment and robotics in smart construction contributes significantly to safety in stealth construction. Autonomous machinery and robots can perform tasks in hazardous or confined spaces, reducing the exposure of human workers to potential risks. Additionally, these technologies are equipped with sensors and advanced navigation systems, allowing them to operate safely and avoid collisions.

Emergency Response Systems (Ma & Wu, 2020; Ramesh et al., 2022)

Smart construction implements advanced emergency response systems that are integral to ensuring safety in stealth construction projects. These systems utilise real-time data from sensors and monitoring devices to detect emergencies such as fires, gas leaks, or structural failures. Automated alerts and notifications enable swift responses, and the integration of smart building materials can enhance the structural integrity of the building, contributing to overall safety.

Collaborative Communication Platforms (Pan & Zhang, 2023; Yu et al., 2022)

Smart construction leverages collaborative platforms that enhance real-time communication among project stakeholders. These platforms facilitate instant communication between workers, supervisors, and emergency responders, ensuring that safety information is relayed promptly. Quick and effective communication is crucial in stealth construction, where coordination and response times are vital in maintaining a safe construction environment.

Safety-Enhancing Wearables (Niknejad et al., 2020)

Wearable technologies are integrated into smart construction practices to enhance safety in stealth construction projects. Smart helmets, vests, and other wearables with sensors can monitor vital signs, detect fatigue, and provide real-time safety alerts. These wearables contribute to the well-being of construction personnel and ensure that safety protocols are followed throughout the construction process.

Construction Duration

In construction, where efficiency and duration are paramount, the acceleration of construction duration becomes a crucial objective. Smart construction emerges as a transformative force, employing innovative technologies and methodologies to streamline construction processes and achieve accelerated timelines. This section details the extensive ways construction duration is achieved through the strategic implementation of smart construction, ensuring that stealth construction projects are discreet and executed with unprecedented efficiency.

BIM for Precise Planning (Kolarić et al., 2022)

Smart construction utilises BIM as a foundational tool for precise planning and coordination. BIM creates a detailed digital representation of the construction project, enabling stakeholders to visualise and analyse every aspect before physical construction begins. This meticulous planning facilitates streamlined workflows, reduces errors, and accelerates decision-making processes, setting the stage for expedited construction in stealth projects.

Prefabrication and Modular Construction (Loo & Wong, 2023)

Prefabrication and modular construction techniques are pivotal in accelerating construction duration within smart construction practices. Components of buildings are manufactured off-site under controlled conditions, ensuring precision and quality. Once on-site, these prefabricated elements can be quickly assembled, significantly reducing construction timelines. This approach aligns seamlessly with stealth construction objectives, allowing for efficient and inconspicuous assembly.

Robotics and Automation (Al-Bashar et al., 2024)

Smart construction leverages robotics and automation to enhance construction duration in stealth projects. Robotic systems can perform repetitive and labour-intensive tasks with precision and duration, reducing the reliance on manual labour. Automated construction processes, such as bricklaying and 3D printing, enable swift progress while maintaining a low profile, contributing to the inconspicuous nature of the construction site.

Advanced Construction Materials (Song et al., 2021)

Implementing advanced materials is a hallmark of smart construction, contributing to efficiency and stealth. High-performance materials, such as self-healing concrete and lightweight composites, not only enhance the durability of structures but also expedite the construction process. These materials often require less time for installation and maintenance, contributing to accelerated construction timelines in stealth projects.

IoT for Real Time Monitoring (Dian et al., 2020)

The IoT is harnessed in smart construction for real-time monitoring and management of construction processes. IoT sensors embedded in construction equipment, materials, and structures provide continuous data streams. This data enables project managers to monitor progress, identify bottlenecks, and make data-driven decisions to optimise construction duration. The seamless integration of IoT technologies ensures that construction activities proceed efficiently in stealth projects.

Digital Twins for Simulation and Optimisation (Kor et al., 2023)

Digital twins, virtual replicas of physical structures or construction processes, are employed in smart construction for simulation and optimisation. Planners can simulate different scenarios, identify potential challenges, and optimise construction sequences by creating a digital twin of a construction project. This virtual planning ensures that the actual construction process unfolds with precision and duration, aligning with the inconspicuous nature of stealth construction.

Drones for Site Surveys and Inspections (Gohari et al., 2022)

Drones are utilised in smart construction for rapid site surveys and inspections, expediting data collection and analysis. Drones with cameras and sensors can quickly assess construction sites, monitor progress, and identify potential issues. This real-time information enables swift decision-making, accelerating the overall construction duration in stealth projects without compromising precision.

Collaborative Construction Software (Onungwa et al., 2021)

Smart construction employs collaborative software that facilitates seamless communication and coordination among project stakeholders. These platforms enable real-time collaboration, document sharing, and streamlined communication, reducing delays caused by miscommunication. The efficient exchange of information contributes to the overall acceleration of construction duration in stealth projects.

AI for Planning and Decision-Making (Pan & Zhang, 2023)

AI algorithms are deployed in smart construction for advanced planning and decision-making. AI analyses vast datasets to optimise construction schedules, allocate resources efficiently, and predict potential challenges. The predictive capabilities of AI contribute to proactive problem-solving, ensuring that construction duration is maintained while adhering to the discreet nature of stealth construction.

Lean Construction Principles (Hamzeh et al., 2021)

Smart construction embraces lean construction principles, emphasising eliminating waste and optimising workflows. By minimising non-value-added activities and streamlining processes, lean construction ensures that resources are utilised efficiently, contributing to accelerated construction timelines. This approach aligns with the inconspicuous goals of stealth construction by promoting a streamlined and efficient construction site.

Economy

In the context of stealth construction, where the objectives include efficiency, inconspicuousness, and resource optimisation, achieving economic benefits is crucial. Smart construction, with its innovative technologies and methodologies, plays a pivotal role in realising economic advantages in stealth construction projects. The extensive way the economy is achieved through the strategic implementation of smart construction ensures that the construction process is discreet and economically efficient.

Precise Planning with BIM (Pan & Zhang, 2023)

Smart construction begins with accurate planning and BIM is a cornerstone of achieving economic efficiency in stealth construction. BIM allows for creating a comprehensive digital model encompassing the entire construction project. This model facilitates accurate cost estimations, resource planning, and project scheduling, ensuring that economic considerations are embedded in the planning phase. By identifying potential challenges and optimising construction sequences, BIM contributes to cost savings and streamlined workflows.

Prefabrication and Modular Construction Techniques (Gallo et al., 2021)

Smart construction emphasises using prefabrication and modular construction techniques to achieve economic benefits in stealth projects. Prefabricated components, manufactured off-site under controlled conditions, lead to cost savings in material usage, labour, and construction time. Additionally, modular construction allows for faster on-site assembly, reducing overall project costs and aligning with stealth construction's discreet and efficient nature.

Optimised Resource Utilisation with Digital Twins (Broo & Schooling, 2023)

Digital twins, virtual replicas of physical structures or construction processes, are employed in smart construction to optimise resource utilisation. By simulating construction scenarios, planners can identify opportunities for resource efficiency and cost savings. This virtual planning with digital twins allows for informed decision-making, minimising waste and enhancing economic performance in stealth construction projects.

Predictive Analytics for Cost Management (Zhu et al., 2022)

Smart construction utilises predictive analytics to manage costs effectively in stealth projects. By analysing historical data and real-time information, predictive analytics algorithms can forecast potential cost overruns and identify areas for cost optimisation. This proactive approach to cost management ensures that stealth construction projects stay within budget constraints while maintaining economic efficiency.

Energy-Efficient Systems for Operational Savings (Wang, 2022)

Beyond the construction phase, smart construction contributes to economic benefits through the integration of energy-efficient systems. Energy-efficient buildings result in long-term operational savings by reducing energy consumption and associated costs. Using bright lighting, HVAC systems, and renewable energy sources ensures that stealth construction projects are discreet and economically sustainable in their ongoing operational phases.

Remote Monitoring and Management (Dian et al., 2020)

Smart construction leverages remote monitoring and management technologies to achieve economic efficiency in stealth construction. Remote sensors and IoT devices allow for real-time monitoring of construction sites, enabling project managers to make informed decisions without needing constant on-site presence. This approach reduces travel costs, enhances efficiency, and contributes economic benefits to stealth construction projects.

Life Cycle Cost Analysis (Panteli et al., 2020)

Smart construction includes life cycle cost analysis as a strategic tool to assess the economic viability of stealth construction projects. This analysis considers upfront construction costs and long-term costs associated with operation, maintenance, and eventual decommissioning. By evaluating the entire life cycle, smart construction ensures that economic considerations are comprehensive and aligned with stealth construction's discreet and efficient goals.

Lean Construction Principles for Waste Reduction (Hamzeh et al., 2021)

Lean construction principles are integral to achieving economic benefits in stealth construction. Lean construction ensures efficient resource utilisation and cost-effective project delivery by minimising waste, optimising workflows, and eliminating non-value-added activities. This approach aligns with the inconspicuous goals of stealth construction by promoting a streamlined and economically efficient construction site.

Blockchain for Transparent and Efficient Transactions (Liu et al., 2023)

Blockchain technology is employed in smart construction to enhance transparency and efficiency in financial transactions. Smart contracts on blockchain platforms facilitate secure and automated payment processes, reducing the risk of disputes and delays. This transparent and efficient financial management contributes to economic benefits in stealth construction projects, ensuring smooth financial transactions throughout the project lifecycle.

Aesthetics

Achieving aesthetic excellence in stealth construction, where the seamless integration of structures with their surroundings is paramount, demands a thoughtful combination of design ingenuity and technological innovation. Smart construction emerges as a transformative force in realising aesthetic goals, ensuring that buildings blend into their environments inconspicuously and exhibit a harmonious and visually appealing presence. This section shows extensive ways in which aesthetic objectives are achieved through the strategic implementation of smart construction, creating a visual collaboration between architecture and nature.

Digital Design and Virtual Prototyping (Rafsanjani & Nabizadeh, 2023)

Smart construction begins with the digital design and virtual prototyping of structures using advanced tools and technologies. Digital design platforms enable architects and designers to create intricate, context-aware designs that consider natural surroundings. Virtual prototyping allows for the simulation of different design elements, ensuring that the final structure aligns seamlessly with the aesthetic goals of stealth construction. This digital approach provides a robust foundation for achieving aesthetic excellence in the built environment.

BIM for Comprehensive Visualisation (Wang & Xie, 2024)

BIM is a powerful tool for comprehensive visualisation in smart construction. BIM platforms create a three-dimensional digital representation of the entire construction project, allowing stakeholders to visualise the structure in its intended environment. This visualisation aids in refining design elements, optimising spatial relationships, and ensuring that the building aligns aesthetically with its surroundings. BIM contributes to the seamless integration of the structure into the natural landscape, an essential aspect of achieving aesthetic goals in stealth construction.

Sustainable and Context-Aware Design (Chen et al., 2022)

Smart construction strongly emphasises sustainable and context-aware design principles. Aesthetic excellence in stealth construction is about blending into the environment and ensuring the design is responsive to ecological considerations. Utilising eco-friendly materials, incorporating green roofs, and embracing sustainable design practices contribute to a visually appealing structure that enhances its aesthetic harmony with the surroundings.

Advanced Materials and Facade Innovations (Kraft et al., 2022)

Implementing advanced materials and facade innovations is instrumental in achieving aesthetic objectives in stealth construction. Smart construction explores materials that not only enhance the structural integrity of the building but also contribute to its visual appeal. Innovations in facades, such as dynamic glass technologies and responsive shading systems, allow for adaptive aesthetics that respond to changing environmental conditions. These advancements ensure that the building's aesthetic qualities remain dynamic and harmonious with the surrounding context.

Intelligent Lighting Design (Cheng et al., 2020)

Aesthetic excellence in stealth construction extends to intelligent lighting design. Smart construction incorporates energy-efficient lighting systems and is strategically designed to enhance the structure's visual impact. Dynamic lighting solutions, responsive to natural light conditions and the time of day, contribute to the overall aesthetic harmony of the building with its surroundings. This attention to lighting design ensures that the structure remains inconspicuous yet visually captivating.

Landscaping Integration and Green Spaces (Huang et al., 2020)

Smart construction integrates landscaping and green spaces as integral components of the aesthetic vision. The seamless transition between built structures and natural landscapes is achieved through thoughtful landscaping design. Green roofs, vertical gardens, and strategically placed vegetation contribute to the visual appeal of the building while enhancing its integration with the environment. These green elements create a tranquil and aesthetically pleasing atmosphere around the structure, aligning with the goals of stealth construction.

AI for Design Optimisation (Panchalingam & Chan, 2021)

AI is leveraged in smart construction for design optimisation, contributing to aesthetic excellence. AI algorithms analyse vast datasets to identify design patterns, material choices, and aesthetic preferences. This data-driven approach optimises design decisions for visual appeal while maintaining stealth construction's discreet and harmonious nature.

Smart Infrastructure and Development for Stealth Construction in the Construction Stage

In stealth construction, the construction stage is the pivotal act where the blueprint transforms into tangible reality. Smart construction methodologies strategically employed during this phase elevate the unnoticeable integration of architectural marvels with their surroundings. Also, BIM becomes the focal point, providing architects and builders with a digital platform to orchestrate a precise construction process. This construction process is visualised through BIM, allowing for precise planning, coordination, and visualisation. Additionally, this digital rehearsal ensures that every element aligns seamlessly with the environment, contributing to the discreet elegance of stealth construction. Furthermore, integrating IoT devices enhances this performance by providing real-time monitoring of construction activities. Safety is heightened, and the construction process becomes an adaptive, responsive entity, ensuring the project remains on course while maintaining a low profile.

Digital twins are essential in stealth construction within the construction stage. These virtual replicas enable architects and builders to simulate and optimise design elements, ensuring the final structure harmonises aesthetically with its natural surroundings, as discussed throughout this chapter. Also, robotics and automation technologies take the stage, executing tasks with precision and efficiency, reducing manual labour, and ensuring the construction process remains inconspicuous. The use of advanced construction materials, integral to smart construction, becomes the materialisation of artistic intent, contributing to structural integrity and aligning with sustainability goals. As the construction site transforms, these materials serve as the building blocks of structures and statements that resonate with the environment and the principles of stealth construction.

In addition, selecting the construction stage as the primary domain for smart construction is deeply rooted in its direct impact on the tangible outcome of a project. BIM, which can guide precise planning and coordination, is indispensable during this phase. Infused with smart construction technologies, the construction process aligns seamlessly with stealth construction's aesthetic and contextual goals. The deployment of advanced materials, automation, and robotics during this stage is akin to the artist's brushstrokes on canvas, influencing the efficiency and outcome of the project. As each beam is laid and every corner takes shape, sustainability principles are embedded in construction.

Moreover, the real-time adaptability afforded by smart construction technologies in the construction stage is a strategic advantage. As vigilant overseers, IoT devices allow for constant monitoring, enabling quick adjustments to the construction process, adherence to safety protocols, and immediate responses to unforeseen challenges. The construction stage, therefore, is not just the act of creation but a dynamic, responsive performance where smart construction technologies ensure that the building emerges unobstructed yet majestically. In this phase, the combination of innovative technologies and architectural vision blossoms into a structure that not only stands as a testament to human ingenuity but also seamlessly integrates into the natural landscape, embodying the principles of stealth construction.

Conclusion

The discussion about using smart construction for stealth projects has showcased technology skills, strategic planning, and environmental consciousness in the complex world of construction advancements. The process involves a complete approach to subtle and sustainable development, from planning before construction begins to building and then managing after construction. Integrating technologies such as BIM, digital twins, IoT, and advanced materials exemplifies a paradigm shift in the construction industry. These technologies, strategically employed, not only ensure precision in planning and execution but also contribute to the aesthetic harmony and energy efficiency of structures, seamlessly blending them into their natural environments. The focus on the construction stage as the epicentre of smart construction activities underscores the transformative impact of technology on the tangible realisation of architectural visions. As the built environment evolves, the principles of uninterrupted integration, energy efficiency, and community engagement emerge as the cornerstones of a new era in construction practices, where smart construction serves as a guiding force towards a future where structures seamlessly coexist with their surroundings, reflecting the harmonious intersection of human ingenuity and environmental stewardship.

References

Ahsan, A. (2024). *Integration of BIM and GIS for smart construction management* [Doctoral dissertation, Politecnico di Torino].

Akhras, G. (2000). Smart materials and smart systems for the future. *Canadian Military Journal*, *1*(3), 25–31.

Al-Bashar, M., Taher, M. A., Islam, M. K., & Ahmed, H. (2024). The impact of advanced robotics and automation on supply chain efficiency in industrial manufacturing: A comparative analysis between the USA and Bangladesh. *Global Mainstream Journal of Business, Economics, Development & Project Management*, *3*(3), 28–41.

Ali, A. O., Elmarghany, M. R., Abdelsalam, M. M., Sabry, M. N., & Hamed, A. M. (2022). Closed-loop home energy management system with renewable energy sources in a smart grid: A comprehensive review. *Journal of Energy Storage*, *50*, 1–24. https://doi.org/10.1016/j.est.2022.104609

Alkhatib, H., Lemarchand, P., Norton, B., & O'Sullivan, D. T. J. (2021). Deployment and control of adaptive building facades for energy generation, thermal insulation, ventilation and daylighting: A review. *Applied Thermal Engineering*, *185*, 1–15. https://doi.org/10.1016/j.applthermaleng.2020.116331

Andersen, A. D., Markard, J., Bauknecht, D., & Korpås, M. (2023). Architectural change in accelerating transitions: Actor preferences, system architectures, and flexibility technologies in the German energy transition. *Energy Research & Social Science*, *97*, 1–22. https://doi.org/10.1016/j.erss.2023.102945

Anwar, A., Mahmood, A. N., & Pickering, M. (2017). Modeling and performance evaluation of stealthy false data injection attacks on smart grid in the presence of corrupted measurements. *Journal of Computer and System Sciences*, *83*(1), 58–72.

Baduge, S. K., Thilakarathna, S., Perera, J. S., Arashpour, M., Sharafi, P., Teodosio, B., Shringi, B., & Mendis, P. (2022). Artificial intelligence and smart vision for building and construction 4.0: Machine and deep learning methods and applications. *Automation in Construction*, *141*, 1–26. https://doi.org/10.1016/j.autcon.2022.104440

Bagherian, M. A., & Mehranzamir, K. (2020). A comprehensive review on renewable energy integration for combined heat and power production. *Energy Conversion and Management*, *224*, 1–38. https://doi.org/10.1016/j.enconman.2020.113454

Brintrup, A., Pak, J., Ratiney, D., Pearce, T., Wichmann, P., Woodall, P., & McFarlane, D. (2020). Supply chain data analytics for predicting supplier disruptions: A case study in complex asset manufacturing. *International Journal of Production Research*, *58*(11), 3330–3341.

Broo, D. G., & Schooling, J. (2023). Digital twins in infrastructure: Definitions, current practices, challenges and strategies. *International Journal of Construction Management*, *23*(7), 1254–1263.

Catalano, C., Meslec, M., Boileau, J., Guarino, R., Aurich, I., Baumann, N., Chartier, F., Dalix, P., Deramond, S., Laube, P., Lee, A., K., Ochsner, P., Pasturel, M., Soret, M., & Moulherat, S. (2021). Smart sustainable cities of the new millennium: towards design for nature. *Circular Economy and Sustainability*, *1*(3), 1053–1086.

Charles, S., & Mishra, P. (2021). A survey of network-on-chip security attacks and countermeasures. *ACM Computing Surveys (CSUR)*, *54*(5), 1–36.

Chen, Y., Huang, D., Liu, Z., Osmani, M., & Demian, P. (2022). Construction 4.0, industry 4.0, and building information modeling (BIM) for sustainable building development within the smart city. *Sustainability*, *14*(16), 1–37. https://doi.org/10.3390/su141610028

Cheng, Y., Fang, C., Yuan, J., & Zhu, L. (2020). Design and application of a smart lighting system based on distributed wireless sensor networks. *Applied Sciences*, *10*(23), 1–21. https://doi.org/10.3390/app10238545

D'Auria, A., Tregua, M., & Vallejo-Martos, M. C. (2018). Modern conceptions of cities as smart and sustainable and their commonalities. *Sustainability*, *10*(8), 1–18. https://doi.org/10.3390/su10082642

Desha, C., Reeve, A., & el Baghdadi, O. (2024). Nature-based design for health and well- being promoting cities. In R. B. Egenhoefer (Ed.), *Routledge handbook of sustainable design* (pp. 424–439). Routledge.

Dian, F. J., Vahidnia, R., & Rahmati, A. (2020). Wearables and the Internet of Things (IoT), applications, opportunities, and challenges: A survey. *IEEE Access*, *8*, 69200–69211.

Ding, J., Mei, H., I, C. L., Zhang, H., & Liu, W. (2020). Frontier progress of unmanned aerial vehicles optical wireless technologies. *Sensors*, *20*(19), 1–35. https://doi.org/10.3390/s20195476

Elghaish, F., Hosseini, M. R., Matarneh, S., Talebi, S., Wu, S., Martek, I., Poshdar, M., & Ghodrati, N. (2021). Blockchain and the 'Internet of Things' for the construction industry: Research trends and opportunities. *Automation in Construction*, *132*, 1–15. https://doi.org/10.1016/j.autcon.2021.103942

Figueiredo, K., Pierott, R., Hammad, A. W., & Haddad, A. (2021). Sustainable material choice for construction projects: A life cycle sustainability assessment framework based on BIM and Fuzzy-AHP. *Building and Environment*, *196*, 1–14. https://doi.org/10.1016/j.buildenv.2021.107805

Gallo, P., Romano, R., & Belardi, E. (2021). Smart green prefabrication: Sustainability performances of industrialized building technologies. *Sustainability*, *13*(9), 1–31. https://doi.org/10.3390/su13094701

Gao, X., Pishdad-Bozorgi, P., Shelden, D. R., & Tang, S. (2021). Internet of things enabled data acquisition framework for smart building applications. *Journal of Construction Engineering and Management*, *147*(2), 1–15. https://doi.org/10.1061/(ASCE)CO.1943-7862.0001983

Getuli, V., Capone, P., Bruttini, A., & Sorbi, T. (2022). A smart objects library for BIM-based construction site and emergency management to support mobile VR safety training experiences. *Construction Innovation*, *22*(3), 504–530.

Gohari, A., Ahmad, A. B., Rahim, R. B. A., Supa'at, A. S. M., Abd Razak, S., & Gismalla, M. S. M. (2022). Involvement of surveillance drones in smart cities: A systematic review. *IEEE Access*, *10*, 56611–56628.

Gondia, A., Siam, A., El-Dakhakhni, W., & Nassar, A. H. (2020). Machine learning algorithms for construction projects delay risk prediction. *Journal of Construction Engineering and Management*, *146*(1), 1–16. https://doi.org/10.1061/(ASCE)CO.1943-7862.0001736

Grover, A. (2021). *Hyperfunctional energy landscapes: Retrofitting public space with renewable energy infrastructure*. https://scholarsbank.uoregon.edu/xmlui/handle/1794/26334

Guesmi, A., Hanif, M. A., Ouni, B., & Shafique, M. (2023). Physical adversarial attacks for camera-based smart systems: Current trends, categorization, applications, research challenges, and future outlook. *IEEE Access*, *11*(2), 109617–109668. https://doi.org/10.1109/ACCESS.2023.3321118.

Hamidavi, T., Abrishami, S., & Hosseini, M. R. (2020). Towards intelligent structural design of buildings: A BIM-based solution. *Journal of Building Engineering*, *32*, 1–15. https://doi.org/10.1016/j.jobe.2020.101685

Hammi, B., Zeadally, S., Khatoun, R., & Nebhen, J. (2022). Survey on smart homes: Vulnerabilities, risks, and countermeasures. *Computers & Security*, *117*, 1–25. https://doi.org/10.1016/j.cose.2022.102677

Hamzeh, F., González, V. A., Alarcon, L. F., & Khalife, S. (2021, July). Lean construction 4.0: Exploring the challenges of development in the AEC industry. In *Proceedings of the 29th annual conference of the international group for lean construction (IGLC), Lima, Peru* (pp. 207–216).

Himeur, Y., Ghanem, K., Alsalemi, A., Bensaali, F., & Amira, A. (2021). Artificial intelligence based anomaly detection of energy consumption in buildings: A review, current trends and new perspectives. *Applied Energy*, *287*, 1–26. https://doi.org/10.1016/j.apenergy.2021.116601

Hu, X., Zhou, Y., Vanhullebusch, S., Mestdagh, R., Cui, Z., & Li, J. (2022). Smart building demolition and waste management frame with image-to-BIM. *Journal of Building Engineering*, *49*, 1–16. https://doi.org/10.1016/j.jobe.2022.104058

Huang, H., Zhang, M., Yu, K., Gao, Y., & Liu, J. (2020). Construction of complex network of green infrastructure in smart city under spatial differentiation of landscape. *Computer Communications*, *154*(2), 380–389.

Ibrahim, F. S. B., Esa, M. B., & Rahman, R. A. (2021). The adoption of IoT in the Malaysian construction industry: Towards construction 4.0. *International Journal of Sustainable Construction Engineering and Technology*, *12*(1), 56–67.

Jafari, M., & Alipour, A. (2021). Review of approaches, opportunities, and future directions for improving aerodynamics of tall buildings with smart facades. *Sustainable Cities and Society*, *72*, 1–20. https://doi.org/10.1016/j.scs.2021.102979

Jiang, Y., Li, M., Guo, D., Wu, W., Zhong, R. Y., & Huang, G. Q. (2022). Digital twin-enabled smart modular integrated construction system for on-site assembly. *Computers in Industry*, *136*, 1–18. https://doi.org/10.1016/j.compind.2021.103594

Jones, K., Mosca, L., Whyte, J., Davies, A., & Glass, J. (2022). Addressing specialization and fragmentation: Product platform development in construction consultancy firms. *Construction Management and Economics*, *40*(11–12), 918–933.

Kabirifar, K., Mojtahedi, M., Wang, C. C., & Tam, T. V. (2020). A conceptual foundation for effective construction and demolition waste management. *Cleaner Engineering and Technology*, *1*, 1–12. https://doi.org/10.1016/j.clet.2020.100019

Khan, A. M., Osińska, M., & Boehlke, J. (2024). Nexus of circular economy and sustainable economic growth for EU countries. In A. Glińska-Neweś & P. Ulkuniemi (Eds.),*The human dimension of the circular economy* (pp. 46–72). Edward Elgar Publishing.

Kolarić, S., Vukomanović, M., & Ramljak, A. (2022). Analyzing the level of detail of construction schedule for enabling site logistics planning (SLP) in the building information modeling (BIM) environment. *Sustainability*, *14*(11), 1–22. https://doi.org/10.3390/su14116701

Kor, M., Yitmen, I., & Alizadehsalehi, S. (2023). An investigation for integration of deep learning and digital twins towards Construction 4.0. *Smart and Sustainable Built Environment*, *12*(3), 461–487.

Korkas, C. D., Terzopoulos, M., Tsaknakis, C., & Kosmatopoulos, E. B. (2022). Nearly optimal demand side management for energy, thermal, EV and storage loads: An Approximate Dynamic Programming approach for smarter buildings. *Energy and Buildings*, *255*, 1–11. https://doi.org/10.1016/j.enbuild.2021.111676

Kraft, R., Kahnt, A., Grauer, O., Thieme, M., Wolz, D. S., Schlüter, D., Tietze, M., Curbach, M., Holschemacher, K., Jäger, H., & Böhm, R. (2022). Advanced carbon reinforced concrete technologies for façade elements of nearly zero-energy buildings. *Materials*, *15*(4), 1–19. https://doi.org/10.3390/ma15041619

Li, Z., Zhang, J., Li, M., Huang, J., & Wang, X. (2020). A review of smart design based on interactive experience in building systems. *Sustainability*, *12*(17), 1–15. https://doi.org/10.3390/su12176760

Liang, C. J., Wang, X., Kamat, V. R., & Menassa, C. C. (2021). Human–robot collaboration in construction: Classification and research trends. *Journal of Construction Engineering and Management*, *147*(10), 1–23. https://doi.org/10.1061/(ASCE)CO.1943-7862.0002154

Liu, H., Han, S., & Zhu, Z. (2023). Blockchain technology toward smart construction: Review and future directions. *Journal of Construction Engineering and Management*, *149*(3), 1–16.

Liu, Q., Shi, R., & Lu, L. (2024). Existing vibration control techniques applied in construction and mechanical engineering. *Mechatronics and Automation Technology*, *1*(1), 599–618. https://doi.org/10.3233/ATDE231155

Loo, B. P., & Wong, R. W. (2023). Towards a conceptual framework of using technology to support smart construction: The case of modular integrated construction (MiC). *Buildings*, *13*(2), 1–16. https://doi.org/10.3390/buildings13020372

Ma, G., & Wu, Z. (2020). BIM-based building fire emergency management: Combining building users' behavior decisions. *Automation in Construction*, *109*, 1–16. https://doi.org/10.1016/j.autcon.2019.102975

Marocco, M., & Garofolo, I. (2021). Integrating disruptive technologies with facilities management: A literature review and future research directions. *Automation in Construction*, *131*, 1–15. https://doi.org/10.1016/j.autcon.2021.103917

Mars, R., & Kohlstedt, K. (2020). *The 99% invisible city: A field guide to the hidden world of everyday design*. Houghton Mifflin.

Merabet, G. H., Essaaidi, M., Talei, H., & Benhaddou, D. (2022). Towards advanced technologies for smart building management: Linking building components and energy use. In A. Tomar, P. H. Nguyen, S. Mishra (Eds.), *Control of smart buildings: An integration to grid and local energy communities* (pp. 179–202). Springer Nature Singapore.

Metallidou, C. K., Psannis, K. E., & Egyptiadou, E. A. (2020). Energy efficiency in smart buildings: IoT approaches. *IEEE Access*, *8*, 63679–63699.

Mohsan, S. A. H., Khan, M. A., Mazinani, A., Alsharif, M. H., & Cho, H. S. (2022). Enabling underwater wireless power transfer towards sixth generation (6g) wireless networks: Opportunities, recent advances, and technical challenges. *Journal of Marine Science and Engineering*, *10*(9), 1–36. https://doi.org/10.3390/jmse10091282

Mun, R. S. (2023). *Strategies for influential interactivity in the physical domain* [Doctoral dissertation, Massachusetts Institute of Technology].

Musarat, M. A., Alaloul, W. S., Khan, A. M., Ayub, S., & Jousseaume, N. (2024). A survey- based approach of framework development for improving the application of Internet of Things in the construction industry of Malaysia. *Results in Engineering*, *21*, 1–18. https://doi.org/10.1016/j.rineng.2024.101823

Nalon, G. H., Santos, R. F., de Lima, G. E. S., Andrade, I. K. R., Pedroti, L. G., Ribeiro, J. C. L., & de Carvalho, J. M. F. (2022). Recycling waste materials to produce self-sensing concretes for smart and sustainable structures: A review. *Construction and Building Materials*, *325*, 1–30. https://doi.org/10.1016/j.conbuildmat.2022.126658

Nazari, S., Borrelli, F., & Stefanopoulou, A. (2020). Electric vehicles for smart buildings: A survey on applications, energy management methods, and battery degradation. *Proceedings of the IEEE*, *109*(6), 1128–1144.

Niknejad, N., Ismail, W. B., Mardani, A., Liao, H., & Ghani, I. (2020). A comprehensive overview of smart wearables: The state-of-the-art literature, recent advances, and future challenges. *Engineering Applications of Artificial Intelligence*, *90*, 1–61. https://doi.org/10.1016/j.engappai.2020.103529

Obaideen, K., Yousef, B. A., AlMallahi, M. N., Tan, Y. C., Mahmoud, M., Jaber, H., & Ramadan, M. (2022). An overview of smart irrigation systems using IoT. *Energy Nexus*, *7*, 1–11. https://doi.org/10.1016/j.nexus.2022.100124

Ogut, O., Tzortzi, N. J., & Bertolin, C. (2022). Vertical green structures to establish sustainable built environment: A systematic market review. *Sustainability*, *14*(19), 1–36. https://doi.org/10.3390/su141912349

Oke, A. E., & Arowoiya, V. A. (2022). An analysis of the application areas of augmented reality technology in the construction industry. *Smart and Sustainable Built Environment*, *11*(4), 1081–1098.

Okpala, I., Nnaji, C., & Karakhan, A. A. (2020). Utilizing emerging technologies for construction safety risk mitigation. *Practice Periodical on Structural Design and Construction*, *25*(2), 1–13. https://doi.org/10.1061/(ASCE)SC.1943-5576.0000468

Oluleye, B. I., Chan, D. W., Saka, A. B., & Olawumi, T. O. (2022). Circular economy research on building construction and demolition waste: A review of current trends and future research directions. *Journal of Cleaner Production*, *357*, 1–18. https://doi.org/10.1016/j.jclepro.2022.131927

Onungwa, I., Olugu-Uduma, N., & Shelden, D. R. (2021). Cloud BIM technology as a means of collaboration and project integration in smart cities. *Sage Open*, *11*(3), 1–9. https://doi.org/10.1177/21582440211033250

Ozarisoy, B. (2022). Energy effectiveness of passive cooling design strategies to reduce the impact of long-term heatwaves on occupants' thermal comfort in Europe: Climate change and mitigation. *Journal of Cleaner Production*, *330*, 1–54. https://doi.org/10.1016/j.jclepro.2021.129675

Pan, Y., & Zhang, L. (2023). Integrating BIM and AI for smart construction management: Current status and future directions. *Archives of Computational Methods in Engineering*, *30*(2), 1081–1110.

Panchalingam, R., & Chan, K. C. (2021). A state-of-the art review on artificial intelligence for Smart Buildings. *Intelligent Buildings International*, *13*(4), 203–226.

Panteli, C., Kylili, A., & Fokaides, P. A. (2020). Building information modelling applications in smart buildings: From design to commissioning and beyond A critical review. *Journal of Cleaner Production*, *265*, 1–13. https://doi.org/10.1016/j.jclepro.2020.121766

Putrada, A. G., Abdurohman, M., Perdana, D., & Nuha, H. H. (2022). Machine learning methods in smart lighting toward achieving user comfort: A survey. *IEEE Access*, *10*, 45137–45178.

Qayyum, F., Jamil, H., & Ali, F. (2023). A review of smart energy management in residential buildings for smart cities. *Energies*, *17*(1), 1–29. https://doi.org/10.3390/en17010083

Quach, S. K. (2022). *Changing business direction towards a more industrialized approach using modular construction*. https://odr.chalmers.se/items/8288cf35-1ec7-4dcb-8e0b-833f3980a329

Rafsanjani, H. N., & Nabizadeh, A. H. (2023). Towards digital architecture, engineering, and construction (AEC) industry through virtual design and construction (VDC) and digital twin. *Energy and Built Environment*, *4*(2), 169–178.

Rahimian, F. P., Goulding, J. S., Abrishami, S., Seyedzadeh, S., & Elghaish, F. (2021). *Industry 4.0 solutions for building design and construction: A paradigm of new opportunities*. Routledge.

Ramesh, G., Logeshwaran, J., & Rajkumar, K. (2022). The smart construction for image pre? processing of mobile robotic systems using neuro fuzzy logical system approach. *NeuroQuantology*, *20*(10), 6354–6367.

Rane, N., Choudhary, S., & Rane, J. (2023). *Artificial Intelligence (AI) and Internet of Things (IoT)-based sensors for monitoring and controlling in architecture, engineering, and construction: Applications, challenges, and opportunities*. https://ssrn.com/abstract=4642197

Sajid, S., Haleem, A., Bahl, S., Javaid, M., Goyal, T., & Mittal, M. (2021). Data science applications for predictive maintenance and materials science in context to Industry 4.0. *Materials today: Proceedings*, *45*, 4898–4905. https://doi.org/10.1016/j.matpr.2021.01.357

Salor, L., C. & Baeza, V. M. (2023). Harnessing the potential of emerging technologies to break down barriers in tactical communications. *Telecom*, *4*(4), 709–731. https://doi.org/10.3390/telecom4040032

Santos, P. G. D. (2020). *Hybrid performance-based wood panels for a smart construction* [Doctoral dissertation, 00500: Universidade de Coimbra].

Sepasgozar, S. M. (2021). Differentiating digital twin from digital shadow: Elucidating a paradigm shift to expedite a smart, sustainable built environment. *Buildings*, *11*(4), 1–16. https://doi.org/10.3390/buildings11040151

Seyedolhosseini, A., Masoumi, N., Modarressi, M., & Karimian, N. (2020). Daylight adaptive smart indoor lighting control method using artificial neural networks. *Journal of Building Engineering*, *29*, 1–10. https://doi.org/10.1016/j.jobe.2019.101141

Shanti, M. Z., Cho, C. S., de Soto, B. G., Byon, Y. J., Yeun, C. Y., & Kim, T. Y. (2022). Real- time monitoring of work-at-height safety hazards in construction sites using drones and deep learning. *Journal of Safety Research*, *83*, 364–370. https://doi.org/10.1016/j.jsr.2022.09.011

Shurrab, H., Jonsson, P., & Johansson, M. I. (2022). Managing complexity through integrative tactical planning in engineer-to-order environments: Insights from four case studies. *Production Planning & Control*, *33*(9–10), 907–924.

Siddiqui, A. A., Ewer, J. A., Lawrence, P. J., Galea, E. R., & Frost, I. R. (2021). Building Information Modelling for performance-based Fire Safety Engineering analysis – A strategy for data sharing. *Journal of Building Engineering*, *42*, 1–14. https://doi.org/10.1016/j.jobe.2021.102794

Sidorova, L. (2022). *Enhancing aesthetics in industrial architecture: Integrating 3D-printed facade structures in the design of an incineration plant in Tromso*. https://www.politesi.polimi.it/handle/10589/214551

Sofi, A., Regita, J. J., Rane, B., & Lau, H. H. (2022). Structural health monitoring using wireless smart sensor network–An overview. *Mechanical Systems and Signal Processing*, *163*, 1–14. https://doi.org/10.1016/j.ymssp.2021.108113

Song, T., Jiang, B., Li, Y., Ji, Z., Zhou, H., Jiang, D., Seok, I., Murugadoss, V., Wen, N., & Colorado, H. (2021). Self-healing materials: A review of recent developments. *ES Materials & Manufacturing*, *14*(2), 1–19.

Taheri, S., Hosseini, P., & Razban, A. (2022). Model predictive control of heating, ventilation, and air conditioning (HVAC) systems: A state-of-the-art review. *Journal of Building Engineering*, *60*, 1–22. https://doi.org/10.1016/j.jobe.2022.105067

Thurman, L. L., Stein, B. A., Beever, E. A., Foden, W., Geange, S. R., Green, N., Gross, J., E., Lawrence, D., J., LeDee, O., Olden, J., D., Thompson, L., M., & Young, B. E. (2020). Persist in place or shift in space? Evaluating the adaptive capacity of species to climate change. *Frontiers in Ecology and the Environment*, *18*(9), 520–528.

Tseng, K. H., Chung, M. Y., Chen, L. H., & Chou, L. A. (2022). A study of green roof and impact on the temperature of buildings using integrated IoT system. *Scientific Reports*, *12*(1), 1–16. https://doi.org/10.1038/s41598-022-20552-6

Umoh, A. A., Adefemi, A., Ibewe, K. I., Etukudoh, E. A., Ilojianya, V. I., & Nwokediegwu, Z. Q. S. (2024). Green architecture and energy efficiency: A review of innovative design and construction techniques. *Engineering Science & Technology Journal*, *5*(1), 185–200.

Vagtholm, R., Matteo, A., Vand, B., & Tupenaite, L. (2023). Evolution and current state of building materials, construction methods, and building regulations in the UK: Implications for sustainable building practices. *Buildings*, *13*(6), 1–37. https://doi.org/10.3390/buildings13061480

Vijayan, D. S., Koda, E., Sivasuriyan, A., Winkler, J., Devarajan, P., Kumar, R. S., Jakimiuk, A., Osinski, P., Podlasek, A., & Vaverková, M. D. (2023). Advancements in solar panel technology in civil engineering for revolutionizing renewable energy solutions – A review. *Energies*, *16*(18), 1–33. https://doi.org/10.3390/en16186579

Vukovic, N., Koriugina, U., Illarionova, D., Pankratova, D., Kiseleva, P., & Gontareva, A. (2021). Towards smart green cities: Analysis of integrated renewable energy use in smart cities. *Strategic Planning for Energy and the Environment*, *40*(1), 75–94. https://doi.org/10.13052/spee1048-5236.4015

Wang, J. (2022). Optimized mathematical model for energy efficient construction management in smart cities using building information modeling. *Strategic Planning for Energy and the Environment*, *41*(1), 61–80. https://doi.org/10.13052/spee1048-5236.4113

Wang, R., Lu, S., & Feng, W. (2020). Impact of adjustment strategies on building design process in different climates oriented by multiple performance. *Applied Energy*, *266*, 1–17.

Wang, X., & Xie, M. (2024). Integration of 3DGIS and BIM and its application in visual detection of concealed facilities. *Geo-Spatial Information Science*, *27*(1), 132–141.

Yu, Y., Yazan, D. M., Junjan, V., & Iacob, M. E. (2022). Circular economy in the construction industry: A review of decision support tools based on information & communication technologies. *Journal of Cleaner Production*, *349*, 1–18. https://doi.org/10.1016/j.jclepro.2022.131335

Zhang, L., & Kim, C. (2023). Chromatics in urban landscapes: Integrating interactive genetic algorithms for sustainable color design in marine cities. *Applied Sciences*, *13*(18), 1–26. https://doi.org/10.3390/app131810306

Zhang, M., Shen, C., He, N., Han, S., Li, Q., Wang, Q., & Guan, X. (2019). False data injection attacks against smart gird state estimation: Construction, detection and defense. *Science China Technological Sciences*, *62*(12), 2077–2087.

Zhang, Q., Yuan, R., Singh, V. P., Xu, C. Y., Fan, K., Shen, Z., Wang, G., & Zhao, J. (2022). Dynamic vulnerability of ecological systems to climate changes across the Qinghai-Tibet Plateau, China. *Ecological Indicators*, *134*, 1–11. https://doi.org/10.1016/j.ecolind.2021.108483

Zhu, H., Hwang, B. G., Ngo, J., & Tan, J. P. S. (2022). Applications of smart technologies in construction project management. *Journal of Construction Engineering and Management*, *148*(4), 1–12. https://doi.org/10.1061/(ASCE)CO.1943-7862.0002260

Zoghi, M., & Kim, S. (2020). Dynamic modeling for life cycle cost analysis of BIM-based construction waste management. *Sustainability*, *12*(6), 1–21. https://doi.org/10.3390/su12062483

Section 3

Stealth (Post-construction): Value Management, Partnering, and Whole Life Cycle

Chapter Eight

Value Management in Stealth Construction

Abstract

The implementation of value management (VM) principles in stealth construction projects is explored comprehensively in the chapter. It elucidated how VM positively influences various facets of construction, including environmental protection, health and safety, project delivery duration, economy, and aesthetics. Applying VM techniques, such as proactive risk management, resource optimisation, and stakeholder collaboration, is essential for achieving project objectives while ensuring sustainability, efficiency, and stakeholder satisfaction. Furthermore, the chapter emphasises VM's benefits, challenges, and relevance across all stages of the construction lifecycle, from pre-construction planning to post-construction evaluation, underscoring its integral role in driving continuous improvement and innovation in the construction industry. Overall, the discourse emphasises the importance of VM in optimising outcomes and maximising value in stealth construction projects.

Keywords: Value management; stealth construction; sustainability; risk management; construction optimisation

Introduction

In construction, Ab-Ghani et al. (2021) stated that value management (VM) is a systematic process to maximise stakeholder value throughout the project lifecycle. It involves identifying and understanding the needs and expectations of stakeholders, defining value objectives, and implementing strategies to achieve them. One of the primary objectives of VM is to optimise the use of resources while delivering projects that meet or exceed quality standards. This approach fosters collaboration among project participants, including clients, designers,

Stealth Construction: Integrating Practices for Resilience and Sustainability, 197–214

Published under exclusive licence by Emerald Publishing Limited
doi:10.1108/978-1-83608-182-120251008

contractors, and end-users, to ensure that decisions are aligned with project objectives and priorities. Venkataraman and Pinto (2023) also added that VM begins with a comprehensive analysis of stakeholders' requirements and preferences at the outset of a construction project. This entails engaging stakeholders in workshops or meetings to elicit their input and understand their needs. Through effective communication and active listening, project teams can identify value drivers and prioritise them based on their importance to stakeholders. This initial phase sets the foundation for establishing clear project objectives and performance criteria to guide decision-making throughout the project lifecycle.

Once project objectives are defined, VM systematically evaluates design alternatives, construction methods, and materials to optimise value. This may include value engineering exercises to identify opportunities to reduce costs without sacrificing performance or quality. Project teams can uncover efficiencies and drive continuous improvement by challenging assumptions and exploring innovative solutions. VM also encompasses risk management strategies to mitigate potential threats to project success and enhance value delivery.

Furthermore, VM requires ongoing monitoring and evaluation throughout construction to ensure that project objectives are met and value maximised. This involves tracking key performance indicators, such as cost, schedule, and quality metrics, and making timely adjustments. Additionally, VM fosters collaboration and innovation, encouraging stakeholders to share insights and best practices to drive value creation. When VM principles are engaged, construction projects can achieve greater efficiency, effectiveness, and stakeholder satisfaction, ultimately delivering outcomes that exceed expectations.

Benefits of Implementing VM in Construction

Oke and Aigbavboa (2017), Kuitert et al. (2024), and Adekunle et al. (2022) have identified that VM in construction offers many benefits that positively impact project outcomes, stakeholders and the broader industry, as summarised in Fig. 5. To begin with, it fosters a culture of collaboration and teamwork among project participants, including clients, designers, contractors, and suppliers. When stakeholders are involved in decision-making, VM ensures that diverse perspectives are considered, leading to more informed and effective decisions. This collaborative approach enhances communication and trust among team members, ultimately facilitating smoother project execution and reducing the likelihood of conflicts or disputes.

Furthermore, VM enables projects to achieve better cost control and optimisation of resources. Through value engineering and value analysis techniques, project teams can identify opportunities to streamline processes, eliminate waste, and reduce unnecessary expenses without compromising quality or performance. When maximising the value derived from every dollar spent, VM helps to minimise project costs and improve overall project profitability. This, in turn, enhances the financial viability of construction projects and increases the return on investment for stakeholders.

Another significant benefit of VM is its ability to enhance project quality and innovation. Prioritising value objectives and performance criteria, project

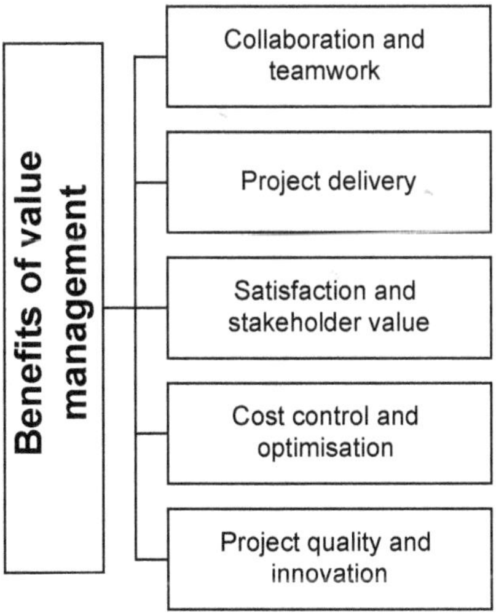

Fig. 5. Benefits of Value Management.

teams are incentivised to seek out innovative solutions and best practices that can deliver superior outcomes. VM encourages continuous improvement and the adoption of new technologies, materials, and construction methods that can drive efficiency, sustainability, and resilience in construction projects. As a result, VM contributes to raising industry standards and advancing state-of-the-art construction practices.

In addition, VM contributes to improved stakeholder satisfaction and stakeholder value. When project decisions are aligned with stakeholder needs and expectations, VM ensures that projects deliver meaningful and valuable outcomes to stakeholders. This increases satisfaction among clients, end-users, and other project stakeholders, enhancing their confidence in the project team and fostering positive relationships. Ultimately, by maximising stakeholder value, VM contributes to the long-term success and reputation of construction projects and the organisations involved.

Challenges in Implementing VM in Construction

VM in construction faces several challenges that can hinder its successful implementation and impact on project outcomes, as stated by Alhumaid et al. (2024), Venkataraman and Pinto (2023), Kineber et al. (2022), Ahmed and Sobuz (2020), and Kannimuthu et al. (2019). One significant challenge is resistance to change among stakeholders. Construction projects often involve numerous stakeholders with varying perspectives, interests, and priorities. Convincing these stakeholders to embrace VM principles and participate actively in the process can be difficult, especially if they are accustomed to traditional project delivery methods. Overcoming resistance to change requires effective communication, stakeholder engagement, and education to demonstrate the benefits of VM and address concerns.

Secondly, inadequate planning and preparation can impede the effectiveness of VM initiatives. VM requires careful planning and coordination to ensure that the right people are involved at the right time and that sufficient resources are allocated to support the process. Also, without proper planning, VM workshops may be poorly attended, key stakeholders may be excluded from discussions, and decisions may be made without adequate consideration of alternatives. To address this challenge, project teams must develop comprehensive VM plans that clearly define objectives, roles and responsibilities, and timelines for implementation. Another challenge is the complexity of construction projects, making it difficult to identify and prioritise value objectives. Construction projects involve numerous interrelated factors, including design requirements, technical specifications, regulatory requirements, and stakeholder preferences. Balancing competing priorities and trade-offs requires careful analysis and decision-making, which can be challenging in the face of uncertainty and ambiguity. VM tools and techniques, such as value engineering and multi-criteria decision analysis, can help project teams navigate this complexity by providing structured approaches to identifying and evaluating value options.

Furthermore, inadequate data and information can hinder VM efforts by limiting the ability to make informed decisions. VM relies on accurate, up-to-date information about project requirements, constraints, and performance metrics. However, construction projects often suffer from data fragmentation, inconsistency, and inaccessibility, making it difficult to assess value and identify improvement opportunities. Addressing this challenge requires investment in data management systems and technologies that enable project teams to collect, analyse, and share information effectively throughout the project lifecycle.

Moreover, time and resource constraints can pose significant challenges to VM in construction. Construction projects are typically subject to tight schedules and budgets, leaving little room for VM activities perceived as time-consuming or costly. Project teams may struggle to allocate sufficient time and resources to VM, leading to rushed decisions, superficial analysis, and missed opportunities for value optimisation. To overcome this challenge, project teams must prioritise VM as an integral part of project planning and execution, allocating adequate resources and integrating VM activities into project workflows. Ultimately, cultural and organisational barriers can hinder the adoption of VM practices within construction organisations. Traditional project delivery methods and organisational structures may reinforce siloed thinking and discourage collaboration and innovation. Changing organisational culture to embrace VM requires strong leadership, effective change management strategies, and investment in training and development.

VM for Stealth Construction

VM for stealth construction involves a strategic approach to maximising the efficiency and effectiveness of construction processes while maintaining value. One key aspect is the integration of advanced technologies at various stages of construction, from planning to execution. For instance, Building Information Modelling (BIM) facilitates precise planning and design. It allows for seamless coordination among different teams, minimising the need for extensive on-site presence and reducing the likelihood of detection. Additionally, adopting prefabrication and modular construction techniques enables components to be assembled off-site, thereby minimising on-site activities and noise, which is crucial for stealth operations.

Another crucial element of VM in stealth construction is the implementation of sustainable practices. Prioritising eco-friendly materials and energy-efficient systems, stealth construction projects can minimise environmental footprint while reducing the detection risk. For example, Amiri et al. (2021) and Marut et al. (2020) highlighted that green building materials such as recycled steel and low-emission concrete contribute to sustainability goals. It enhances the project's stealth capabilities by reducing the need for frequent deliveries and on-site fabrication. Furthermore, integrating renewable energy sources such as solar panels or wind turbines can provide on-site power without relying on conspicuous external sources, further enhancing the project's covert profile.

Moreover, VM in stealth construction extends beyond physical aspects, including risk mitigation and security measures. Tahmasebi (2024) and Alqudhaibi et al. (2023) illustrated that advanced risk assessment techniques, such as predictive analytics and scenario modelling, can identify potential vulnerabilities and threats early in the project lifecycle, allowing for proactive countermeasures. Additionally, stringent security protocols, including access controls and surveillance systems, help safeguard the construction site from unauthorised access and potential espionage. When these measures are integrated into the project's VM framework, stealth construction endeavours can ensure operational efficiency, confidentiality, and security, which are essential for maintaining a low profile in sensitive or high-risk environments.

When incorporating VM into the construction process to achieve stealth construction goals, it is necessary to consider several aspects of the project. These include shaping the building's cross-section, handling construction visibility, managing energy transmission, and deploying countermeasures. The following sections offer an in-depth exploration of each of these vital elements.

Building Cross-Section

VM plays a critical role in ensuring that building cross-section development aligns with the objectives of stealth construction. The shape of the building is a fundamental aspect that directly impacts its stealth capabilities. Architects and engineers can optimise the building's design through VM to minimise its visual and radar signatures. This may involve incorporating angular or irregular shapes that scatter incoming waves, making the structure less detectable by surveillance technologies, as Ma et al. (2001), Golato et al. (2014), and Khan et al. (2023) stipulated in their studies. Additionally, VM facilitates the integration of features such as facades with non-reflective materials or coatings that absorb or diffuse electromagnetic waves, further enhancing the building's stealth characteristics.

Internally, VM drives the selection of construction techniques and materials contributing to the building's stealth profile. For instance, Bertram et al. (2019) and

Jin et al. (2020) postulated that modular construction methods allow for assembling critical components off-site, reducing on-site activities and noise levels and thus enhancing covert operations. Similarly, the strategic use of sound-absorbing materials and acoustic insulation helps dampen noise generated during construction, minimising the risk of detection. Optimising internal construction processes through VM, stealth construction projects can achieve greater efficiency while maintaining a low profile.

Material usage is another critical aspect of building cross-section development in stealth construction, and VM plays a crucial role in selecting appropriate materials that balance stealth requirements with structural integrity and cost considerations. For example, Rwahwire and Ssebagala (2023) and Wu (2020) explained in their studies that using composite materials with radar-absorbing properties can help reduce the building's radar cross-section, making it less detectable to radar surveillance. Likewise, Hons (2019) added that lightweight materials such as carbon fibre or advanced polymers may minimise the building's overall weight, reducing the risk of structural vibrations that could betray its presence. Through VM, construction teams can evaluate various material options and make informed decisions that optimise the building's stealth characteristics without compromising its performance or budgetary constraints.

Handling Visibility

VM is integral to ensuring that handling invisibility aspects aligns with the principles of stealth construction. One crucial factor is the colour and texture of the construction, which can significantly affect its visibility. Through VM, construction teams can select colours and textures that blend seamlessly with the surrounding environment, reducing the structure's visual footprint (Al-Kodmany, 2023; Naqvi, 2023). For instance, using earth tones or camouflage patterns can help conceal the building from aerial or ground-based surveillance. Additionally, incorporating textured surfaces that break up light reflections can enhance invisibility by reducing specular highlights that might give away the building's presence.

Location plays a pivotal role in handling invisibility, and VM enables careful consideration of site selection to minimise exposure and maximise concealment. Rippa (2020) and Barua (2023) identified that VM can ensure that the building remains hidden from prying eyes by strategically positioning the construction site in areas with natural barriers or existing infrastructure, such as dense foliage or urban landscapes. Moreover, choosing locations with minimal traffic or human activity reduces the likelihood of detection and enhances the project's overall stealth capabilities. Through VM, construction teams can analyse various site options and make informed decisions that optimise invisibility while mitigating logistical challenges and risks.

Furthermore, interaction with the environment is another critical aspect of handling invisibility in stealth construction, and VM facilitates the implementation of measures to minimise the building's environmental impact and interaction. For example, Al-Kodmany (2023) and Ofori et al. (2023) stated that using sustainable landscaping techniques to integrate the structure into its surroundings helps disguise its presence and reduce its ecological footprint. Similarly, incorporating passive design strategies, such as orientation for optimal solar exposure and natural ventilation, enhances energy efficiency and contributes to the building's seamless integration with the environment. When environmentally sensitive approaches are optimised through VM, stealth construction projects can achieve invisibility while fostering harmony with the surrounding ecosystem.

Managing Energy Transmission

VM is essential in ensuring that energy transmission management aligns with the objectives of stealth construction. Rathor and Saxena (2020) noted that energy management is a crucial aspect that optimises energy resource consumption and distribution throughout the construction process. Through VM, construction teams can implement energy-efficient practices and technologies, such as Light-emitting Diode (LED) lighting, smart Heating, Ventilation, and Air Conditioning (HVAC) systems, and energy-efficient appliances, to minimise energy consumption and reduce the project's environmental footprint. Additionally, VM facilitates the integration of automated energy management systems that monitor and control energy usage in real-time, ensuring optimal performance while minimising detection risks.

Climate action planning is another essential component of managing energy transmission in stealth construction, and VM enables the development and implementation of strategies to mitigate climate change impacts. According to Ghosh et al. (2020), VM ensures that stealth construction projects contribute to global efforts to combat climate change by prioritising sustainable practices, such as reducing carbon footprint and mitigating greenhouse gas emissions. This may involve incorporating carbon offset initiatives, implementing green building certifications, or adopting renewable energy solutions to reduce reliance on fossil fuels and minimise environmental impact.

Furthermore, energy efficiency is a fundamental consideration in managing energy transmission for stealth construction, and VM drives the adoption of technologies and practices that maximise energy efficiency while maintaining operational effectiveness. Kamel and Memari (2022) and Li et al. (2020) explained that through VM practice, construction teams can optimise building envelopes, improve insulation, and utilise energy-efficient equipment and systems to minimise energy loss and enhance overall performance. Additionally, VM facilitates the implementation of energy-efficient construction techniques, such as passive solar design and daylighting strategies, to harness natural energy sources and reduce reliance on external power sources. By prioritising energy efficiency through VM, stealth construction projects can minimise their energy footprint and enhance operational resilience in remote or sensitive environments.

Renewable energy plays a crucial role in managing energy transmission for stealth construction, and VM enables the integration of renewable energy sources, such as solar, wind, and geothermal, to power construction operations and supports long-term sustainability goals (Al-Shetwi, 2022; Khan & Al-Ghamdi, 2021). Through VM, construction teams can assess the feasibility of renewable energy technologies, evaluate cost-effectiveness, and develop implementation plans that maximise energy generation while minimising environmental impact. Additionally, VM facilitates the integration of energy storage solutions, such as batteries and hydrogen fuel cells, to store excess energy and ensure uninterrupted power supply during stealth construction operations. By leveraging renewable energy through VM, stealth construction projects can reduce their reliance on traditional energy sources, enhance operational efficiency, and minimise their environmental footprint.

Deploying Countermeasures

VM is critical in ensuring that deploying countermeasures aligns with the objectives of stealth construction. Banerjee and Nayaka (2022) illustrated that one key aspect is implementing a functional construction system integrating security features seamlessly into the building's design and operations. Through VM, construction teams can identify and prioritise security requirements, such as access control systems, surveillance cameras, and intrusion detection sensors, and integrate them into the construction process from the outset. This proactive approach ensures that security measures are effectively incorporated into the building's infrastructure, minimising vulnerabilities and enhancing its ability to thwart potential threats.

Furthermore, proactive practices are essential in deploying countermeasures for stealth construction, and VM enables construction teams to adopt proactive strategies to mitigate risks and enhance security. Pargaonkar (2023) stated that by conducting comprehensive risk assessments and threat analyses, VM facilitates the identification of potential vulnerabilities early in the project lifecycle. This allows construction teams to implement pre-emptive measures, such as strengthening perimeter defences, enhancing cybersecurity protocols, and training personnel in security procedures to reduce the likelihood of security breaches. Additionally, VM supports the development of contingency plans and emergency response protocols to ensure swift and effective responses to security incidents, further enhancing the project's resilience against threats.

Abidi et al. (2022) mentioned that predictive planning can be another critical aspect of deploying countermeasures in stealth construction, and VM enables construction teams to leverage predictive analytics and modelling tools to anticipate and mitigate security risks. Through VM, construction teams can analyse historical data, simulate potential scenarios, and identify emerging threats to inform proactive security strategies. This may involve deploying advanced surveillance technologies, such as artificial intelligence and machine learning algorithms, to detect suspicious behaviour and anomalies in real-time, enabling rapid response and intervention. Additionally, VM supports the integration of predictive planning into construction scheduling and resource allocation, ensuring that security measures are effectively implemented without disrupting project timelines or budget constraints. By embracing predictive planning through VM, stealth construction projects can stay ahead of evolving security threats and maintain a strong defence posture throughout the project lifecycle.

VM Towards Stealth Construction

VM for stealth construction embodies a strategic approach to optimising construction processes while prioritising discretion and covert operations. In an era where security concerns and the need for confidentiality are paramount, integrating VM principles becomes imperative. This approach involves proactive planning, meticulous risk assessment, and the strategic deployment of resources to ensure that construction projects remain undetected or inconspicuous. When value is managed in every construction phase, from planning and design to execution and operation, stakeholders can achieve operational efficiency without compromising security or confidentiality.

At the heart of VM for stealth construction lies the seamless integration of advanced technologies across all aspects of the project. From sophisticated surveillance systems to state-of-the-art materials and construction techniques, leveraging cutting-edge innovations is essential to maintaining a low profile. When incorporated into these technologies, construction teams can enhance the project's stealth capabilities while optimising resource utilisation and minimising risks. Moreover, VM ensures that technological solutions are carefully tailored to meet stealth construction's unique requirements and challenges, allowing for agile responses to emerging threats and evolving security needs.

Furthermore, VM in stealth construction extends beyond the physical aspects to encompass strategic decision-making and risk mitigation. Through proactive VM practices, construction teams can identify potential vulnerabilities and security risks early in the project lifecycle, enabling them to implement pre-emptive measures to address these concerns. This may involve deploying countermeasures, such as enhanced security protocols, access controls, and surveillance systems, to safeguard construction sites and personnel from external threats. Additionally, VM facilitates the development of contingency plans and emergency response strategies to ensure resilience against unforeseen events, thereby safeguarding project timelines and objectives.

From the target point of view of VM towards stealth construction, we have coined five (5) central sustainable principles in environmental protection, health and safety, project delivery duration, economy, and aesthetics. These principles focus on integrating stealth construction technologies into every stage of the construction process, from the beginning to the end and throughout the project's lifespan.

Environmental Protection

Environmental protection in stealth construction is a critical aspect that can be effectively achieved by implementing VM principles. Integrating VM into the construction process and stealth construction projects can minimise their environmental impact and promote sustainability in several ways. VM contributes to environmental protection in stealth construction by selecting materials and construction methods that prioritise sustainability. Jalaei et al. (2021) and Backes and Traverso (2021) identified that VM techniques, such as life cycle costing and environmental impact assessments, enable project teams to evaluate the environmental consequences of different material choices and construction practices. By considering resource consumption, carbon emissions, and waste generation, stakeholders can identify environmentally friendly alternatives that minimise negative environmental impacts. This may involve using recycled or renewable materials, adopting energy-efficient construction techniques, and implementing green building standards such as Leadership in Energy and Environmental Design (LEED) or Building Research Establishment Environmental Assessment Methodology (BREEAM).

Furthermore, VM facilitates optimising energy performance and resource efficiency in stealth construction projects. Ebirim et al. (2024) discussed that when energy requirements and consumption patterns are systematically analysed, project teams can identify opportunities to reduce energy usage and improve operational efficiency throughout the project lifecycle. This may include incorporating passive design strategies to optimise natural lighting and ventilation, installing energy-efficient systems and appliances, and integrating renewable energy technologies such as solar panels or geothermal heating. When energy efficiency and resource conservation are prioritised, VM helps minimise the ecological footprint of stealth construction projects and promotes environmental sustainability.

Moreover, VM supports the implementation of environmental protection measures during the construction and operation phases of stealth projects. This includes strategies to minimise pollution, mitigate environmental disturbances, and protect natural habitats and ecosystems (Baloch et al., 2023). During construction, VM techniques can be used to develop comprehensive environmental management plans that outline measures to prevent soil erosion, manage stormwater runoff, and minimise air and noise pollution. Additionally, VM supports integrating sustainable landscaping practices and habitat restoration efforts to mitigate the project's ecological impact. Throughout the operation phase, VM enables ongoing monitoring and maintenance activities to ensure that environmental protection measures remain effective and compliance with environmental regulations is maintained.

Implementing VM in stealth construction projects is crucial in promoting environmental protection and sustainability. VM enables stakeholders to achieve environmentally responsible outcomes without compromising project objectives when this approach is fostered in decision-making that considers environmental impacts alongside economic and social factors. Through careful planning, strategic decision-making, and ongoing monitoring and evaluation, VM helps minimise the environmental footprint of stealth construction projects. It contributes to the preservation and conservation of natural resources for future generations.

Health and Safety

Implementing VM in stealth construction can actively contribute to achieving health and safety goals by prioritising measures that mitigate risks and ensure the well-being of workers and stakeholders throughout the project lifecycle, as Argaw et al. (2020) and Kordi et al. (2021) discussed. To begin with, VM enables proactive identification and mitigation of health and safety hazards associated with stealth construction projects. When stakeholders are involved early in risk assessments and value engineering workshops, project teams can identify potential hazards and develop effective control measures to prevent accidents and injuries. This proactive approach to risk management ensures that health and safety considerations are integrated into project planning and decision-making from the outset, minimising the likelihood of incidents occurring during construction.

Moreover, VM facilitates optimising construction processes to enhance worker safety and well-being. Thakur et al. (2023) and Jeong and Yoon (2016) highlighted that through value analysis and process improvement initiatives, project teams can streamline workflows, eliminate inefficiencies, and reduce exposure to hazardous conditions. This may involve redesigning work sequences, implementing ergonomic improvements, and investing in safety equipment and training programs to protect workers from occupational hazards. By continuously seeking opportunities to enhance safety performance, VM fosters a culture of safety excellence within the project team and promotes a safer working environment for all personnel involved.

Furthermore, VM supports adopting innovative technologies and practices that improve health and safety outcomes in stealth construction projects. Tender et al. (2022) discussed that when advancements in construction automation, robotics, and digitalisation are leveraged, project teams can minimise the need for manual labour in hazardous environments and reduce the risk of accidents and injuries. Additionally, VM encourages using virtual reality simulations and safety planning tools to identify and address potential safety issues before they arise, enhancing hazard awareness and risk mitigation efforts. It facilitates effective communication and collaboration among project stakeholders, which is essential for ensuring health and safety in stealth construction projects. By promoting open dialogue and information sharing, VM helps align stakeholders' expectations and priorities regarding health and safety, fostering a collective commitment to achieving safety excellence. This collaborative approach enables stakeholders to coordinate efforts, address emerging safety concerns, and implement corrective actions promptly, thereby minimising risks and promoting a safe working environment for everyone involved.

In summary, implementing VM actively contributes to achieving health and safety objectives in stealth construction by proactively identifying and mitigating hazards, optimising construction processes, adopting innovative technologies, and fostering collaboration among stakeholders. Prioritising health and safety throughout the project lifecycle, VM ensures that workers and stakeholders remain protected from harm and that projects are completed safely and successfully.

Project Delivery Duration

Implementing VM in stealth construction is pivotal in achieving efficient project delivery durations by streamlining processes, optimising resource allocation, and minimising delays throughout the project lifecycle. To begin with, Spellacy et al. (2021) stated that VM contributes to shortening project delivery durations by identifying and eliminating inefficiencies in the construction process. Through value engineering workshops and analysis, project teams can identify redundant activities, streamline workflows, and optimise project schedules to reduce unnecessary delays. This proactive approach to process optimisation ensures that construction activities are executed promptly and efficiently, minimising the project's overall duration.

Moreover, Venkataraman and Pinto (2023) and Ebirim et al. (2024) highlighted that VM optimises resource allocation to expedite project delivery durations. When critical tasks and allocating resources are prioritised effectively, project teams can ensure that human resources, materials, and equipment are utilised efficiently to meet project milestones and deadlines. This may involve reallocating resources from non-critical activities to accelerate progress on critical path tasks, implementing just-in-time delivery systems to minimise inventory and storage costs, and leveraging economies of scale to negotiate favourable terms with suppliers. When resource allocation is optimised, VM enables projects to be completed faster and cost-effectively.

Furthermore, according to Kineber et al. (2021), VM supports identifying and mitigating risks that could cause project delivery delays. Through rigorous risk assessment and management processes, project teams can anticipate potential obstacles and develop contingency plans to address them proactively. This may involve identifying alternative suppliers or subcontractors to mitigate supply chain disruptions, implementing robust project controls and monitoring systems to track progress and identify deviations from the schedule, and establishing clear communication channels to address issues and resolve conflicts on time. When risks are managed, VM helps prevent delays and ensures that projects are delivered on schedule. Additionally, VM fosters collaboration and communication among project stakeholders, essential for minimising delays and expediting project delivery durations (Fulford & Standing, 2014). By promoting transparency and accountability, VM creates a culture of teamwork and cooperation among project participants, enabling them to work together towards common goals and overcome challenges collectively. This collaborative approach facilitates timely decision-making, reduces misunderstandings and conflicts, and fosters a sense of ownership and commitment to project success. VM enhances project efficiency and accelerates delivery durations by promoting effective collaboration.

In summary, implementing VM actively contributes to achieving efficient project delivery durations in stealth construction by streamlining processes, optimising resource allocation, managing risks, and fostering collaboration among stakeholders. When proactively addressing inefficiencies, optimising resource utilisation, and promoting effective communication and teamwork, VM ensures that projects are completed faster and cost-effectively, ultimately enhancing project success and stakeholder satisfaction.

Economy

Implementing VM in stealth construction actively contributes to achieving economy by maximising resource utilisation efficiency, optimising cost-effectiveness, and minimising wastage throughout the project lifecycle. VM is crucial in maximising resource utilisation efficiency in stealth construction projects. When systematically analysing project requirements and identifying opportunities for optimisation, project teams can ensure that resources such as materials, human resources, and equipment are utilised to their fullest potential (Ammirato et al., 2023). This may involve value engineering exercises to identify cost-effective alternatives, lean construction practices to minimise waste and inefficiencies, and strategic resource allocation to prioritise critical project activities. Through proactive resource management, VM enables projects to achieve economic growth by maximising the value of every dollar spent.

Moreover, VM facilitates the optimisation of cost-effectiveness in stealth construction projects. Bode and Ragab (2024) stated that by scrutinising project costs and identifying opportunities for cost reduction, project teams can minimise expenses without compromising quality or performance. This may involve conducting cost-benefit analyses to evaluate the economic feasibility of different options, negotiating favourable terms with suppliers and subcontractors, and implementing value-based procurement strategies to maximise value for money. Additionally, VM encourages innovation and creativity in cost management, leading to the adoption of novel solutions and technologies that offer cost savings and efficiencies.

Furthermore, VM supports the minimisation of wastage in stealth construction projects, which is essential for achieving economy. Noorzai (2023) and Karaz et al. (2021) reiterated that when lean construction principles and practices are implemented, project teams can identify and eliminate sources of waste throughout the project lifecycle. This may involve reducing excess inventory, optimising material usage, and streamlining construction processes to minimise downtime and rework. Additionally, VM encourages adopting sustainable practices that minimise environmental impact and resource consumption, further contributing to the overall economy of the project. When waste reduction is prioritised, VM ensures that resources are utilised efficiently and responsibly, maximising value and minimising costs.

Additionally, VM fosters collaboration and communication among project stakeholders, essential for achieving economy in stealth construction projects. When transparency and accountability are implemented and promoted, VM creates a culture of teamwork and cooperation among project participants, enabling them to work together towards common goals and identify cost savings and efficiency opportunities. This collaborative approach facilitates sharing best practices, lessons learned, and innovative ideas, leading to continuous improvement and optimisation of project performance.

In summary, implementing VM actively contributes to achieving economy in stealth construction by maximising resource efficiency, optimising cost-effectiveness, minimising wastage, and fostering stakeholder collaboration. From prioritising value for money, minimising costs, and maximising efficiency, VM ensures that projects achieve economy by delivering high-quality outcomes that meet or exceed stakeholders' expectations at the lowest possible cost.

Aesthetics

Incorporating VM into stealth construction actively achieves aesthetics by ensuring that the design and execution of the project align with aesthetic objectives while maintaining the project's covert nature. VM enables stakeholders to prioritise aesthetic considerations from the project's inception (Zimina et al., 2012). When architects, designers, and other stakeholders are involved in value engineering workshops, project teams identify aesthetic goals and develop strategies to achieve them within the constraints of stealth construction requirements. This proactive approach ensures that aesthetic considerations integrate into the project's design and planning stages, laying the groundwork for a visually appealing outcome.

Moreover, VM supports the selection of materials, finishes, and architectural elements that enhance the aesthetic appeal of stealth construction projects. Venkataraman and Pinto (2023) and Hosseinpour et al. (2022) identified that through value analysis and cost-benefit assessments, project teams evaluate different options and identify those that best balance aesthetics with practicality and functionality. This may involve choosing materials that mimic natural surroundings or blend seamlessly into the built environment, selecting finishes that complement the project's surroundings, and incorporating architectural elements that enhance visual interest while maintaining efficiency.

Furthermore, VM facilitates optimising design and construction processes to achieve aesthetic objectives efficiently. Best and De Valence (2002) and Wong et al. (2009) illustrated that by streamlining workflows, eliminating inefficiencies, and leveraging innovative technologies, project teams realise aesthetic goals while minimising costs and maximising efficiency. This may involve using BIM software to visualise design concepts and identify potential conflicts or inconsistencies, implementing prefabrication techniques to achieve precision and consistency in construction, and adopting sustainable practices that enhance the project's aesthetic appeal while reducing environmental impact. Through strategic planning and execution, VM enables stakeholders to achieve desired aesthetics without compromising project objectives or timelines.

VM for Stealth Construction from the Post-construction Stage

VM is crucial in optimising outcomes and maximising stakeholder value in the post-construction stage of stealth construction projects. In this stage, one key aspect of VM is evaluating project performance and compliance with project objectives. Project teams can identify success and improvement areas through thorough assessment and analysis of key performance indicators such as cost, schedule, quality, and stakeholder satisfaction. This information is invaluable for informing future projects and refining processes to achieve even better outcomes.

Additionally, VM in the post-construction stage involves the implementation of lessons learned and best practices identified during the project. By documenting successes and challenges encountered throughout the construction process, project teams can develop a repository of knowledge that can be leveraged to improve future projects. This may include refining design and construction processes, optimising resource allocation, and enhancing risk management strategies. When lessons learned are applied, VM ensures that each successive project builds upon past successes, driving continuous improvement and innovation in stealth construction practices.

Furthermore, VM in the post-construction stage extends to the ongoing maintenance and operation of completed projects. When proactive maintenance programs and monitoring systems are established, project teams can ensure that assets remain in optimal condition and continue to deliver value over the long term. This may involve implementing preventive maintenance schedules, conducting regular inspections, and promptly addressing any issues or deficiencies. Also, prioritising proactive maintenance asset management and VM helps preserve the integrity and functionality of completed projects, maximising their lifespan and value for stakeholders.

The focus of VM in the post-construction stage shows the need to assess the actual performance and outcomes of the project once it has been completed. While VM principles are essential throughout the entire project lifecycle, including the pre-construction and construction stages, it is in the post-construction stage where the project's tangible results become evident. This is the stage where stakeholders can evaluate whether the project has met its objectives, delivered the expected value, and identified areas for improvement. By conducting thorough post-construction evaluations and applying lessons learned to future projects, VM ensures that the benefits of value optimisation are realised and sustained over the long term.

Conclusion

Implementing VM in stealth construction projects is crucial for achieving optimal outcomes across various aspects such as environmental protection, health and safety, project delivery duration, economy, and aesthetics. When VM principles are integrated into each stage of the construction process, from pre-construction planning to post-construction evaluation, project teams can maximise efficiency, minimise risks, and deliver projects that meet or exceed stakeholder expectations. Through proactively identifying opportunities, strategic decision-making, and continuous improvement efforts, VM ensures that stealth construction projects fulfil their covert objectives and contribute to sustainability, safety, cost-effectiveness, and visual appeal. By prioritising value throughout the project lifecycle, stakeholders can achieve success in stealth construction projects while simultaneously promoting long-term value creation and innovation in the construction industry.

References

Ab-Ghani, R., Zakaria, N., & Ye, K. M. (2021). Rethinking value management (VM) integration within the strategic phase of the construction project value chain. *International Journal of Sustainable Construction Engineering and Technology*, *12*(5), 80–93.

Abidi, M. H., Mohammed, M. K., & Alkhalefah, H. (2022). Predictive maintenance planning for industry 4.0 using machine learning for sustainable manufacturing. *Sustainability*, *14*(6), 1–27.

Adekunle, P., Aigbavboa, C., Akinradewo, O., Oke, A., & Aghimien, D. (2022). Construction information management: Benefits to the construction industry. *Sustainability*, *14*(18), 1–21.

Ahmed, S., & Sobuz, M. H. R. (2020). Challenges of implementing lean construction in the construction industry in Bangladesh. *Smart and Sustainable Built Environment*, *9*(2), 174–207.

Alhumaid, A. M., Bin Mahmoud, A. A., & Almohsen, A. S. (2024). Value engineering adoption's barriers and solutions: The case of Saudi Arabia's construction industry. *Buildings*, *14*(4), 1–19.

Al-Kodmany, K. (2023). Greenery-covered tall buildings: A review. *Buildings, 13*(9), 1–53. https://doi.org/10.3390/buildings13092362

Alqudhaibi, A., Albarrak, M., Aloseel, A., Jagtap, S., & Salonitis, K. (2023). Predicting cybersecurity threats in critical infrastructure for industry 4.0: A proactive approach based on attacker motivations. *Sensors, 23*(9), 1–17.

Al-Shetwi, A. Q. (2022). Sustainable development of renewable energy integrated power sector: Trends, environmental impacts, and recent challenges. *Science of The Total Environment, 822*, 1–18. https://doi.org/10.1016/j.scitotenv.2022.153645

Amiri, A., Emami, N., Ottelin, J., Sorvari, J., Marteinsson, B., Heinonen, J., & Junnila, S. (2021). Embodied emissions of buildings – A forgotten factor in green building certificates. *Energy and Buildings, 241*, 1–12. https://doi.org/10.1016/j.enbuild.2021.110962

Ammirato, S., Felicetti, A. M., Linzalone, R., Corvello, V., & Kumar, S. (2023). Still our most important asset: A systematic review on human resource management in the midst of the fourth industrial revolution. *Journal of Innovation & Knowledge, 8*(3), 1–14. https://doi.org/10.1016/j.jik.2023.100403

Argaw, S. T., Troncoso-Pastoriza, J. R., Lacey, D., Florin, M. V., Calcavecchia, F., Anderson, D., Burleson, W., Vogel, J., O'Leary, C., Eshaya-Chauvin, B., & Flahault, A. (2020). Cybersecurity of Hospitals: Discussing the challenges and working towards mitigating the risks. *BMC Medical Informatics and Decision Making, 20*(1), 1–10.

Backes, J. G., & Traverso, M. (2021). Application of life cycle sustainability assessment in the construction sector: A systematic literature review. *Processes, 9*(7), 1–31.

Baloch, Q. B., Shah, S. N., Iqbal, N., Sheeraz, M., Asadullah, M., Mahar, S., & Khan, A. U. (2023). Impact of tourism development upon environmental sustainability: A suggested framework for sustainable ecotourism. *Environmental Science and Pollution Research, 30*(3), 5917–5930.

Banerjee, A., & Nayaka, R. R. (2022). A comprehensive overview on BIM-integrated cyber physical system architectures and practices in the architecture, engineering and construction industry. *Construction Innovation, 22*(4), 727–748.

Barua, M. (2023). *Lively cities: Reconfiguring urban ecology*. University of Minnesota Press.

Bertram, N., Fuchs, S., Mischke, J., Palter, R., Strube, G., & Woetzel, J. (2019). Modular construction: From projects to products. *McKinsey & Company: Capital Projects & Infrastructure, 1*(1), 1–34.

Best, R., & De Valence, G. (Eds.). (2002). *Design and construction: Building in value*. Routledge.

Bode, A. T., & Ragab, Z. B. (2024). Adoption of project management strategies and minimization of time and cost in project execution. *American Journal of Economics and Business Management, 7*(3), 1–23.

Ebirim, W., Ninduwezuor-Ehiobu, N., Usman, F. O., Olu-lawal, K. A., Ani, E. C., & Montero, D. J. P. (2024). Project management strategies for accelerating energy efficiency in HVAC systems amidst climate change. *International Journal of Management & Entrepreneurship Research, 6*(3), 512–525.

Fulford, R., & Standing, C. (2014). Construction industry productivity and the potential for collaborative practice. *International Journal of Project Management, 32*(2), 315–326.

Ghosh, A., Misra, S., Bhattacharyya, R., Sarkar, A., Singh, A. K., Tyagi, V. C., Kumar, R., V., & Meena, V. S. (2020). Agriculture, dairy and fishery farming practices and greenhouse gas emission footprint: A strategic appraisal for mitigation. *Environmental Science and Pollution Research, 27*, 10160–10184. https://doi.org/10.1007/s11356-020-07949-4

Golato, A., Demirli, R., & Santhanam, S. (2014). Lamb wave scattering by a symmetric pair of surface-breaking cracks in a plate. *Wave Motion, 51*(8), 1349–1363.

Hons, E. M. B. (2019). *Improving the performance of thermoplastic composite structural joints*. https://pureadmin.qub.ac.uk/ws/portalfiles/portal/483312894/Improving_the_performance_of_thermoplastic_composit

Hosseinpour, N., Kazemi, F., & Mahdizadeh, H. (2022). A cost-benefit analysis of applying urban agriculture in sustainable park design. *Land Use Policy, 112*, 1–11. https://doi.org/10.1016/j.landusepol.2021.105834

Jalaei, F., Zoghi, M., & Khoshand, A. (2021). Life cycle environmental impact assessment to manage and optimize construction waste using Building Information Modeling (BIM). *International Journal of Construction Management*, *21*(8), 784–801.

Jeong, B. K., & Yoon, T. E. (2016). Improving IT process management through value stream mapping approach: A case study. *JISTEM-Journal of Information Systems and Technology Management*, *13*(1), 389–404.

Jin, R., Hong, J., & Zuo, J. (2020). Environmental performance of off-site constructed facilities: A critical review. *Energy and buildings*, *207*, 1–9. https://doi.org/10.1016/j.enbuild.2019.109567

Kamel, E., & Memari, A. M. (2022). Residential building envelope energy retrofit methods, simulation tools, and example projects: A review of the literature. *Buildings*, *12*(7), 1–28.

Kannimuthu, M., Raphael, B., Palaneeswaran, E., & Kuppuswamy, A. (2019). Optimizing time, cost and quality in multi-mode resource-constrained project scheduling. *Built Environment Project and Asset Management*, *9*(1), 44–63.

Karaz, M., Teixeira, J. C., & Rahla, K. M. (2021). Construction and demolition waste – A shift toward lean construction and building information model. In *Sustainability and automation in smart constructions: Proceedings of the International Conference on Automation Innovation in Construction (CIAC-2019), Leiria, Portugal* (pp. 51–58). Springer International Publishing.

Khan, S. A., & Al-Ghamdi, S. G. (2021). Renewable and integrated renewable energy systems for buildings and their environmental and socio-economic sustainability assessment. In *Energy systems evaluation (Volume 1) sustainability assessment* (pp. 127–144). Springer International Publishing.

Khan, M. S., Shakoor, R. A., Fayyaz, O., & Ahmed, E. M. (2023). A focused review on techniques for achieving cloaking effects with metamaterials. *Optik – International Journal for Light and Electron Optics*, *297*, 1–20. https://doi.org/10.1016/j.ijleo.2023.171575

Kineber, A. F., Othman, I., Oke, A. E., Chileshe, N., & Buniya, M. K. (2021). Impact of value management on building projects success: Structural equation modeling approach. *Journal of Construction Engineering and Management*, *147*(4), 1–15. https://doi.org/10.1061/(ASCE)CO.1943-7862.0002026

Kineber, A. F., Siddharth, S., Chileshe, N., Alsolami, B., & Hamed, M. M. (2022). Addressing of value management implementation barriers within the Indian construction industry: A PLS-SEM approach. *Sustainability*, *14*(24), 1–24.

Kordi, N. E., Belayutham, S., & Che Ibrahim, C. K. I. (2021). Mapping of social sustainability attributes to stakeholders' involvement in construction project life cycle. *Construction Management and Economics*, *39*(6), 513–532.

Kuitert, L., Willems, J., & Volker, L. (2024). Value integration in multi-functional urban projects: A value driven perspective on sustainability transitions. *Construction Management and Economics*, *42*(2), 182–198.

Li, H. X., Li, Y., Jiang, B., Zhang, L., Wu, X., & Lin, J. (2020). Energy performance optimisation of building envelope retrofit through integrated orthogonal arrays with data envelopment analysis. *Renewable Energy*, *149*, 1414–1423. https://doi.org/10.1016/j.renene.2019.10.143

Ma, Z., Merkus, H. G., & Scarlett, B. (2001). Extending laser diffraction for particle shape characterization: Technical aspects and application. *Powder Technology*, *118*(1–2), 180–187.

Marut, J. J., Alaezi, J. O., & Obeka, I. C. (2020). A review of alternative building materials for sustainable construction towards sustainable development. *Journal of Modern Materials*, *7*(1), 68–78.

Naqvi, S. M. S. (2023). *Maximizing green space in a building complex through alternative landscape design elements* [Doctoral dissertation, Guru Gobind Singh Indraprastha University].

Noorzai, E. (2023). Evaluating lean techniques to improve success factors in the construction phase. *Construction Innovation*, *23*(3), 622–639.

Ofori, E. K., Li, J., Radmehr, R., Zhang, J., & Shayanmehr, S. (2023). Environmental consequences of ISO 14001 in European economies amidst structural change and technology innovation: Insights from green governance dynamism. *Journal of Cleaner Production*, *411*, 1–16. https://doi.org/10.1016/j.jclepro.2023.137301

Oke, A. E., & Aigbavboa, C. O. (2017). *Sustainable value management for construction projects* (pp. 75–86). Springer.

Pargaonkar, S. (2023). Advancements in security testing: A comprehensive review of methodologies and emerging trends in software quality engineering. *International Journal of Science and Research (IJSR)*, *12*(9), 61–66.

Rathor, S. K., & Saxena, D. (2020). Energy management system for smart grid: An overview and key issues. *International Journal of Energy Research*, *44*(6), 4067–4109.

Rippa, A. (2020). *Borderland infrastructures: Trade, development, and control in western China* (p. 307). Amsterdam University Press.

Rwahwire, S., & Ssebagala, I. (2023). Radar cross-section reducing metamaterials. *Electromagnetic Metamaterials: Properties and Applications*, *2*(3), 341–362. https://doi.org/10.1002/9781394167074.ch16

Spellacy, J., Edwards, D. J., Roberts, C. J., Hayhow, S., & Shelbourn, M. (2021). An investigation into the role of the quantity surveyor in the value management workshop process. *Journal of Engineering, Design and Technology*, *19*(2), 423–445.

Tahmasebi, M. (2024). Beyond defense: Proactive approaches to disaster recovery and threat intelligence in modern enterprises. *Journal of Information Security*, *15*(2), 106–133.

Tender, M., Couto, J. P., Gibb, A., Fuller, P., & Yeomans, S. (2022). Emerging technologies for health, safety and well-being in construction industry. *Industry 4.0 for the Built Environment: Methodologies, Technologies and Skills*, *1*(1), 369–390. https://doi.org/10.1007/978-3-030-82430-3_16

Thakur, V., Akerele, O. A., & Randell, E. (2023). Lean and Six Sigma as continuous quality improvement frameworks in the clinical diagnostic laboratory. *Critical Reviews in Clinical Laboratory Sciences*, *60*(1), 63–81.

Venkataraman, R. R., & Pinto, J. K. (2023). *Cost and value management in projects*. John Wiley & Sons.

Wong, F. W., Lam, P. T., & Chan, E. H. (2009). Optimising design objectives using the Balanced Scorecard approach. *Design Studies*, *30*(4), 369–392.

Wu, J. (2020). *Advanced metric wave radar*. Springer.

Zimina, D., Ballard, G., & Pasquire, C. (2012). Target value design: Using collaboration and a lean approach to reduce construction cost. *Construction Management and Economics*, *30*(5), 383–398.

Chapter Nine

Partnering in Stealth Construction

Abstract

This chapter delved into the multifaceted landscape of construction partnering, exploring its key aspects, promotion factors, and associated challenges. It examined how partnering principles are applied in various construction contexts, including stealth construction, where integration of advanced technologies and collaborative practices is pivotal. Moreover, it highlighted the significance of partnering in addressing crucial considerations such as environmental protection, health and safety, project delivery duration, aesthetics, and economy during and after the construction phase. Through collaborative efforts and shared responsibilities, construction partnering emerges as a cornerstone for achieving excellence and sustainability in the built environment.

Keywords: Construction partnering; stealth construction; health and safety; project optimisation; collaboration

Introduction

Partnering in construction involves collaborative efforts among various stakeholders within the industry to achieve common goals and objectives. Mudzvokorwa et al. (2020) stated that it entails fostering strong relationships between contractors, subcontractors, architects, engineers, and clients, among others, to enhance project outcomes. Through partnering, stakeholders work closely together from project inception to completion, sharing responsibilities, risks, and rewards. This approach emphasises open communication, mutual trust, and respect among all parties involved. Leveraging each other's strengths and expertise, partnering in construction aims to streamline processes, improve efficiency, and ultimately deliver high-quality projects on time and within budget. Furthermore, partnering encourages innovation and creativity as stakeholders collaborate to overcome challenges and find solutions

Stealth Construction: Integrating Practices for Resilience and Sustainability, 215–233

Published under exclusive licence by Emerald Publishing Limited
doi:10.1108/978-1-83608-182-120251009

that meet project requirements and exceed expectations. Thus, partnering in construction fosters a culture of teamwork and cooperation, leading to successful project delivery and long-term relationships built on trust and mutual benefit.

Key Aspects of Partnering in Construction

According to Koolwijk et al. (2022) and Evans et al. (2020), the key aspects of partnering in construction revolve around collaboration, communication, mutual trust, and shared objectives. To begin with, *collaboration* is fundamental to partnering in construction. This involves stakeholders actively working together, sharing resources, and aligning their efforts towards common goals. Whether it is contractors, subcontractors, architects, engineers, or clients, everyone involved in the project plays a crucial role in its success. Through fostering a collaborative environment, partnering enables stakeholders to leverage their expertise and experience to overcome challenges and achieve optimal outcomes.

Furthermore, *communication* is another essential aspect of partnering in construction. Clear and open communication channels are established among all parties involved in the project, ensuring that information flows smoothly and efficiently. This includes regular meetings, progress updates, and transparent discussions about project milestones, timelines, and potential issues. Effective communication helps prevent misunderstandings, minimises delays, and fosters a sense of accountability among stakeholders. In addition to communication, *mutual trust* is a cornerstone of partnering in construction. Building and maintaining trust among stakeholders is essential for successful collaboration. Trust is earned through consistent actions, reliability, and integrity. When stakeholders trust each other, they are more likely to share information, take calculated risks, and confidently work towards common objectives. This creates a supportive and cohesive working environment where everyone feels valued and respected. Additionally, *shared objectives* are a vital aspect of partnering in construction. All stakeholders align their interests and goals to prioritise the project's success. This involves setting clear objectives, defining expectations, and establishing measurable outcomes everyone strives to achieve. When united in pursuing common goals, stakeholders can effectively coordinate their efforts, make informed decisions, and navigate challenges together.

Factors that Promote Partnering Among Construction Professionals

To achieve the set goal of a project right from planning to execution, the professionals involved must be in unison about what is needed and necessary for the project to achieve quality and standard. Cutting across every construction phase, the following factors are discussed as a pathway towards achieving project delivery of cost, quality, and duration, as summarised in Table 3.

Clear Communication Channels (Aripin et al., 2023)

Establishing open and transparent lines of communication is paramount in promoting partnering among construction professionals. When communication

Table 3. Factors That Promote Partnering Among Construction Professionals.

Factors Promoting Partnership	**Author(s)**
Clear communication channels	Aripin et al. (2023)
Mutual respect and trust	Gunduz and Abdi (2020)
Shared goals and objectives	Deep et al. (2021)
Effective conflict resolution mechanisms	Phillips-Alonge (2017)
Continuous collaboration	Aghimien et al. (2022)
Transparent decision-making processes	Li et al. (2005) and Evans et al. (2020)
Recognition of individual contributions	Köhler et al. (2022)
Commitment to continuous improvement	Pozzi et al. (2023)
Establishment of formal partnerships	Dewulf and Kadefors (2012)
Emphasis on building long-term relationships	Ali and Haapasalo (2023)

flows freely, professionals can easily exchange ideas, share pertinent information, and address concerns on time. Whether through regular team meetings, project management software, or digital communication platforms, ensuring that all stakeholders have access to the necessary channels facilitates effective collaboration and decision-making throughout the project lifecycle.

Mutual Respect and Trust (Gunduz & Abdi, 2020)

Cultivating a culture of mutual respect and trust is essential for fostering strong partnerships among construction professionals. When individuals feel valued and respected for their expertise and contributions, they are more likely to collaborate openly and constructively. Building trust among team members encourages honest communication, promotes teamwork, and fosters a supportive work environment where everyone feels empowered to contribute their ideas and insights without fear of judgement or criticism.

Shared Goals and Objectives (Deep et al., 2021)

Aligning interests and objectives is critical for promoting partnering among construction professionals. By clearly defining shared goals and objectives at the outset of a project, all stakeholders can understand what needs to be accomplished and work together towards expected outcomes. Whether delivering a project on time and within budget, achieving sustainability targets, or exceeding client expectations, having a shared vision helps unite professionals from different disciplines and backgrounds, driving collaboration and cooperation throughout the project lifecycle.

Effective Conflict Resolution Mechanisms (Phillips-Alonge, 2017)

Conflict is inevitable in any collaborative endeavour, but effective conflict resolution mechanisms can help prevent disputes from escalating and damaging relationships among construction professionals. Through facilitated discussions, mediation, or arbitration, having a structured approach to resolving conflicts ensures that issues are addressed promptly and constructively, allowing the project to move forward without unnecessary delays or disruptions.

Continuous Collaboration (Aghimien et al., 2022)

Encouraging continuous collaboration among construction professionals is critical to promoting partnering and maximising project success. Rather than working in isolation, professionals should actively seek opportunities to collaborate and leverage each other's expertise throughout the project lifecycle. Through cross-functional teams, interdisciplinary workshops, or joint problem-solving sessions, fostering a culture of collaboration encourages innovation, drives efficiency, and leads to better outcomes for all stakeholders involved.

Transparent Decision-Making Processes (Evans et al., 2020; Li et al., 2005)

Involving all stakeholders in decision-making processes and providing visibility into critical decisions are essential for promoting partnering among construction professionals. When individuals have a voice in the decision-making process and understand the rationale behind key decisions, they are likelier to buy into the project vision and feel invested in its success. With considerations through regular updates, decision logs, or stakeholder consultations, ensuring transparency in decision-making promotes trust, accountability, and alignment among team members.

Recognition of Individual Contributions (Köhler et al., 2022)

Recognising and celebrating the contributions of individual professionals within the team is essential for promoting partnering and fostering a positive work environment. This could be through formal recognition programs, team awards, or simple expressions of gratitude; acknowledging the efforts and achievements of team members boosts morale, motivates individuals to perform at their best, and strengthens bonds among team members.

Commitment to Continuous Improvement (Pozzi et al., 2023)

Embracing a continuous learning and improvement culture is essential for promoting partnering among construction professionals. Implementing ongoing training and development programs, lessons-learned sessions, or post-project reviews encourages professionals to reflect on their experiences, learn from their mistakes, share best practices, promote innovation, drive excellence, and ensure that lessons are applied to future projects.

Establishment of Formal Partnerships (Dewulf & Kadefors, 2012)

Formalising partnerships through contractual agreements, alliances, or joint ventures can help strengthen relationships among construction professionals and provide a framework for collaboration. This can be done through strategic alliances with preferred suppliers, joint ventures with complementary firms, or long-term partnerships with key clients. Establishing formal partnerships creates opportunities for mutual growth, shared risk, and collaborative innovation, ultimately leading to better outcomes for all parties involved.

Emphasis on Building Long-Term Relationships (Ali & Haapasalo, 2023)

Prioritising the development of long-term relationships based on trust, respect, and mutual benefit is essential for promoting partnering among construction professionals. Rather than focusing solely on short-term gains or transactions, professionals should invest in building meaningful relationships with their counterparts, clients, and other stakeholders. Whether through regular communication, networking events, or collaborative initiatives, emphasising the importance of long-term relationships fosters loyalty, trust, and goodwill, paving the way for future collaboration and success.

Challenges in Construction Partnering

While partnering in construction offers numerous benefits, it also comes with its fair share of challenges that can hinder its successful implementation. Saraji et al. (2021) illustrated that one of the primary challenges is overcoming resistance to change. Adversarial relationships and a lack of stakeholder trust have traditionally characterised the construction industry. Therefore, shifting to a collaborative approach requires a cultural shift and a willingness to embrace new working methods. Resistance to change can manifest at various levels, from frontline workers accustomed to traditional project delivery methods to senior management hesitant to relinquish control over decision-making processes. This challenge requires effective change management strategies, including education, training, and ongoing communication to foster buy-in and support for partnering initiatives.

Another significant challenge in construction partnering is managing conflicting priorities among stakeholders (Adekola et al., 2020). Construction projects typically involve multiple stakeholders with diverse interests, including clients, contractors, subcontractors, architects, engineers, and regulatory authorities. These stakeholders may have competing priorities, such as cost, schedule, quality, and safety, leading to conflicts and disagreements if not properly managed. Balancing these conflicting priorities requires effective leadership, negotiation skills, and a collaborative approach to decision-making. Establishing clear communication channels and mechanisms for resolving conflicts promptly and constructively is essential for maintaining alignment and harmony among stakeholders throughout the project lifecycle.

Furthermore, Odeyemi et al. (2024) and Evans et al. (2020) identified that addressing cultural differences among stakeholders is a common challenge in construction partnering. Construction projects often involve diverse teams with different organisational cultures, work practices, and communication styles. These cultural differences can lead to misunderstandings, miscommunication, and friction, undermining collaboration and teamwork. Overcoming this challenge requires building cultural awareness, fostering respect for diversity, and promoting inclusive practices that value and leverage all stakeholders' unique perspectives and contributions. Establishing common goals and objectives and facilitating team-building activities and cultural sensitivity training can help bridge cultural divides and promote a cohesive working environment conducive to successful partnering.

Implementing effective collaboration tools and technologies is another challenge in construction partnering (Porwal & Hewage, 2013). While advancements in digital technology have revolutionised many aspects of the construction industry, adopting and integrating these tools into partnering initiatives can be complex and challenging. Issues such as interoperability, data security, and user adoption can pose significant barriers to successful implementation. Additionally, not all stakeholders may have access to or be proficient in using these tools, further complicating the collaboration process. Overcoming this challenge requires careful planning, investment in training and support, and collaboration with technology providers to ensure that the tools and technologies chosen align with the specific needs and capabilities of the project team.

Partnering in Stealth Construction

Stealth construction represents a groundbreaking approach that seamlessly integrates advanced technologies throughout the various phases of construction, enhancing efficiency and effectiveness in project delivery. This innovative practice leverages cutting-edge technologies like Building Information Modelling (BIM), drones, augmented reality (AR), and advanced data analytics to streamline processes and optimise outcomes. From project planning and design to construction and maintenance, stealth construction harnesses the power of technology to improve collaboration, communication, and decision-making among stakeholders.

One of the key aspects of stealth construction is its ability to minimise disruptions and maximise productivity through innovative construction practices. Anant and Charpe (2022) and Jayawardana et al. (2023) mentioned that stealth construction reduces on-site construction time when adopting off-site fabrication, modular construction, and prefabricated components and minimises the impact on surrounding communities and environments. This approach accelerates project delivery and enhances safety and quality by providing controlled environments for construction activities.

Furthermore, stealth construction enables real-time monitoring and control of construction processes, allowing project teams to identify and address issues before they escalate proactively. Through the integration of sensors, Internet of Things (IoT) devices, and advanced analytics, stakeholders can monitor progress, track resources, and optimise workflows in real time, leading to improved project outcomes and cost savings. Additionally, Zainuddin et al. (2024) and Alizadehsalehi and Hadavi (2023) discussed that by leveraging AR and virtual reality (VR) technologies, stakeholders can visualise construction plans and identify potential clashes or conflicts, facilitating informed decision-making and minimising rework. Another significant aspect of stealth construction is its focus on sustainability and environmental stewardship. By incorporating green building practices, renewable energy systems, and eco-friendly materials, stealth construction reduces carbon footprint and promotes environmental sustainability. This aligns with the growing demand for sustainable construction solutions and helps project teams meet regulatory requirements and achieve green building certifications.

To achieve the above, the chapter considered four aspects of the construction that must be considered. These include a cross-section of the building, energy transmission, visibility, and countermeasures. These sections combine to bring a form of resilience and sustainability to the project as it moves from the initial to the completion stages of construction. What partnering in stealth construction brings to projects is discussed below.

A Cross-Section of the Building

Construction partnering plays an adhesive role in realising a cross-section of the building in the context of stealth construction. Evans et al. (2020) and Lahdenperä (2012) illustrated that this innovative construction approach emphasises the integration of advanced technologies and collaborative practices to optimise project outcomes. For this study, a cross-section of the building, defined around *the shape of the building, internal construction, material usage, and other factors*, represents a vertical slice through the structure and summarises various construction elements, including structural components, mechanical systems, and architectural features. Through construction partnering, stakeholders work together to ensure that each cross-section aspect is seamlessly integrated and optimised for efficiency, functionality, and sustainability.

In terms of the shape of the building, construction partnering facilitates the exploration and selection of architectural designs that meet aesthetic objectives and align with functional requirements and sustainability goals (Goubran & Cucuzzella, 2019; Rippmann, 2016). Stakeholders collaborate to evaluate various design options, considering building orientation, envelope design, and passive solar strategies to maximise energy efficiency and occupant comfort. Partnering brings together architects, engineers, contractors, and other experts to ensure the building's shape is designed to use less energy, have a smaller environmental footprint, and perform better overall.

Furthermore, Hughes et al. (2012) and Zakaria et al. (2018) mentioned that internal construction within the cross-section of the building is another crucial aspect addressed through construction partnering. Stakeholders can work collaboratively to design and implement efficient structural systems, mechanical and electrical infrastructure, and interior finishes that meet project requirements while adhering to budgetary and schedule constraints. Through partnering, construction professionals can leverage BIM and other advanced technologies to visualise and coordinate internal construction elements, ensuring seamless integration and minimising conflicts or clashes during construction. Additionally, partnering enables stakeholders to explore innovative construction methods, such as off-site fabrication and modular construction, to enhance productivity, quality, and safety while reducing construction timelines and on-site disruptions.

In addition, material usage is a key consideration in the cross-section of the building, and construction partnering facilitates the selection and integration of sustainable building materials and systems. Stakeholders collaborate to identify materials that minimise environmental impact, promote resource efficiency, and enhance indoor environmental quality, as illustrated by Obiuto et al. (2024). From low-emission finishes and recycled content materials to energy-efficient insulation and water-saving fixtures, partnering enables informed decision-making regarding material selection to optimise performance and reduce lifecycle costs. Moreover, partnering fosters collaboration with suppliers and manufacturers to ensure the availability of sustainable materials and support the implementation of green building practices throughout the construction process.

Other factors addressed through construction partnering concerning the cross-section of the building include accessibility, safety, and resilience. Stakeholders collaborate to design and implement features that enhance accessibility for all occupants, including individuals with challenges, while ensuring compliance with regulatory requirements and industry standards. Additionally, partnering facilitates the integration of robust safety measures and emergency systems to protect workers and occupants during construction and throughout the building's lifecycle. Moreover, partnering enables stakeholders to assess and address resilience considerations, such as climate change adaptation, natural disaster preparedness, and long-term durability, to ensure that the building can withstand future challenges and remain functional and resilient.

Energy Transmission

Construction partnering is vital in shaping the approach towards energy transmission in stealth construction. When stakeholders such as architects, engineers, contractors, and energy specialists are involved, partnering ensures a comprehensive and integrated approach to incorporating *renewable energy, climate action planning, energy management, and energy efficiency* strategies into the construction process.

Mihailova et al. (2022) stated that renewable energy integration is a key focus area addressed through construction partnering. Collaborating with renewable energy experts, stakeholders can assess the feasibility and potential of incorporating renewable energy sources such as solar, wind, and geothermal into the building design. Through partnering, stakeholders can explore innovative technologies and financing options to optimise the use of renewable energy, reducing reliance on fossil fuels and lowering carbon emissions over the building's lifecycle.

Climate action planning is another critical aspect addressed through construction partnering. Stakeholders collaborate to develop comprehensive climate action plans that outline strategies for mitigating climate change impacts and enhancing resilience in the built environment. Enwin and Ikiriko (2024) highlighted that by integrating climate-responsive design principles and adaptive strategies into the construction process, partnering ensures that buildings are better equipped to withstand extreme weather events, temperature fluctuations, and other climate-related risks.

Furthermore, energy management is a core component of stealth construction facilitated through partnership. Digitemie and Ekemezie (2024) stated that by leveraging advanced technologies such as Building Energy Management Systems (BEMS) and smart metres, stakeholders can monitor and optimise energy consumption in real time, identifying opportunities for energy savings and operational efficiency improvements. Through partnering, stakeholders can develop customised energy management plans tailored to the specific needs and objectives of the building, maximising energy performance and reducing operating costs.

Additionally, Manfren et al. (2021) demonstrated that energy efficiency is a central focus area addressed through construction partnering. When collaborating on selecting and integrating energy-efficient building materials, systems, and technologies, stakeholders can optimise energy performance and reduce energy waste throughout the building's lifecycle. Through partnering, stakeholders can conduct energy modelling, performance simulations, and lifecycle cost analyses to identify cost-effective energy efficiency measures and design strategies. Additionally, partnering enables stakeholders to implement best practices such as passive design strategies, daylighting, and natural ventilation to enhance energy efficiency and occupant comfort further.

Visibility

Construction partnering plays a crucial role in shaping the visibility aspects of construction projects within stealth construction. When architects, designers, contractors, and environmental specialists are engaged, partnering ensures that factors such as the *colour and texture of the construction, its location, interaction with the environment, and other visibility-related considerations* are carefully considered and integrated into the project.

The colour and texture of the construction are key elements addressed through construction partnering. Zhang and Kim (2023) and Kosori⊠ et al. (2021) asserted that stakeholders collaborate to select materials and finishes that blend harmoniously with the surrounding environment while also enhancing the aesthetic appeal of the building. When leveraging the collective expertise of architects and designers, partnering ensures that the colour palette and texture of the construction complement the natural landscape or urban context, creating visually pleasing and cohesively built environments.

Location is another critical aspect addressed through construction partnering. Stakeholders work together to evaluate potential construction sites and assess their suitability regarding visibility, accessibility, and environmental impact (Baah et al., 2022). When conducting thorough site analyses and environmental assessments, partnering enables informed decision-making regarding the optimal location for the construction project. Additionally, partnering facilitates collaboration with local communities and stakeholders to address concerns and ensure the project aligns with community values and priorities.

Interaction with the environment is a central focus area addressed through construction partnering. As Wuni and Shen (2022) discussed, stakeholders collaborate to minimise the environmental footprint of the construction project and mitigate its impact on the surrounding ecosystem. By incorporating sustainable design principles, such as green roofs, rain gardens, and permeable paving, partnering ensures that the construction interacts harmoniously with the natural environment, enhancing biodiversity, improving air and water quality, and mitigating urban heat island effects.

Other visibility-related factors, such as signage, lighting, and landscaping, are also addressed through construction partnering. Stakeholders collaborate to develop signage and wayfinding systems that enhance the construction site's visibility and usability while promoting safety and accessibility. Additionally, partnering facilitates the integration of energy-efficient lighting and landscaping features that enhance the visual appeal of the construction project while minimising energy consumption and environmental impact.

In conclusion, construction partnering is crucial in shaping the visibility of construction projects within stealth construction. Fostering collaboration, innovation, and informed decision-making among stakeholders, partnering ensures that factors such as the colour and texture of the construction, its location, interaction with the environment, and other visibility-related considerations are carefully considered and integrated into the project. This collaborative approach enhances the aesthetic appeal and functionality of the built environment and contributes to creating sustainable and resilient communities.

Countermeasures

Construction partnering is essential in developing and implementing countermeasures in stealth construction. When actively engaging architects, engineers, contractors, and security specialists, partnering ensures that *functional construction systems, tactics such as construction methods, and other relevant factors* are carefully considered and integrated into the project to enhance security and resilience.

Functional construction systems are a key focus area addressed through construction partnering. Pargoo and Ilbeigi (2023) described that stakeholders collaborate to design and implement systems that countermeasures against potential security threats and risks. This may include integrating access control systems, surveillance cameras, intrusion detection sensors, and other security technologies into the building's design. When leveraging the collective expertise of security specialists and construction professionals, partnering ensures that functional construction systems are seamlessly integrated into the project to provide robust protection against unauthorised access, intrusion, and other security breaches.

Tactics, such as construction methods, are another critical aspect addressed through construction partnering. Adekola et al. (2020) and Cantelmi et al. (2021) discussed that stakeholders work together to develop construction methodologies that enhance security and resilience while minimising vulnerabilities and risks. This may involve using blast-resistant materials, reinforced structural systems, and other construction techniques to withstand various hazards, including natural disasters, terrorist attacks, and other security threats. Collaborating on selecting and implementing tactical construction methods ensures the project is built to withstand potential threats and protect occupants and assets.

Other factors related to countermeasures, such as site layout, emergency preparedness, and risk management strategies, are also addressed through construction partnering. Stakeholders collaborate to develop site plans that optimise security and access control while minimising potential vulnerabilities. This may include placing barriers, bollards, and other physical security measures to deter unauthorised access and mitigate vehicle-borne threats. Additionally, partnering facilitates the development of emergency response plans and procedures to ensure that the project is prepared to respond effectively to security incidents, natural disasters, and other emergencies.

Fostering collaboration, innovation, and informed decision-making among stakeholders and partnering ensures that functional construction systems, tactics such as construction methods, and other relevant factors are carefully considered and integrated into the project to enhance security and resilience. This collaborative approach protects occupants and assets and creates safer, more secure built environments for current and future generations.

Partnering Towards Stealth Construction

Partnering towards stealth construction marks a paradigm shift in the construction industry, where collaborative endeavours among diverse stakeholders' usher in a new era of innovation and efficiency. This transformative approach summarises a multifaceted commitment to integrating advanced technologies, leveraging cutting-edge practices, and prioritising security, sustainability, and resilience throughout every phase of the building project. Through active engagement and synergistic collaboration, architects, engineers, contractors, security specialists, and other key players work in unison to chart a course towards construction excellence that meets and exceeds the evolving demands of the modern built environment.

At the forefront of partnering towards stealth construction lies a steadfast dedication to enhancing security measures. Mishra et al. (2020) and George et al. (2023) explained that this entails carefully examining potential vulnerabilities and threats and strategically integrating sophisticated security systems, robust access controls, and resilient construction techniques. Partnering the collective expertise of security specialists and construction professionals ensures that buildings are fortified against various risks, ranging from natural disasters to deliberate sabotage. Through proactive risk management and vigilant oversight, partnering endeavours aim to instil a sense of safety and security, safeguarding occupants, assets, and critical infrastructure with unwavering diligence.

Furthermore, partnering towards stealth construction embodies a holistic commitment to sustainability and resilience, echoing the imperatives of the 21st century (Bibri, 2022; Krantz, 2020). By embracing green building principles, harnessing renewable energy sources, and integrating resilient design features, stakeholders collaborate to minimise environmental impact, optimise energy efficiency, and enhance structural integrity. This collaborative effort extends beyond compliance with regulatory standards, transcending conventional boundaries to embrace a holistic sustainability vision encompassing social, economic, and environmental considerations. Through the adoption of innovative construction techniques, materials, and technologies, partnering initiatives pave the way for the creation of buildings that not only withstand the test of time but also serve as beacons of sustainability, resilience, and innovation in an ever-changing world.

Considering partnering towards achieving stealth construction, we have identified five (5) central sustainable principles centred on environmental protection, health and safety, project delivery duration, economy, and aesthetics. These principles emphasise the integration of stealth construction technologies across all stages of the construction process, from inception to completion and throughout the project's whole life cycle.

Environmental Protection

In stealth construction, environmental protection is achieved through a concerted effort among stakeholders facilitated by construction partnering. To begin with, Jellinek et al. (2021) illustrated that partnering enables collaborative decision-making processes that prioritise environmental considerations from the project's inception. Architects, engineers, contractors, and environmental specialists work together to assess the environmental impact of the construction project and develop strategies to minimise harm to the surrounding ecosystem. This may involve conducting environmental impact assessments, identifying sensitive habitats or ecosystems, and implementing measures to mitigate potential adverse effects on air and water quality, wildlife habitats, and biodiversity.

Furthermore, Yang et al. (2021) and Umoh et al. (2024) added that construction partnering fosters the integration of sustainable building practices and green technologies into the construction process. When leveraging stakeholders' collective expertise, partnering initiatives promote eco-friendly materials, energy-efficient systems, and renewable energy sources to minimise resource consumption and reduce carbon emissions. Through collaborative planning and design, stakeholders explore innovative solutions such as passive design strategies, green roofs, and rainwater harvesting systems to enhance environmental performance and promote sustainability. Additionally, partnering facilitates the adoption of construction techniques that minimise waste generation, optimise material use, and encourage recycling and reuse, further reducing the project's environmental footprint.

In addition, construction partnering encourages ongoing monitoring and evaluation of environmental performance throughout the project lifecycle. McGrath et al. (2021) discussed that by establishing transparent communication channels and data-sharing mechanisms, stakeholders can track key environmental indicators, assess progress towards sustainability goals, and identify opportunities for improvement. This collaborative approach enables continuous learning and adaptation, allowing stakeholders to implement corrective actions and refine strategies to enhance environmental protection. Additionally, partnering facilitates engagement with regulatory agencies, local communities, and environmental advocacy groups, fostering transparency, accountability, and social responsibility in environmental management practices. Through construction partnering, environmental protection becomes a shared responsibility, with stakeholders working together to minimise environmental impact and create healthier, more sustainable built environments for current and future generations.

Health and Safety

Health and safety in stealth construction are prioritised and achieved through robust collaboration and partnership among various stakeholders involved in the project. Construction partnering fosters a proactive approach to health and safety, beginning with the early planning stages (Akinlolu et al., 2022; Tayeh et al., 2020). Architects, engineers, contractors, and health and safety professionals work together to identify potential hazards, assess risks, and develop comprehensive safety plans tailored to the project's needs. Stakeholders and partnering initiatives ensure that health and safety considerations are integrated into every aspect of the construction process.

One key aspect of achieving health and safety in stealth construction through construction partnering is the implementation of rigorous safety protocols and procedures. Cordeiro et al. (2022) and Osei-Asibey et al. (2021) stated that stakeholders collaborate to establish clear guidelines and standards for safe work practices, personal protective equipment (PPE) requirements, and emergency response protocols. Through ongoing communication and training initiatives, partnering initiatives promote a culture of safety awareness and accountability among all project participants. Regular safety audits and inspections are conducted to monitor compliance with established protocols and identify areas for improvement, ensuring that health and safety remain top priorities throughout the project lifecycle.

Furthermore, construction partnering facilitates the adoption of innovative technologies and practices to enhance health and safety in stealth construction projects. Shafiq and Afzal (2020) and Alizadehsalehi et al. (2020) described that by embracing digital tools like BIM, VR, and drones, stakeholders can conduct virtual safety assessments, identify potential hazards, and simulate emergency scenarios in a controlled environment. Additionally, partnering initiatives promote using modular construction techniques, prefabricated components, and off-site fabrication to minimise on-site hazards, reduce exposure to construction-related risks, and enhance overall safety performance. Through collaborative efforts and a commitment to continuous improvement, construction partnering ensures that health and safety remain paramount in stealth construction projects, protecting the well-being of workers, occupants, and the broader community.

Project Delivery Duration

In stealth construction, achieving optimal project duration results from strategic collaboration and partnership facilitated by construction partnering. From the project's inception, Ghanbari et al. (2023) and Harit and Judson (2023) reiterated that stakeholders such as architects, engineers, contractors, and project managers work together to develop comprehensive project plans and schedules that prioritise efficiency and minimise delays. Through open communication channels and regular coordination meetings, partnering initiatives enable stakeholders to identify critical path activities, allocate resources effectively, and anticipate potential challenges before they arise. By bringing together all team members with diverse expertise and experience, partnering ensures that project durations are carefully planned and managed to meet or exceed established timelines.

Furthermore, construction partnering promotes the adoption of innovative construction methods and technologies to streamline project delivery and accelerate construction schedules. Stakeholders collaborate to explore prefabrication, modular construction, and other off-site fabrication techniques to expedite construction and reduce on-site labour requirements (Langston & Zhang, 2021; Wu et al., 2021). When digital tools such as BIM, project management software, and real-time monitoring systems are embraced, partnering initiatives facilitate efficient coordination and communication among project teams, enabling proactive problem-solving and timely decision-making. Through collaborative efforts and a commitment to continuous improvement, construction partnering empowers stakeholders to optimise project durations, delivering high-quality construction projects on time and within budget.

Aesthetics and Economy

Achieving a harmonious balance between aesthetics and economy in stealth construction is a multifaceted endeavour that relies on collaborative efforts and strategic partnerships facilitated by construction partnering. Khan and McNally (2023) and Saad et al. (2024) mentioned that architects, engineers, contractors, and other stakeholders collaborate closely to develop innovative design solutions that enhance the built environment's visual appeal while optimising cost-effectiveness. Partnering initiatives among professionals ensure that aesthetic considerations are seamlessly integrated into the project's design and construction process without compromising economic viability.

Lu et al. (2022) and Salam et al. (2024) stated that construction partnering fosters an iterative design approach emphasising value engineering and cost optimisation to achieve aesthetic objectives within budgetary constraints. Through collaborative brainstorming sessions and design charrettes, stakeholders explore creative design alternatives, material selections, and construction techniques that maximise aesthetic impact while minimising costs. Participating in value engineering exercises analysing lifecycle costs and partnering initiatives empowers stakeholders to make informed decisions that balance aesthetics and cost-effectiveness, maximising the project's value for everyone involved.

Furthermore, construction partnering promotes adopting sustainable building practices and innovative technologies that enhance aesthetics and economy in stealth construction projects. Stakeholders collaborate to identify eco-friendly materials, energy-efficient systems, and resilient design features that not only improve the visual appeal of the building but also contribute to long-term cost savings and environmental sustainability (Alahira et al., 2024). By embracing green building principles and incorporating sustainable design strategies from the outset, partnering initiatives ensure that aesthetic considerations are aligned with economic and environmental goals, resulting in visually striking and financially prudent buildings.

Through effective communication, collaboration, and shared decision-making, construction partnering empowers stakeholders to achieve a harmonious synthesis of aesthetics and economy in stealth construction projects. When a collaborative culture that values creativity, innovation, and cost-consciousness is initiated in construction, partnering initiatives enable stakeholders to realise their aesthetic vision while maximising the project's economic value. Through a holistic approach that integrates design excellence, cost-effectiveness, and sustainability, construction partnering ensures that stealth construction projects deliver exceptional aesthetic outcomes that are both visually captivating and financially sustainable.

Partnering for Stealth Construction from the Post-construction Stage

In the post-construction stage of stealth construction, partnership plays a crucial role in ensuring the long-term success and sustainability of the built environment. While the physical construction may be complete, the collaboration among stakeholders remains essential for the building's ongoing maintenance, operation, and optimisation. Architects, engineers, contractors, facility managers, and building owners collaborate to address maintenance needs, monitor performance metrics, and implement upgrades or modifications as necessary. This ongoing partnership ensures that the building meets evolving needs and remains resilient and functional.

Furthermore, partnership in the post-construction stage extends beyond the immediate stakeholders to include the broader community and environment. Stakeholders collaborate to assess the building's impact on its surroundings, implement measures to mitigate environmental impact and engage with the community to address any concerns or issues that may arise. In the post-construction stage, when stakeholders work together and communicate openly, it helps build trust and transparency. This teamwork promotes sustainable practices that benefit the people using the building and the larger community.

The post-construction stage is particularly well-suited for partnership in stealth construction because it focuses on long-term sustainability and resilience. Unlike the pre-construction and construction stages, which primarily focus on planning and implementation, the post-construction stage emphasises ongoing maintenance, operation, and optimisation of the building. Partnership in this stage allows stakeholders to address emerging challenges, optimise performance, and adapt to changing needs and conditions over time. Finally, collaboration after construction helps keep the building valuable for its users and the community over time.

Conclusion

Collaboration through construction partnering emerges as a cornerstone for achieving multifaceted goals spanning security, sustainability, efficiency, and aesthetic appeal in stealth construction. Through active engagement of architects, engineers, contractors, and specialists, construction partnering ensures seamless integration of advanced technologies, innovative practices, and robust strategies across various project phases. From meticulous planning to post-construction stages, partnering fosters proactive approaches to address challenges, optimise resources, and maximise value, ultimately yielding buildings prioritising safety, environmental responsibility, and long-term functionality. Furthermore, partnership in stealth construction extends far beyond traditional project management paradigms, transcending silos to embrace a holistic vision of construction excellence. When communication, collaboration, and shared responsibility are emphasised, partnering initiatives enable stakeholders to navigate complexities, innovate solutions, and foster resilience in the built environment. Through a collective commitment to sustainability, efficiency, and community engagement, construction partnering ensures that stealth construction projects meet and exceed expectations, serving as enduring symbols of progress and stewardship in the modern built environment.

References

Adekola, J., Fischbacher-Smith, D., & Fischbacher-Smith, M. (2020). Inherent complexities of a multi-stakeholder approach to building community resilience. *International Journal of Disaster Risk Science*, *11*(1), 32–45.

Aghimien, D., Aigbavboa, C., Oke, A., Thwala, W., & Moripe, P. (2022). Digitalization of construction organisations – A case for digital partnering. *International Journal of Construction Management*, *22*(10), 1950–1959.

Akinlolu, M., Haupt, T. C., Edwards, D. J., & Simpeh, F. (2022). A bibliometric review of the status and emerging research trends in construction safety management technologies. *International Journal of Construction Management*, *22*(14), 2699–2711.

Alahira, J., Ninduwezuor-Ehiobu, N., Olu-Lawal, K. A., Ani, E. C., & Ejibe, I. (2024). Eco-innovative graphic design practices: Leveraging fine arts to enhance sustainability in industrial design. *Engineering Science & Technology Journal*, *5*(3), 783–793.

Ali, F., & Haapasalo, H. (2023). Development levels of stakeholder relationships in collaborative projects: Challenges and preconditions. *International Journal of Managing Projects in Business*, *16*(8), 58–76.

Alizadehsalehi, S., & Hadavi, A. (2023). Synergies of lean, BIM, and extended reality (LBX) for project delivery management. *Sustainability*, *15*(6), 1–27.

Alizadehsalehi, S., Hadavi, A., & Huang, J. C. (2020). From BIM to extended reality in AEC industry. *Automation in Construction*, *116*, 1–13. https://doi.org/10.1016/j.autcon.2020.103254

Anant, K. M., & Charpe, P. S. (2022, April). Analysis of seismic forces for earthquake-resistant constructions. In J. K. Deka, P. S. Robi & B. Sharma (Eds.), *International conference on emerging global trends in engineering and technology* (pp. 145–163). Springer Nature Singapore.

Aripin, Z., Mulyani, S. R., & Haryaman, A. (2023). Marketing strategy in project sustainability management efforts in extractive industries: Building a reciprocity framework for community engagement. *Kriez Academy: Journal of Development & Community Service*, *1*(1), 25–38.

Baah, C., Acquah, I. S. K., & Ofori, D. (2022). Exploring the influence of supply chain collaboration on supply chain visibility, stakeholder trust, environmental and financial performances: A partial least square approach. *Benchmarking: An International Journal*, *29*(1), 172–193.

Bibri, S. E. (2022). The social shaping of the metaverse as an alternative to the imaginaries of data-driven smart cities: A study in science, technology, and society. *Smart Cities*, *5*(3), 832–874.

Cantelmi, R., Di Gravio, G., & Patriarca, R. (2021). Reviewing qualitative research approaches in the context of critical infrastructure resilience. *Environment Systems and Decisions*, *41*(3), 341–376.

Cordeiro, L., Gnatta, J. R., Ciofi-Silva, C. L., Price, A., de Oliveira, N. A., Almeida, R. M., Mainardi, G., M., Srinivas, S., Chan, W., Levin, A., S., & Padoveze, M. C. (2022). Personal protective equipment implementation in healthcare: A scoping review. *American Journal of Infection Control*, *50*(8), 898–905.

Deep, S., Gajendran, T., & Jefferies, M. (2021). A systematic review of 'enablers of collaboration' among the participants in construction projects. *International Journal of Construction Management*, *21*(9), 919–931.

Dewulf, G., & Kadefors, A. (2012). Collaboration in public construction – Contractual incentives, partnering schemes and trust. *Engineering Project Organization Journal*, *2*(4), 240–250.

Digitemie, W. N., & Ekemezie, I. O. (2024). A comprehensive review of Building Energy Management Systems (BEMS) for improved efficiency. *World Journal of Advanced Research & Reviews*, *21*(3), 829–841.

Enwin, A. D., & Ikiriko, T. D. (2024). Resilient and regenerative sustainable urban housing solutions for Nigeria. *World Journal of Advanced Research & Reviews*, *21*(2), 1078 1099.

Evans, M., Farrell, P., Elbeltagi, E., Mashali, A., & Elhendawi, A. (2020). Influence of partnering agreements associated with BIM adoption on stakeholder's behaviour in construction mega-projects. *International Journal of BIM & Engineering Science*, *3*(1), 1–20.

George, A. S., George, A. H., & Baskar, T. (2023). Digitally immune systems: Building robust defences in the age of cyber threats. *Partners Universal International Innovation Journal*, *1*(4), 155–172.

Ghanbari, M., Zolfaghari, D., & Yadegari, Z. (2023). Mitigating construction delays in Iran: An empirical evaluation of building information modeling and integrated project delivery. *Journal of Engineering Management & Systems Engineering*, *2*(3), 170–179.

Goubran, S., & Cucuzzella, C. (2019). Integrating the sustainable development goals in building projects. *Journal of Sustainability Research*, *1*(1), 1–43.

Gunduz, M., & Abdi, E. A. (2020). Motivational factors and challenges of cooperative partnerships between contractors in the construction industry. *Journal of Management in Engineering*, *36*(4), 1–10.

Harit, M., & Judson, L. (2023). Mitigation strategies for delay factors in healthcare projects. *Journal of Building Construction*, *5*(1), 1–16.

Hughes, D., Williams, T., & Ren, Z. (2012). Differing perspectives on collaboration in construction. *Construction Innovation*, *12*(3), 355–368.

Jayawardana, J., Jayasinghe, J. A. S. C., Sandanayake, M., Kulatunga, A. K., & Zhang, G. (2023). Prefabricated construction in Sri Lanka: A proposed adoption strategy and a pilot case study from sustainability perspective. *Engineering Journal of Institution & Engineering Sri Lanka*, *56*(1), 71–80. http://doi.org/10.4038/engineer.v56i1.7562

Jellinek, S., Lloyd, S., Catterall, C., & Sato, C. (2021). Facilitating collaborations between researchers and practitioners in ecosystem management and restoration. *Ecological Management & Restoration*, *22*(2), 208–213.

Khan, M., & McNally, C. (2023). A holistic review on the contribution of civil engineers for driving sustainable concrete construction in the built environment. *Developments in the Built Environment*, *16*(1), 1–15. https://doi.org/10.1016/j.dibe.2023.100273

Köhler, J., Sönnichsen, S. D., & Beske-Jansen, P. (2022). Towards a collaboration framework for circular economy: The role of dynamic capabilities and open innovation. *Business Strategy & the Environment*, *31*(6), 2700–2713.

Koolwijk, J., van Oel, C., & Bel, M. (2022). The interplay between financial rules, trust and power in strategic partnerships in the construction industry. *Engineering, Construction & Architectural Management*, *29*(3), 1089–1108.

Kosorić, V., Lau, S. K., Tablada, A., Bieri, M., & M. Nobre, A. (2021). A holistic strategy for successful photovoltaic (PV) implementation in Singapore's built environment. *Sustainability*, *13*(11), 1–35. https://doi.org/10.3390/su13116452

Krantz, D. (2020). Solving problems like Maria: A case study and review of collaborative hurricane-resilient solar energy and autogestión in Puerto Rico. *Journal of Sustainability Research*, *3*(1), 1–89. https://doi.org/10.20900/jsr20210004

Lahdenperä, P. (2012). Making sense of the multi-party contractual arrangements of project partnering, project alliancing and integrated project delivery. *Construction Management and Economics*, *30*(1), 57–79.

Langston, C., & Zhang, W. (2021). DfMA: Towards an integrated strategy for a more productive and sustainable construction industry in Australia. *Sustainability*, *13*(16), 1–21.

Li, B., Akintoye, A., Edwards, P. J., & Hardcastle, C. (2005). Critical success factors for PPP/PFI projects in the UK construction industry. *Construction Management & Economics*, *23*(5), 459–471.

Lu, Y., Sood, T., Chang, R., & Liao, L. (2022). Factors impacting integrated design process of net zero energy buildings: An integrated framework. *International Journal of Construction Management*, *22*(9), 1700–1712.

Manfren, M., Sibilla, M., & Tronchin, L. (2021). Energy modelling and analytics in the built environment – A review of their role for energy transitions in the construction sector. *Energies*, *14*(3), 1–29.

McGrath, P., McCarthy, L., Marshall, D., & Rehme, J. (2021). Tools and technologies of transparency in sustainable global supply chains. *California Management Review*, *64*(1), 67–89.

Mihailova, D., Schubert, I., Burger, P., & Fritz, M. M. (2022). Exploring modes of sustainable value co-creation in renewable energy communities. *Journal of Cleaner Production*, *330*, 1–11. https://doi.org/10.1016/j.jclepro.2021.129917

Mishra, S., Anderson, K., Miller, B., Boyer, K., & Warren, A. (2020). Microgrid resilience: A holistic approach for assessing threats, identifying vulnerabilities, and designing corresponding mitigation strategies. *Applied Energy*, *264*, 1–17. https://doi.org/10.1016/j.apenergy.2020.114726

Mudzvokorwa, T., Mwiya, B., & Mwanaumo, E. M. (2020). Improving the contractor-subcontractor relationship through partnering on construction projects in Zambia. *Journal of Construction Engineering & Project Management*, *10*(1), 1–15.

Obiuto, N. C., Ebirim, W., Ninduwezuor-Ehiobu, N., Ani, E. C., Olu-lawal, K. A., & Ugwuanyi, E. D. (2024). Integrating sustainability into HVAC project management: Challenges and opportunities. *Engineering Science & Technology Journal*, *5*(3), 873–887.

Odeyemi, O., Oyewole, A. T., Adeoye, O. B., Ofodile, O. C., Addy, W. A., Okoye, C. C., & Ololade, Y. J. (2024). Entrepreneurship in Africa: A review of growth and challenges. *International Journal of Management & Entrepreneurship Research*, *6*(3), 608–622.

Osei-Asibey, D., Ayarkwa, J., Adinyira, E., Acheampong, A., & Amoah, P. (2021). Roles and responsibilities of stakeholders towards ensuring health and safety at construction site. *Journal of Building Construction & Planning Research*, *9*(1), 90–114.

Pargoo, S. N., & Ilbeigi, M. (2023). A scoping review for cybersecurity in the construction industry. *Journal of Management in Engineering*, *39*(2), 1–13. https://doi.org/10.1061/JMENEA.MEENG-5034

Phillips-Alonge, O. K. (2017). *The influence of partnering on the occurrence and cost of construction conflicts and disputes* [Doctoral dissertation, Northcentral University].

Porwal, A., & Hewage, K. N. (2013). Building Information Modeling (BIM) partnering framework for public construction projects. *Automation in Construction*, *31*(1), 204–214.

Pozzi, R., Rossi, T., & Secchi, R. (2023). Industry 4.0 technologies: Critical success factors for implementation and improvements in manufacturing companies. *Production Planning & Control*, *34*(2), 139–158.

Rippmann, M. (2016). *Funicular Shell Design: Geometric approaches to form finding and fabrication of discrete funicular structures* [Doctoral dissertation, ETH Zurich].

Saad, S., Haris, M., Ammad, S., & Rasheed, K. (2024). AI-assisted building design. In S. Saad, S. Ammad & K. Rasheed (Eds.), *AI in material science* (pp. 143–168). CRC Press.

Salam, M., Killen, C., & Forsythe, P. (2024). Assessing interdisciplinary collaboration in the detailed design phase of construction projects: Applying practice-based inter-organisational theories. *International Journal of Construction Management*, *2*(3), 1–10. https://doi.org/10.1080/15623599.2024.2313820

Saraji, K. M., Streimikiene, D., & Kyriakopoulos, G. L. (2021). Fermatean fuzzy CRITIC-COPRAS method for evaluating the challenges to industry 4.0 adoption for a sustainable digital transformation. *Sustainability*, *13*(17), 1–20.

Shafiq, M. T., & Afzal, M. (2020). Potential of virtual design construction technologies to improve job-site safety in Gulf Corporation Council. *Sustainability*, *12*(9), 1–21.

Tayeh, B. A., Yaghi, R. O., & Abu Aisheh, Y. I. (2020). Project manager interventions in occupational health and safety during the pre-construction phase in the Gaza Strip. *The Open Civil Engineering Journal*, *14*(1), 20–30. https://doi.org/10.2174/1874149502014010020

Umoh, A. A., Adefemi, A., Ibewe, K. I., Etukudoh, E. A., Ilojianya, V. I., & Nwokediegwu, Z. Q. S. (2024). Green architecture and energy efficiency: A review of innovative design and construction techniques. *Engineering Science & Technology Journal*, *5*(1), 185–200.

Wu, Z., Luo, L., Li, H., Wang, Y., Bi, G., & Antwi-Afari, M. F. (2021). An analysis on promoting prefabrication implementation in construction industry towards sustainability. *International Journal of Environmental Research & Public Health*, *18*(21), 1–21.

Wuni, I. Y., & Shen, G. Q. (2022). Developing critical success factors for integrating circular economy into modular construction projects in Hong Kong. *Sustainable Production & Consumption*, *29*, 574–587.

Yang, Z., Chen, H., Mi, L., Li, P., & Qi, K. (2021). Green building technologies adoption process in China: How environmental policies are reshaping the decision-making among alliance-based construction enterprises. *Sustainable Cities & Society*, *73*, 1–15. https://doi.org/10.1016/j.scs.2021.103122

Zainuddin, N. S., Alias, A. H., Abu Bakar, N., & Haron, N. A. (2024). Unlocking the potential of BIM and VR integration to address construction challenges in Malaysia's building industry: A literature review. *Journal of Sustainable Civil Engineering & Technology (JSCET)*, *3*(1), 10–38.

Zakaria, S., Gajendran, T., Skitmore, M., & Brewer, G. (2018). Key factors influencing the decision to adopt industrialised building systems technology in the Malaysian construction industry: An inter-project perspective. *Architectural Engineering & Design Management*, *14*(1–2), 27–45.

Zhang, L., & Kim, C. (2023). Chromatics in urban landscapes: Integrating interactive genetic algorithms for sustainable color design in marine cities. *Applied Sciences*, *13*(18), 1–26.

Chapter Ten

Whole Life Cycle in Stealth Construction

Abstract

The chapter discussed the comprehensive integration of whole life cycle (WLC) principles in construction, mainly focusing on its application in stealth construction. It outlined the challenges of implementing WLC practices, emphasising the need for proactive planning and meticulous execution. The study highlighted key aspects of the WLC in stealth construction, including considerations for building design, energy transmission, visibility management, and security countermeasures. Additionally, it underscores the importance of addressing environmental protection, health and safety, project delivery duration, economy, and aesthetics throughout the construction process to ensure the development of resilient, sustainable, and visually appealing structures that meet the needs of present and future generations.

Keywords: Stealth construction; whole life cycle; sustainable construction; project delivery; stealth

Introduction

Asante et al. (2022) stated that the whole life cycle (WLC) in construction encompasses a series of orchestrated phases, each crucial in bringing a structure from concept to completion. It commences with the inception phase, where architects, engineers, and clients collaborate to outline project objectives, assess feasibility, and establish design criteria. During this phase, critical decisions are made regarding budgetary constraints, site selection, and initial concept designs. Stakeholders meticulously analyse the environmental impact, regulatory requirements, and the project's overall viability. This phase sets the foundation for the construction process, influencing subsequent stages profoundly.

Stealth Construction: Integrating Practices for Resilience and Sustainability, 235–257

Published under exclusive licence by Emerald Publishing Limited
doi:10.1108/978-1-83608-182-120251010

Following inception, the design phase takes centre stage, characterised by intense creativity and meticulous planning. Rane et al. (2023) added that architects and engineers translate conceptual ideas into tangible blueprints and construction documents, incorporating structural, mechanical, and electrical specifications. Furthermore, advanced computer-aided design (CAD) technologies streamline this process, allowing for precise modelling and virtual simulations to optimise functionality and efficiency. Concurrently, stakeholders refine project scopes, finalise material selections, and navigate regulatory compliance. The design phase serves as the blueprint for construction, guiding subsequent activities and ensuring alignment with project objectives.

As design solidifies, the construction phase unfolds, marking the tangible manifestation of meticulous planning and collaborative efforts. Nizma et al. (2024) asserted that skilled labourers, contractors, and project managers work synergistically to execute construction activities according to established timelines and specifications. This phase encompasses site preparation, foundation laying, structural erection, and installation of mechanical and electrical systems. Also, adherence to stringent quality control measures and safety protocols is paramount, mitigating risks and ensuring compliance with industry standards. Effective communication and coordination among stakeholders are imperative, facilitating seamless progress and addressing unforeseen challenges proactively.

Upon completion of construction, the project transitions into the operation and maintenance phase, where the constructed facility assumes its intended function and undergoes routine upkeep to preserve longevity and functionality, as stated by Ramakrishnan et al. (2021) and Harris et al. (2021). Richerson (2022) emphasised that facility managers oversee day-to-day operations, implement maintenance protocols, and manage occupant needs. Regular inspections, repairs, and upgrades are conducted to safeguard against wear and tear, technological obsolescence, and evolving regulatory requirements. With this, sustainable practices are increasingly integrated, optimising energy efficiency and minimising environmental impact throughout the facility's life cycle. The operation and maintenance phase ensures the continued functionality and value of the constructed asset, fostering a legacy of durability and sustainability in the built environment.

WLC Principles in Construction

In achieving the construction WLC, the following phases are considered: inception, design, construction, and operation and maintenance.

Inception Phase (Bahramian & Yetilmezsoy, 2020; Dawood & Vukovic, 2015)

- Stakeholders from various disciplines, including architects, engineers, developers, and community representatives, come together to deliberate on the project's overarching vision and goals, emphasising the importance of sustainability from the outset. Through collaborative workshops and brainstorming sessions, they articulate a shared understanding of the project's purpose, scope, and desired outcomes, considering environmental and social dimensions.

- Feasibility studies are conducted comprehensively, encompassing site conditions, regulatory requirements, market demand, and financial viability. Environmental assessments delve into the ecological sensitivity of the proposed site, identifying potential risks and opportunities for conservation and mitigation measures.
- Budgetary constraints are carefully weighed against sustainability objectives, focusing on long-term value creation and return on investment. Through cost-benefit analysis and scenario planning, stakeholders explore different pathways to achieving sustainability targets within the project's financial parameters.
- Site selection criteria are established based on a comprehensive evaluation of potential locations, considering factors such as access to transportation networks, availability of renewable resources, and compatibility with surrounding land uses. Environmental impact assessments guide decision-making, highlighting areas of concern and informing strategies for minimising the project's ecological footprint.
- Community engagement activities are conducted to solicit input from residents, businesses, and organisations, fostering a sense of ownership and inclusion in the project development process. Through public forums, surveys, and focus groups, stakeholders gather valuable insights into community needs, preferences, and concerns, which are integrated into the project's design and planning.

Design Phase (Kovacic & Zoller, 2015; Roberts et al., 2020)

- Sustainable design principles are woven into the fabric of the project from its conceptualisation, focusing on creating aesthetically pleasing, environmentally responsible, and socially responsive buildings. Architects explore innovative design strategies, such as passive solar design, natural ventilation, and green roofs, to optimise energy performance and reduce reliance on mechanical systems.
- Material selections are guided by a commitment to sustainability, emphasising sourcing renewable, recycled, and locally sourced materials whenever possible. Life cycle assessments inform decisions about material procurement, evaluating factors such as embodied energy, carbon footprint, and end-of-life recyclability.
- Building technologies are deployed strategically to enhance energy efficiency, occupant comfort, and operational performance. From energy-efficient lighting systems to smart building controls, technology solutions are tailored to meet the specific needs of each project while minimising environmental impact.
- Collaboration between design disciplines is facilitated through integrated design charrettes and multidisciplinary workshops, fostering cross-functional dialogue and creativity. Engineers work closely with architects to optimise building systems for energy efficiency, structural integrity, and resilience to climate change.
- Stakeholder engagement continues throughout the design phase, with regular progress updates, design reviews, and feedback sessions ensuring that sustainability goals remain front and centre. Client input is solicited at critical milestones, allowing for adjustments and refinements to the design based on evolving needs and priorities.

Construction Phase (Buyle et al., 2013; Hasik et al., 2019)

- Sustainable construction practices are implemented on the ground, guided by detailed construction documents and specifications that reflect the project's sustainability goals and requirements. Site managers and contractors collaborate closely to develop construction schedules that minimise disruption to local ecosystems and communities.
- Quality control measures are rigorously enforced to ensure that materials are sourced responsibly, installed correctly, and meet specified performance criteria. Regular inspections and testing protocols are conducted to verify compliance with environmental standards and building codes.
- Worker safety is prioritised through comprehensive training programs, safety protocols, and ongoing supervision. Health and safety practices are integrated into daily work routines, with regular toolbox talks and safety meetings fostering a culture of accountability and awareness.
- Environmental monitoring is conducted throughout the construction phase to track key indicators such as air and water quality, noise levels, and waste generation. Mitigation measures are implemented to address any adverse impacts on the surrounding environment, including erosion control, dust suppression, and stormwater management.
- Collaboration with suppliers and subcontractors extends beyond the construction site, with efforts to source sustainable materials, equipment, and services from certified vendors. Supply chain transparency and accountability are prioritised, with suppliers required to adhere to strict environmental and social standards.

Operation and Maintenance Phase (Liu et al., 2020; Lu et al., 2020)

- Comprehensive maintenance plans are developed based on manufacturer recommendations, industry best practices, and lessons learned from the construction phase. Preventive maintenance schedules ensure that building systems and components remain in optimal condition, minimising downtime and repair costs.
- Energy and water conservation measures are integrated into daily operations, with building managers implementing strategies to reduce consumption, optimise efficiency, and track performance metrics. Smart building technologies, such as automated lighting and Heating, Ventilation, and Air Conditioning (HVAC) controls, are used to monitor energy usage in real-time and identify opportunities for improvement.
- Regular inspections and audits are conducted to assess the condition of building systems and identify potential issues before they escalate. Building occupants are encouraged to report maintenance concerns promptly, with clear channels for communication and feedback, ensuring that problems are addressed on time.
- Sustainability education and outreach efforts are ongoing, with building managers and occupant representatives organising workshops, seminars, and training sessions to promote sustainable behaviours and practices. From energy conservation tips to waste reduction strategies, occupants are empowered to reduce the building's environmental footprint actively.

- Performance tracking and reporting mechanisms are established to monitor progress towards sustainability goals and demonstrate accountability to stakeholders. Key performance indicators are tracked over time, allowing for benchmarking against industry standards and continuous improvement initiatives.

Achieving WLC in Construction

WLC sustainability in construction demands a comprehensive approach that extends beyond the completion of a project. It necessitates careful consideration of environmental, social, and economic factors throughout a building's life, as summarised in Table 4. To commence this endeavour, Samarasekara et al. (2024) and Thanu (2022) mentioned that stakeholders engage in rigorous planning and analysis during the inception phase. Conducting thorough feasibility studies, assessing potential environmental impacts, and setting clear sustainability goals lays the groundwork for a project prioritising longevity and efficiency. Jiang et al. (2023) added that the active involvement of architects, engineers, clients, and community members ensures that diverse perspectives are integrated into the decision-making process, fostering holistic solutions that address multifaceted challenges.

Furthermore, during the design phase, Ho and Iyer-Raniga (2020) explained the pursuit of WLC sustainability manifesting through innovative design strategies and material selections to minimise environmental footprint and optimise resource efficiency. Architects leverage sustainable design principles such as passive solar techniques, efficient building orientation, and renewable materials to enhance energy performance and reduce operational costs over the building's lifespan. Integrating advanced building technologies, such as smart HVAC systems and energy-efficient lighting, improves operational efficiency and occupant comfort. Concurrently, lifecycle assessments evaluate the environmental impacts of design choices, guiding decisions that promote long-term sustainability.

In addition, as construction commences, the focus shifts to implementing sustainable practices on the ground. De Luca et al. (2021) and Ghaffar et al. (2020) mentioned that contractors employ construction techniques that minimise waste generation, maximise recycling opportunities, and reduce energy consumption during the building process. Adherence to stringent quality control measures ensures that materials are installed correctly, enhancing durability and minimising the need for future repairs. Moreover, fostering a culture of safety and well-being among workers ensures compliance with regulations and contributes to the project's social sustainability. Effective project management and communication facilitate stakeholder collaboration, enabling timely resolution of issues and adherence to project schedules.

Finally, upon completion, the commitment to WLC sustainability persists through proactive operation and maintenance practices. Studies by Amaral et al. (2020), Burroughs and Hansen (2020), and Himeur et al. (2023) showed that facility managers implement strategies to optimise energy and water usage, reduce waste generation, and enhance indoor air quality. Regular inspections and preventive maintenance activities are conducted to identify and address potential

Table 4. WLC in Construction.

Authors	Achieving WLC
Thanu (2022) and Samarasekara et al. (2024)	Planning and analysis
Ho and Iyer-Raniga (2020)	Design strategies and material selections
Ghaffar et al. (2020) and De Luca et al. (2021)	Quality control measures
Amaral et al. (2020), Burroughs and Hansen (2020), and Himeur et al. (2023)	Inspections and preventive maintenance activities

issues before they escalate, prolonging the lifespan of building systems and components. Additionally, engaging occupants in sustainability initiatives fosters a culture of environmental stewardship, promoting responsible usage and behaviour. Continuous monitoring and performance tracking allow for ongoing optimisation, ensuring the building remains efficient and resilient throughout its lifecycle. By embracing WLC sustainability principles from inception to operation, stakeholders can create buildings that meet immediate needs and contribute to a sustainable future for future generations.

Challenges in Implementing WLC Principles in Construction

Implementing WLC principles in construction presents many challenges that demand innovative solutions and collaborative efforts from stakeholders. De Toni and Pessot (2021) highlighted that one significant hurdle arises from the complexity and interconnectedness of various project phases and disciplines. Coordinating diverse stakeholders, including architects, engineers, contractors, and developers, requires effective communication and alignment of objectives throughout the project lifecycle. Also, conflicting priorities and divergent interests may impede progress, necessitating compromise and consensus-building to achieve sustainable outcomes. Moreover, integrating sustainability into traditional project delivery methods poses challenges, as existing processes and workflows may not adequately address long-term environmental, social, and economic impacts.

Another challenge lies in the availability and accessibility of data and tools necessary for informed decision-making, as addressed by Barbhuiya and Das (2023). Conducting comprehensive lifecycle assessments and evaluating design choices' environmental and economic impacts requires access to reliable data and sophisticated analysis tools. However, such resources may be limited or inaccessible, particularly for smaller firms or projects with constrained budgets. Additionally, the complexity of assessing lifecycle impacts across multiple dimensions, including energy consumption, carbon emissions, and social equity, must be clarified for the decision-making process. Bridging these knowledge gaps and providing accessible tools and resources is essential for empowering stakeholders to make informed, sustainable decisions.

Furthermore, the fragmented nature of the construction industry presents further challenges to implementing WLC principles. With various stakeholders often operating in fractions and pursuing disparate objectives, achieving sustainability goals can be challenging (Maxner et al., 2022). Lack of coordination and collaboration among stakeholders may result in low outcomes and missed partnership opportunities. Addressing this challenge requires fostering a culture of collaboration and knowledge-sharing across disciplines and organisational boundaries. Encouraging interdisciplinary teamwork and incentivising cooperation can facilitate the integration of diverse perspectives and expertise, leading to more holistic and sustainable solutions.

Moreover, Pórólfsdóttir et al. (2023) added that financial constraints and market dynamics present significant barriers to implementing WLC principles in construction. While sustainable design and construction practices may yield long-term cost savings and benefits, the upfront investment required to adopt these practices can be prohibitive for some stakeholders. Additionally, market pressures and short-term profit motives may incentivise prioritising cost and schedule considerations over sustainability. Overcoming these challenges requires demonstrating the long-term value proposition of WLC sustainability through tangible benefits such as energy savings, operational efficiency, and enhanced resilience. Moreover, leveraging financial incentives, regulatory mechanisms, and market drivers can encourage investment in sustainable construction practices and incentivise adoption across the industry.

WLC for Stealth Construction

Stealth construction represents a cutting-edge approach that integrates advanced technologies and innovative practices across all phases of the construction process, from inception to completion. This innovative methodology emphasises discretion, efficiency, and sustainability, harnessing the power of emerging technologies to streamline operations and minimise environmental impact. Stealth construction relies on a seamless fusion of digital tools, automation, and prefabrication techniques to optimise every aspect of the construction lifecycle. To begin with, during the inception phase, stealth construction leverages predictive analytics and machine learning algorithms to assess project feasibility and forecast potential challenges. Analysing vast datasets and historical trends, stakeholders gain valuable insights into market dynamics, regulatory requirements, and resource availability, enabling informed decision-making and risk mitigation strategies. Moreover, virtual reality (VR) and augmented reality (AR) technologies are utilised to visualise project concepts and engage stakeholders in immersive design experiences, facilitating collaboration and consensus-building (Delgado et al., 2020).

Furthermore, as the project progresses to the design phase, stealth construction embraces Building Information Modelling (BIM) as a central tool for coordination and optimisation (Johansson & Roupé, 2024). It enables multidisciplinary teams to collaborate in real-time, digitally simulating building performance and identifying opportunities for efficiency gains. Advanced parametric design software empowers architects and engineers to explore complex geometries and optimise structural configurations, while generative design algorithms generate countless design iterations to uncover innovative solutions. Additionally, prefabrication and modular construction techniques are employed to streamline construction processes, minimise waste, and accelerate project timelines. Furthermore, stealth construction relies on state-of-the-art robotics, drones, and autonomous vehicles to execute tasks precisely and efficiently throughout construction. Robotics automate repetitive tasks such as bricklaying, concrete pouring, and steel fabrication, enhancing productivity and safety on-site. Jacob-Loyola et al. (2021) added that drones conduct aerial surveys, monitor progress, and perform site inspections, providing real-time data to project teams and enabling proactive decision-making – autonomous vehicles transport materials and equipment, optimising logistics and reducing carbon emissions associated with traditional transportation methods.

In addition, stealth construction harnesses technology to enhance building performance and occupant comfort in the operation and maintenance phase. Broday and Gameiro-da Silva (2023) stated that Internet of Things (IoT) sensors collect real-time data on energy usage, indoor air quality, and equipment performance, enabling predictive maintenance and optimising building operations. Smart building systems adjust environmental conditions dynamically, maximising energy efficiency and reducing operational costs. Furthermore, machine learning algorithms analyse historical data to identify patterns and trends, informing strategies for continuous improvement and long-term sustainability. In essence, stealth construction represents a paradigm shift in the construction industry, where technology catalyses innovation and sustainability across the entire building lifecycle. By embracing digitalisation, automation, and prefabrication, stakeholders can unlock new possibilities for efficiency, resilience, and environmental stewardship, ushering in a new era of construction excellence.

Towards the WLC for stealth construction, there are considerations on the cross-section of the building, energy transmission, visibility, and countermeasures that the construction professionals consider as they strive towards a stealthy infrastructure. These considerations are expatiated below.

Building Cross Section

Throughout the entire life cycle of a construction project, the building cross-section plays a crucial role in shaping the efficiency, functionality, and sustainability of the built environment in stealth construction. Each aspect of the building cross-section, including its shape, internal construction, material usage, and other factors, is meticulously considered and strategically integrated to optimise performance and minimise environmental impact.

Çok (2022) illustrated that the shape of the building cross-section is a fundamental consideration in stealth construction, influencing factors such as energy efficiency, natural ventilation, and aesthetic appeal. By employing sleek and streamlined profiles, such as aerodynamic curves or angular geometries, buildings can minimise air resistance and reduce energy consumption associated with heating, cooling, and ventilation systems. Additionally, strategically placing windows and openings within the cross-section maximises natural daylight penetration while minimising solar heat gain, enhancing indoor comfort and reducing reliance on artificial lighting and HVAC systems.

Furthermore, internally, Broyles et al. (2022) and Chen et al. (2023) mentioned that the construction of the building cross-section is designed to optimise spatial utilisation, structural integrity, and occupant comfort. In stealth construction, internal partitions, columns, and load-bearing elements are strategically positioned to maximise floor area and create flexible, adaptable spaces that accommodate diverse uses and occupancy patterns. Prefabricated building systems and modular construction techniques are often employed to expedite assembly and minimise on-site labour requirements, reducing construction time and costs. Moreover, innovative construction methods, such as cross-laminated timber (CLT) or engineered steel structures, offer lightweight, durable solutions that enhance structural resilience and minimise embodied carbon emissions.

In addition, Elalaouy et al. (2024) illustrated that material usage is crucial within the building cross-section and likewise in stealth construction, where sustainability and resource efficiency are paramount. High-performance building materials, such as recycled steel, engineered wood, and low-emissivity glass, are selected for their durability, thermal insulation properties, and environmental sustainability. Advanced insulation and air sealing techniques are employed to minimise heat loss and air infiltration, enhancing building envelope performance and reducing energy consumption (Fawaier & Bokor, 2022). Additionally, renewable materials and bio-based products, such as bamboo flooring or cellulose insulation, are utilised to further reduce the building's carbon footprint and promote ecological balance.

Other factors influencing the building cross-section in stealth construction include accessibility, adaptability, and resilience to external forces. Burns et al. (2023) and Svensson (2020) stated that universal design principles are integrated into the cross-sectional layout to ensure equitable access for all occupants, including individuals with disabilities or mobility impairments. Furthermore, the cross-section is designed to accommodate future modifications and expansions, allowing the building to evolve and adapt to changing needs and technological advancements. Resilience measures, such as reinforced foundations, impact-resistant glazing, and redundant systems, are implemented to mitigate risks from natural disasters, climate change, and other external threats, safeguarding occupants and assets.

Visibility

Throughout the entire life cycle of a construction project, visibility plays a crucial role in stealth construction, influencing factors such as the colour and texture of the construction, its location, interaction with the environment, and other vital aspects. Visibility refers to how buildings are perceived within their surroundings, and in stealth construction, careful consideration is given to minimise visual impact while optimising functionality, efficiency, and sustainability.

The colour and texture of construction materials are primary considerations in stealth construction, as they significantly impact visibility (Xu et al., 2023).

Light-coloured and reflective surfaces, such as white or light-coloured facades, can blend into the surrounding environment, reducing visual contrast and minimising the building's visibility. Conversely, darker colours and rough textures may stand out more prominently, drawing attention and disrupting the visual harmony of the landscape. In stealth construction, materials with muted tones, earthy hues, or natural textures are often chosen to blend seamlessly with the environment, ensuring minimal visual disruption.

Location is another critical factor in managing visibility in stealth construction, as illustrated by Keeley et al. (2022). Buildings in sensitive or scenic areas, such as natural reserves, historic districts, or cultural landmarks, require special attention to minimise visual impact and preserve the integrity of the surroundings. In such locations, low-profile or underground construction techniques may be employed to reduce the visible footprint of the building and maintain unobstructed views. Additionally, strategic site selection and orientation can optimise natural screening from existing vegetation or topographic features, mitigating visual impact.

Xu et al. (2023) and Nuswantoro and Richter (2024) highlighted that the interaction between buildings and their environment also influences visibility in stealth construction. Passive design strategies, such as building orientation, landscaping, and architectural features, are utilised to harmonise with the natural surroundings and minimise visual intrusion. Building elements such as green roofs, living walls, and vegetated buffers create a seamless transition between the built environment and the natural landscape, enhancing visual aesthetics and ecological resilience. Moreover, attention to detail in architectural design, such as scale, proportion, and massing, ensures that buildings complement rather than dominate the visual context.

Dijokiene et al. (2022) and Serra et al. (2021) reiterated that building height, form, and silhouette impact visibility in stealth construction. Low-profile structures with simple, hidden forms tend to blend more seamlessly into the landscape, while tall or angular buildings may disrupt visual continuity and skyline views. In response, setback requirements, height restrictions, and building setbacks are enforced to maintain visual harmony and preserve scenic vistas. Additionally, architectural elements such as fenestration patterns, rooflines, and building setbacks are carefully designed to minimise visual clutter and enhance overall aesthetics.

Energy Transmission

From the entire life cycle of a construction project, SaberiKamarposhti et al. (2024) and Esposito et al. (2024) identified energy transmission as a central consideration in stealth construction, encompassing renewable energy integration, climate action planning, energy management, and energy efficiency initiatives. Each aspect of energy transmission is strategically addressed to optimise energy performance, reduce carbon emissions, and promote sustainability across all phases of the construction process. Furthermore, renewable energy integration is a cornerstone of energy transmission in stealth construction, where stakeholders prioritise adopting clean and renewable energy sources to power buildings and reducing reliance on fossil fuels. Marszal et al. (2012) and Zhang et al. (2022)

added that solar photovoltaic (PV) systems, wind turbines, and geothermal heat pumps are commonly deployed to generate on-site renewable energy, providing a reliable and sustainable power source for heating, cooling, and electricity needs. Through strategic planning and design, buildings are equipped with renewable energy systems that seamlessly integrate with architectural elements and building systems, maximising energy production and minimising visual impact.

Furthermore, climate action planning is another critical component of energy transmission in stealth construction, where stakeholders develop comprehensive strategies to mitigate climate change and adapt to its impacts. Wernersson et al. (2024) identified that climate action plans encompass a range of initiatives, including greenhouse gas (GHG) emissions reduction targets, energy efficiency measures, and resilience planning. In stealth construction, buildings are designed and constructed to exceed energy efficiency standards and achieve net-zero or net-positive energy performance, effectively reducing carbon emissions and mitigating the building sector's contribution to climate change.

Additionally, energy management is integral to optimising energy transmission in stealth construction, where stakeholders implement robust monitoring, measurement, and control systems to optimise energy usage and performance. As Francisco et al. (2020) and Ahmad and Zhang (2021) put it, advanced building automation systems, smart metres, and real-time energy monitoring software enable stakeholders to track energy consumption patterns, identify inefficiencies, and implement targeted interventions to reduce energy waste. Demand response strategies, load shifting, and peak shaving techniques are employed to manage energy demand effectively, minimise grid reliance, and optimise operational costs.

Energy efficiency is a fundamental principle in energy transmission in stealth construction, where stakeholders prioritise building envelope performance, HVAC system efficiency, and lighting optimisation to minimise energy consumption and maximise operational efficiency. High-performance building materials, such as insulated concrete forms (ICFs), triple-glazed windows, and cool roofs, are selected to enhance thermal insulation and reduce heat transfer through the building envelope (Hachem-Vermette, 2020). Wang et al. (2024) added that energy-efficient HVAC systems, including variable refrigerant flow (VRF) systems, heat pumps, and radiant heating, are installed to provide comfortable indoor environments while minimising energy demand. Additionally, advanced lighting controls, occupancy sensors, and daylight harvesting technologies are deployed to optimise lighting levels and minimise electricity usage.

Countermeasures

Countermeasures play a vital role in stealth construction, encompassing functional construction systems, tactical methods of construction, and other important factors. Countermeasures in stealth construction are designed to enhance security, minimise risks, and optimise efficiency while maintaining discretion and confidentiality. Firstly, Habbal et al. (2024) and Nova (2022) discussed that functional construction systems are integral to countermeasures in stealth construction, where stakeholders prioritise implementing robust and adaptive systems to address security, safety, and operational requirements. Also, advanced access control systems, surveillance cameras, and intrusion detection systems are deployed to safeguard construction sites and prevent unauthorised access. Additionally, fire suppression systems, emergency lighting, and evacuation procedures are established to ensure the safety of workers and mitigate risks during construction activities. Furthermore, resilient building systems, including redundant power supplies, backup generators, and data storage, are integrated to maintain operational continuity and minimise disruptions in emergencies or security incidents.

Secondly, Vijayan et al. (2023) and Yadav and Bhatnagar (2024) discussed how tactical construction methods can be strategically employed in stealth construction to minimise visibility, optimise efficiency, and mitigate environmental impact. Zhou et al. (2021) added that prefabrication and modular construction techniques streamline assembly processes, reduce on-site labour requirements, and accelerate project timelines. By prefabricating building components off-site, stakeholders can minimise disruption to surrounding areas and maintain a low profile during construction. Additionally, construction phasing and sequencing are carefully planned to minimise noise, dust, and other nuisances that could attract unwanted attention or compromise confidentiality. Moreover, construction activities are scheduled to coincide with low-traffic periods and minimise interference with surrounding activities to maintain discretion and minimise disruptions.

Other factors influencing countermeasures in stealth construction include site security, information protection, and stakeholder collaboration. Site security measures, such as perimeter fencing, security patrols, and access control checkpoints, are implemented to deter trespassing, vandalism, and theft, protecting valuable assets and sensitive information. Information protection protocols, including data encryption, secure communication channels, and access controls, are established to safeguard confidential project information and intellectual property. Additionally, stakeholder collaboration and communication are essential for effective risk management and contingency planning, ensuring all parties are aligned on security protocols, emergency procedures, and response strategies.

WLC Towards Stealth Construction

The journey towards achieving stealth construction spans the entire life cycle of a construction project, beginning with relatable planning and strategising in the inception phase. Clear objectives are established, encompassing functional requirements, security considerations, and sustainability goals. Through thorough feasibility studies and risk assessments, potential challenges are identified and addressed proactively, laying a solid foundation for stealth construction practices.

In the design phase, the principles of stealth construction are intricately woven into every aspect of architectural and engineering processes. Advanced technologies and innovative strategies are leveraged to minimise visual impact, optimise energy efficiency, and enhance security. Building layouts are carefully crafted to blend seamlessly with the surrounding environment, while materials and finishes are selected to reduce reflectivity and minimise visibility. Sustainable design principles are integrated to mitigate environmental impact and promote long-term resilience.

During the construction phase, the vision of stealth construction is realised through meticulous execution and attention to detail. Advanced construction techniques, such as prefabrication and modular construction, are employed to minimise on-site disturbances and maintain a low profile. Also, construction activities are coordinated to minimise noise, dust, and disruptions, while robust security measures are implemented to safeguard the site and protect against unauthorised access or information breaches. Through collaboration and communication among stakeholders, the objectives of stealth construction are upheld, creating buildings that meet functional requirements while prioritising discretion, security, and sustainability.

We have identified five (5) central sustainable principles considering the entire project life cycle towards achieving stealth construction. These principles centre on environmental protection, health and safety, project delivery duration, economy, and aesthetics. They emphasise the benefits of integrating stealth practices across all stages of the construction process, from inception to completion and throughout the project's lifespan.

Environmental Protection

Environmental protection in stealth construction is achieved through the comprehensive implementation of WLC practices prioritising sustainability, resource conservation, and ecosystem preservation. From the initial planning stages to the ongoing maintenance of the built environment, a multifaceted approach is adopted to minimise environmental impact and promote ecological resilience, which is in correlation to Omole et al. (2024) study on sustainable urban design.

During the inception phase, environmental protection begins with site selection and assessment. Moore-O'Leary et al. (2017) described that stakeholders can carefully evaluate potential sites based on ecological sensitivity, habitat preservation, and proximity to natural resources. Environmental impact assessments are conducted to identify potential risks and develop mitigation strategies, ensuring that the chosen site minimises disruption to local ecosystems and habitats. Additionally, sustainability goals and objectives are established, guiding decision-making processes throughout the project's life cycle.

In the design phase, environmental protection is integrated into every aspect of architectural and engineering processes. Negi (2021) added that sustainable design principles are prioritised, including passive heating and cooling strategies, energy-efficient building envelopes, and water conservation measures. Building materials with low environmental impact, such as recycled content, rapidly renewable resources, and locally sourced materials, are selected to reduce embodied carbon emissions and minimise ecological footprint. Furthermore, green building certifications are pursued to validate environmental performance and demonstrate commitment to sustainability. Furthermore, environmental protection is upheld throughout the construction phase through responsible construction practices and adherence to environmental regulations, as highlighted by Ajibike et al. (2023).

Construction activities are planned and executed to minimise disturbance to natural habitats, reduce soil erosion, and prevent pollution of air and water resources. Waste management plans are implemented to minimise construction waste and maximise recycling and reuse of materials. Additionally, construction equipment and machinery are selected for their energy efficiency and emissions control, reducing environmental impact.

Environmental protection continues through sustainable building management practices in the operation and maintenance phase. Etukudoh (2024) described that energy and water conservation measures are implemented to optimise operational efficiency and reduce resource consumption. Building systems are monitored and maintained to ensure optimal performance and minimise environmental impact. Furthermore, green cleaning practices, waste reduction initiatives, and indoor air quality management programs are implemented to promote occupant health and well-being while minimising environmental impact.

Health and Safety

In stealth construction, ensuring health and safety involves a comprehensive approach that integrates WLC practices to mitigate risks, protect workers, and promote a safe working environment from project inception to completion. When health and safety considerations are prioritised at every stage of the construction process, stakeholders can minimise accidents, injurics, and occupational hazards while fostering a culture of safety and well-being.

Health and safety considerations are embedded into project planning and risk management strategies during the inception phase (Harris et al., 2021; Lingard & Wakefield, 2019). Stakeholders thoroughly assess potential health and safety hazards associated with site conditions, project scope, and regulatory requirements. Also, risk mitigation plans are developed to address identified hazards and establish hazard identification, reporting, and response protocols. Health and safety objectives are also based on clear goals and performance metrics to guide decision-making throughout the project's life cycle.

In the design phase, health and safety are integrated into architectural and engineering processes to create an environment prioritising occupant well-being and safety (Engineer et al., 2021). Building designs incorporate universal design principles to ensure accessibility and inclusivity for all occupants, including those with disabilities or mobility limitations. Furthermore, building layouts are optimised to minimise ergonomic risks and improve workflow efficiency for construction workers. Passive design strategies, such as natural ventilation and daylighting, enhance indoor air quality and promote occupant comfort while reducing reliance on mechanical systems.

Health and safety practices are implemented throughout construction to mitigate risks and protect workers from potential hazards. Comprehensive health and safety plans are developed and communicated to all project stakeholders, outlining roles, responsibilities, hazard control and mitigation protocols. Choudhry et al. (2008) and Goh and Goh (2016) added that site-specific safety measures, such as personal protective equipment (PPE), fall protection systems, and emergency response procedures, are implemented to safeguard workers and minimise the risk of accidents or injuries. Regular safety inspections and audits are conducted to monitor compliance with health and safety regulations and identify areas for improvement.

Health and safety practices are prioritised in the operation and maintenance phase to ensure the ongoing safety and well-being of building occupants and maintenance personnel. Building systems are maintained regularly to prevent malfunctions and hazards, while emergency response plans are reviewed and updated to address changing conditions and occupancy levels. Additionally, occupant training programs educate building users about safety protocols, emergency procedures, and hazard awareness, empowering them to contribute to a safer and healthier built environment.

Overall, achieving health and safety in stealth construction requires a proactive and collaborative approach that integrates WLC practices to identify, assess, and mitigate risks throughout the project's life cycle. By prioritising health and safety considerations from inception to operation, stakeholders can create a built environment that promotes the well-being of all occupants and workers while minimising accidents, injuries, and occupational hazards.

Project Delivery Duration

Achieving efficient project delivery duration in stealth construction entails a detailed approach that integrates WLC practices to optimise timelines, streamline processes, and minimise delays from project inception through completion. When efficiency and coordination are prioritised at every stage of the construction process, stakeholders can expedite project timelines while maintaining quality and adhering to project objectives. During the inception phase, project delivery duration is influenced by careful planning, feasibility studies, and risk assessments. Scoggins et al. (2022) and Ebirim et al. (2024) suggested that stakeholders establish clear project objectives, scope, and timelines, considering site conditions, regulatory requirements, and resource availability. Comprehensive project planning and scheduling tools are utilised to develop realistic timelines and identify critical path activities. Moreover, risk mitigation and contingency planning strategies are designed to address potential delays and disruptions, ensuring proactive management of project delivery duration.

In the design phase, project delivery duration is optimised through efficient collaboration and decision-making processes. Integrated project delivery (IPD) methodologies are adopted to foster collaboration among multidisciplinary teams, including architects, engineers, contractors, and subcontractors. BIM technologies are leveraged to streamline design coordination, identify clashes, and resolve conflicts before construction begins. Additionally, value engineering exercises are conducted to identify cost-saving opportunities and optimise design solutions without compromising quality or performance.

Aslam et al. (2021) illustrated that project delivery duration is achieved throughout the construction phase through strategic planning, resource management, and efficient execution of construction activities. Lean construction principles are applied to minimise waste, optimise workflows, and eliminate non-value-added activities. Prefabrication and modular construction techniques accelerate construction timelines, reduce on-site labour requirements, and mitigate weather-related delays.

Moreover, just-in-time delivery practices and advanced logistics planning optimise material and equipment procurement, minimising inventory costs and delays.

Project delivery duration is sustained through proactive maintenance and asset management practices in the operation and maintenance phase. Weeks and Leite (2021) added that building management systems (BMS) and computerised maintenance management systems (CMMS) are implemented to monitor building performance, schedule preventive maintenance tasks, and track asset lifecycles. Additionally, predictive maintenance techniques, such as condition-based monitoring and predictive analytics, are utilised to anticipate equipment failures and address maintenance issues before they escalate, minimising downtime and optimising operational efficiency.

Economy

Achieving economic efficiency in stealth construction involves strategically implementing WLC practices to optimise costs, maximise value, and minimise financial risks from project inception to completion. When this approach is adopted, which considers the entire life cycle of the construction project, stakeholders can make informed decisions that prioritise cost-effectiveness while maintaining quality, sustainability, and project objectives.

Economic efficiency begins with comprehensive project planning and budgeting during the inception phase. Obiuto et al. (2024) maintained that stakeholders conduct thorough feasibility studies and financial analyses to assess project viability, identify potential cost drivers, and establish realistic budgets and economic targets. Cost estimation techniques, such as parametric estimating and benchmarking, are utilised to forecast project costs accurately and allocate resources efficiently. Moreover, risk management strategies are developed to identify and mitigate financial risks, ensuring proactive project economics management.

In the design phase, economic efficiency is achieved through value engineering and cost optimisation strategies. Fernandes et al. (2020) illustrated that designers collaborate closely with stakeholders to identify cost savings and value enhancement opportunities without compromising project quality or performance. Also, value engineering exercises are conducted to analyse design alternatives, optimise material selections, and streamline construction processes to achieve cost-effective solutions. Sustainable design principles are also integrated to minimise life cycle costs, reduce operational expenses, and enhance long-term value. Furthermore, economic efficiency is upheld throughout the construction phase through efficient resource management, procurement strategies, and cost-control measures. Mellado and Lou (2020) stated that lean construction principles are applied to minimise waste, optimise workflows, and improve productivity, reducing construction costs and enhancing project efficiency. Strategic procurement practices, such as competitive bidding, negotiated contracts, and strategic sourcing, are employed to obtain competitive pricing and maximise value from subcontractors and suppliers. Furthermore, robust cost monitoring and control systems are implemented to track project expenditures, identify variances, and implement corrective actions to keep the project within budget.

Economic efficiency is sustained through proactive asset management and lifecycle cost analysis in the operation and maintenance phase. BMS and CMMS optimise operational efficiency, minimise maintenance costs, and extend asset lifespan. Additionally, lifecycle cost analysis techniques are employed to assess the total cost of ownership, including acquisition, operation, maintenance, and disposal costs, enabling stakeholders to make informed decisions that maximise long-term value and minimise lifecycle costs.

Aesthetics

Achieving aesthetic excellence in stealth construction involves a holistic approach that integrates WLC practices to create visually harmonious, culturally sensitive, and contextually appropriate built environments from project inception to completion. By prioritising design innovation, architectural integrity, and environmental sensitivity, stakeholders can create buildings that blend seamlessly with their surroundings and inspire and enhance the built environment.

Bao et al. (2022) stated that aesthetics is considered from the outset during the inception phase, with stakeholders collaborating to establish design objectives, vision statements, and aesthetic goals. Also, site analysis and contextual studies are conducted to understand the project site's cultural, historical, and environmental context, informing design decisions and guiding aesthetic strategies. Furthermore, stakeholder engagement processes are employed to solicit input from the community, ensuring that the design reflects local preferences, values, and aspirations.

In the design phase, aesthetics is brought to life through innovative design solutions, sensitive material selections, and meticulous attention to detail (Almusaed et al., 2021). Designers leverage advanced technologies, such as parametric modelling and digital simulations, to explore design possibilities and optimise architectural forms. Also, building materials are carefully chosen for their aesthetic qualities, durability, and sustainability, emphasising natural and locally sourced materials that complement the surrounding environment. Moreover, passive design strategies, such as orientation, massing, and shading, enhance visual appeal while optimising energy performance and occupant comfort.

Throughout the construction phase, aesthetics is upheld through quality craftsmanship, meticulous execution, and adherence to design intent. Skilled artisans and craftsmen collaborate closely with architects and designers to realise intricate details and finishes, ensuring that the built environment reflects the intended aesthetic vision (Deamer & Bernstein, 2010; Naboni & Paoletti, 2015). Additionally, construction activities are coordinated to minimise visual disruption and maintain the integrity of the surrounding landscape. Sustainable construction practices, such as waste reduction and material recycling, minimise environmental impact while enhancing aesthetic quality.

According to Tu (2020) and Fusco-Girard and Vecco (2021), aesthetics is preserved through ongoing maintenance, adaptive reuse, and landscape management practices in operation and maintenance. Building facades and finishes are regularly cleaned and maintained to protect their appearance and extend their lifespan. Furthermore, adaptive reuse strategies are employed to repurpose existing structures and integrate them into the built environment, enhancing visual diversity and preserving architectural heritage. Landscaping and site design are continually managed to improve visual aesthetics, promote biodiversity, and create welcoming outdoor spaces for occupants and visitors.

Overall, achieving aesthetic excellence in stealth construction requires a collaborative and multidisciplinary approach that integrates WLC practices to create buildings that inspire, enrich, and contribute positively to the built environment. By prioritising design innovation, architectural integrity, and environmental sensitivity throughout the project's life cycle, stakeholders can create timeless and visually stunning built environments that resonate with people and communities for generations.

WLC for Stealth Construction from the Post-construction Stage

In the post-construction stage of stealth construction, the focus shifts from the physical construction process to the ongoing management, maintenance, and optimisation of the built environment. WLC practices play a crucial role during this phase, ensuring that the building continues to perform optimally and meet the evolving needs of its occupants and stakeholders. This stage encompasses a range of activities, including facility management, operations, maintenance, renovations, and eventual decommissioning or repurposing of the building. Furthermore, during the post-construction stage, WLC practices are implemented to maximise the building's longevity, efficiency, and sustainability. Comprehensive facility management strategies are developed to monitor and optimise building performance, including energy usage, indoor air quality, and occupant comfort. Regular maintenance programs are established to address wear and tear, prevent deterioration, and extend the lifespan of building systems and components. Additionally, renovations and upgrades may be undertaken to modernise facilities, incorporate new technologies, or adapt to changing user needs and preferences.

Ultimately, the post-construction stage is particularly well-suited for implementing WLC practices because it provides an opportunity to leverage insights gained from the construction process and adjust strategies accordingly. By assessing building systems' performance, identifying improvement areas, and implementing enhancements, stakeholders can optimise the building's efficiency, functionality, and sustainability over time. Moreover, WLC practices emphasise the importance of ongoing monitoring, evaluation, and adaptation, aligning closely with the iterative nature of facility management and operations in the post-construction stage.

Conclusion

Throughout the discussion, we explored the multifaceted approach of integrating WLC practices in stealth construction, encompassing environmental protection, health and safety, project delivery duration, economy, and aesthetics. From the inception to the post-construction stages, stakeholders prioritise sustainability, efficiency, and quality, embedding these principles into every construction process.

When advanced technologies, innovative strategies, and collaborative efforts are leveraged, stealth construction projects strive to minimise environmental impact, enhance safety, optimise timelines, manage costs effectively, and create visually harmonious built environments. Stealth construction helps build strong, eco-friendly, and attractive buildings that meet current and future needs by planning, being careful during construction, and keeping up with maintenance over time.

References

Ahmad, T., & Zhang, D. (2021). Using the internet of things in smart energy systems and networks. *Sustainable Cities and Society*, *68*, 1–22. https://doi.org/10.1016/j.scs.2021.102783

Ajibike, W. A., Adeleke, A. Q., Mohamad, F., Bamgbade, J. A., & Moshood, T. D. (2023). The impacts of social responsibility on the environmental sustainability performance of the Malaysian construction industry. *International Journal of Construction Management*, *23*(5), 780–789.

Almusaed, A., Yitmen, I., Almsaad, A., Akiner, İ., & Akiner, M. E. (2021). Coherent investigation on a smart kinetic wooden façade based on material passport concepts and environmental profile inquiry. *Materials*, *14*(14), 1–22.

Amaral, R. E., Brito, J., Buckman, M., Drake, E., Ilatova, E., Rice, P., Sabbagh, C., Voronkin, S., & Abraham, Y. S. (2020). Waste management and operational energy for sustainable buildings: A review. *Sustainability*, *12*(13), 1–21. https://doi.org/10.3390/su12135337

Asante, R., Faibil, D., Agyemang, M., & Khan, S. A. (2022). Life cycle stage practices and strategies for circular economy: Assessment in construction and demolition industry of an emerging economy. *Environmental Science and Pollution Research*, *29*(54), 82110–82121.

Aslam, M., Gao, Z., & Smith, G. (2021). Integrated implementation of virtual design and construction (VDC) and lean project delivery system (LPDS). *Journal of Building Engineering*, *39*, 1–13. https://doi.org/10.1016/j.jobe.2021.102252

Bahramian, M., & Yetilmezsoy, K. (2020). Life cycle assessment of the building industry: An overview of two decades of research (1995–2018). *Energy and Buildings*, *219*, 1–26. https://doi.org/10.1016/j.enbuild.2020.109917

Bao, Z., Laovisutthichai, V., Tan, T., Wang, Q., & Lu, W. (2022). Design for manufacture and assembly (DfMA) enablers for offsite interior design and construction. *Building Research & Information*, *50*(3), 325–338.

Barbhuiya, S., & Das, B. B. (2023). Life cycle assessment of construction materials: Methodologies, applications and future directions for sustainable decision-making. *Case Studies in Construction Materials*, *19*, 1–33. https://doi.org/10.1016/j.cscm.2023.e02326

Broday, E. E., & Gameiro-da Silva, M. C. (2023). The role of Internet of Things (IoT) in the assessment and communication of indoor environmental quality (IEQ) in buildings: A review. *Smart and Sustainable Built Environment*, *12*(3), 584–606.

Broyles, J. M., Brown, N. C., Hartwell, A. J., Alvarez, E. G., Ismail, M. A., Norford, L. K., & Mueller, C. T. (2022). Shape optimization of a concrete floor systems for sustainability, acoustical, and thermal objectives. In M. F. Hvejsel & P. J. Cruz (Eds.), *Structures and architecture – A viable urban perspective?* (pp. 1121–1128). CRC Press.

Burns, S. P., Mendonca, R. J., & Smith, R. O. (2023). Accessibility of public buildings in the United States: A cross-sectional survey. *Disability & Society*, 1–16. https://doi.org/10.1080/09687599.2023.2239996

Burroughs, H. E., & Hansen, S. J. (2020). *Managing indoor air quality*. River Publishers.

Buyle, M., Braet, J., & Audenaert, A. (2013). Life cycle assessment in the construction sector: A review. *Renewable and Sustainable Energy Reviews*, *26*, 379–388. https://doi.org/10.1016/j.rser.2013.05.001
Chen, T., Tai, K. F., Raharjo, G. P., Heng, C. K., & Leow, S. W. (2023). A novel design approach to prefabricated BIPV walls for multi-storey buildings. *Journal of Building Engineering*, *63*, 1–20. https://doi.org/10.1016/j.jobe.2022.105469
Choudhry, R. M., Fang, D., & Ahmed, S. M. (2008). Safety management in construction: Best practices in Hong Kong. *Journal of Professional Issues in Engineering Education and Practice*, *134*(1), 20–32.
Çok, B. (2022). Introduction modeling of deaf facades in the design of the facades of buildings with the index to renewable energy generation. *AIS-Architecture Image Studies*, *3*(1), 8–33.
Dawood, N., & Vukovic, V. (2015, April). Whole lifecycle information flow underpinned by BIM: Technology, process, policy and people. In *2nd International Conference on Civil and Building Engineering Informatics*, 22 April to 24 April, Tokyo, Japan.
De Luca, A., Chen, L., & Gharehbaghi, K. (2021). Sustainable utilization of recycled aggregates: Robust construction and demolition waste reduction strategies. *International Journal of Building Pathology and Adaptation*, *39*(4), 666–682.
De Toni, A. F., & Pessot, E. (2021). Investigating organisational learning to master project complexity: An embedded case study. *Journal of Business Research*, *129*, 541–554. https://doi.org/10.1016/j.jbusres.2020.03.027
Deamer, P., & Bernstein, P. (2010). *Building (in) the future: Recasting labor in architecture*. Princeton Architectural Press.
Delgado, J. M. D., Oyedele, L., Demian, P., & Beach, T. (2020). A research agenda for augmented and virtual reality in architecture, engineering and construction. *Advanced Engineering Informatics*, *45*, 1–21. https://doi.org/10.1016/j.aei.2020.101122
Dijokiene, D., Alistratovaite-Kurtinaitiene, I., & Cirtautas, M. (2022). Building height regulation: Is it still relevant for the 21st-century city? *28th International Seminar on Urban Form ISUF2021: Urban form and the sustainable and prosperous cities*, 29th June – 3rd July 2021, pp. 1120–1125, Glasgow, Scotland.
Ebirim, W., Usman, F. O., Montero, D. J. P., Ninduwezuor-Ehiobu, N., Olu-lawal, K. A., & Ani, E. C. (2024). Project management strategies for implementing energy-efficient cooling solutions in emerging data center markets. *World Journal of Advanced Research and Reviews*, *21*(2), 1802–1809.
Elalaouy, O., El Ghzaoui, M., & Foshi, J. (2024). Enhancing antenna performance: A comprehensive review of metamaterial utilization. *Materials Science and Engineering*, *304*, 1–24. https://doi.org/10.1016/j.mseb.2024.117382
Engineer, A., Gualano, R. J., Crocker, R. L., Smith, J. L., Maizes, V., Weil, A., & Sternberg, E. M. (2021). An integrative health framework for wellbeing in the built environment. *Building and Environment*, *205*, 1–15. https://doi.org/10.1016/j.buildenv.2021.108253
Esposito, P., Marrasso, E., Martone, C., Pallotta, G., Roselli, C., Sasso, M., & Tufo, M. (2024). A roadmap for the implementation of a renewable energy community. *Heliyon*, *10*(7), 1–25.
Etukudoh, E. A. (2024). Theoretical frameworks of ecopfm predictive maintenance (ecopfm) predictive maintenance system. *Engineering Science & Technology Journal*, *5*(3), 913–923.
Fawaier, M., & Bokor, B. (2022). Dynamic insulation systems of building envelopes: A review. *Energy and Buildings*, *270*, 1–20. https://doi.org/10.1016/j.enbuild.2022.112268
Fernandes, G., O'Sullivan, D., Pinto, E. B., Araújo, M., & Machado, R. J. (2020). Value of project management in university–industry R&D collaborations. *International Journal of Managing Projects in Business*, *13*(4), 819–843.
Francisco, A., Mohammadi, N., & Taylor, J. E. (2020). Smart city digital twin–enabled energy management: Toward real-time urban building energy benchmarking.

Journal of Management in Engineering, *36*(2), 1–11. https://doi.org/10.1061/(ASCE) ME.1943-5479.000074

Fusco-Girard, L., & Vecco, M. (2021). The "intrinsic value" of cultural heritage as driver for circular human-centered adaptive reuse. *Sustainability*, *13*(6), 1–28.

Ghaffar, S. H., Burman, M., & Braimah, N. (2020). Pathways to circular construction: An integrated management of construction and demolition waste for resource recovery. *Journal of Cleaner Production*, *244*, 1–9. https://doi.org/10.1016/j.jclepro.2019.118710

Goh, Y. M., & Goh, W. M. (2016). Investigating the effectiveness of fall prevention plan and success factors for program-based safety interventions. *Safety Science*, *87*, 186–194.

Habbal, A., Ali, M. K., & Abuzaraida, M. A. (2024). Artificial intelligence trust, risk and security management (AI TRiSM): Frameworks, applications, challenges and future research directions. *Expert Systems with Applications*, *240*, 1–14. https://doi.org/10.1016/j.eswa.2023.122442

Hachem-Vermette, C. (2020). Selected high-performance building envelopes. *Solar Buildings and Neighborhoods: Design Considerations for High Energy Performance*, *1*(1), 67–100. https://doi.org/10.1007/978-3-030-47016-6_3

Harris, F., McCaffer, R., Baldwin, A., & Edum-Fotwe, F. (2021). *Modern construction management*. John Wiley & Sons.

Hasik, V., Escott, E., Bates, R., Carlisle, S., Faircloth, B., & Bilec, M. M. (2019). Comparative whole-building life cycle assessment of renovation and new construction. *Building and Environment*, *161*, 1–10. https://doi.org/10.1016/j.buildenv.2019.106218

Himeur, Y., Elnour, M., Fadli, F., Meskin, N., Petri, I., Rezgui, Y., Bensaali, F., & Amira, A. (2023). AI- big data analytics for building automation and management systems: A survey, actual challenges and future perspectives. *Artificial Intelligence Review*, *56*(6), 4929–5021.

Ho, T. K. O., & Iyer-Raniga, U. (2020). Life cycle costing: Evaluate sustainability outcomes for building and construction sector. In *industry, innovation and infrastructure, encyclopedia of the UN sustainable development goals* (pp. 1–11). Springer.

Jacob-Loyola, N., Muñoz-La Rivera, F., Herrera, R. F., & Atencio, E. (2021). Unmanned aerial vehicles (UAVs) for physical progress monitoring of construction. *Sensors*, *21*(12), 1–27. https://doi.org/10.3390/s21124227

Jiang, A., Ao, Y., Yang, R., & Wang, T. (2023). Rural infrastructure lifecycle inclusiveness impact path analysis: Combining logical framework and structural equation modeling. *Advances in Civil Engineering*, *1*(1), 1–14. https://doi.org/10.1155/2023/8092064

Johansson, M., & Roupé, M. (2024). Real-world applications of BIM and immersive VR in construction. *Automation in Construction*, *158*, 1–21. https://doi.org/10.1016/j.autcon.2023.105233

Keeley, A. R., Komatsubara, K., & Managi, S. (2022). The value of invisibility: Factors affecting social acceptance of renewable energy. *Energy Sources, Part B: Economics, Planning, and Policy*, *17*(1), 1–20. https://doi.org/10.1080/15567249.2021.1983891

Kovacic, I., & Zoller, V. (2015). Building life cycle optimization tools for early design phases. *Energy*, *92*, 409–419. https://doi.org/10.1016/j.energy.2015.03.027

Lingard, H., & Wakefield, R. (2019). *Integrating work health and safety into construction project management*. John Wiley & Sons.

Liu, Y., Zhang, Y., Ren, S., Yang, M., Wang, Y., & Huisingh, D. (2020). How can smart technologies contribute to sustainable product lifecycle management? *Journal of Cleaner Production*, *249*, 1–9. https://doi.org/10.1016/j.scitotenv.2020.137870

Lu, Q., Xie, X., Parlikad, A. K., & Schooling, J. M. (2020). Digital twin-enabled anomaly detection for built asset monitoring in operation and maintenance. *Automation in Construction*, *118*, 1–16. https://doi.org/10.1016/j.autcon.2020.103277

Marszal, A. J., Heiselberg, P., Jensen, R. L., & Nørgaard, J. (2012). On-site or off-site renewable energy supply options? Life cycle cost analysis of a net zero energy building in Denmark. *Renewable Energy*, *44*, 154–165. https://doi.org/10.1016/j.renene.2012.01.079

Maxner, T., Dalla Chiara, G., & Goodchild, A. (2022). Identifying the challenges to sustainable urban last-mile deliveries: Perspectives from public and private stakeholders. *Sustainability*, *14*(8), 1–21. https://doi.org/10.3390/su14084701

Mellado, F., & Lou, E. C. (2020). Building information modelling, lean and sustainability: An integration framework to promote performance improvements in the construction industry. *Sustainable Cities and Society*, *61*, 1–13. https://doi.org/10.1016/j.scs.2020.102355

Moore-O'Leary, K. A., Hernandez, R. R., Johnston, D. S., Abella, S. R., Tanner, K. E., Swanson, A. C., Kreitler, J., & Lovich, J. E. (2017). Sustainability of utility-scale solar energy – critical ecological concepts. *Frontiers in Ecology and the Environment*, *15*(7), 385–394.

Naboni, R., & Paoletti, I. (2015). *Advanced customization in architectural design and construction*. Springer International Publishing.

Negi, A. (2021). Green buildings: Strategies for energy-efficient and eco-friendly designs. *Mathematical Statistician and Engineering Applications*, *70*(1), 683–689.

Nizma, C., Bangun, R., Benhur, B., Cahyoginarti, C., & Zuardi, M. (2024). The role of organizational structure in project management. *Journal Syntax Transformation*, *5*(2), 590–597.

Nova, K. (2022). Security and resilience in sustainable smart cities through cyber threat intelligence. *International Journal of Information and Cybersecurity*, *6*(1), 21–42.

Nuswantoro, B., & Richter, A. (2024). Avatars as visibility artifacts in digital work. *Thirty-Second European Conference on Information Systems (ECIS 2024)*, May 1, pp. 1–17, Paphos, Cyprus.

Obiuto, N. C., Ebirim, W., Ninduwezuor-Ehiobu, N., Ani, E. C., Olu-lawal, K. A., & Ugwuanyi, E. D. (2024). Integrating sustainability into HVAC project management: Challenges and opportunities. *Engineering Science & Technology Journal*, *5*(3), 873–887.

Omole, F. O., Olajiga, O. K., & Olatunde, T. M. (2024). Sustainable urban design: A review of eco-friendly building practices and community impact. *Engineering Science & Technology Journal*, *5*(3), 1020–1030.

Pórólfsdóttir, E., Árnadóttir, Á., & Heinonen, J. (2023). Net zero emission buildings: A review of academic literature and national roadmaps. *Environmental Research: Infrastructure and Sustainability*, *3*(4), 1–27. https://doi.org/10.1088/2634-4505/ad0e80

Ramakrishnan, K., O'Reilly, K., & Budds, J. (2021). The temporal fragility of infrastructure: Theorizing decay, maintenance, and repair. *Environment and Planning E: Nature and Space*, *4*(3), 674–695.

Rane, N., Choudhary, S., & Rane, J. (2023). Leading-edge technologies for architectural design: A comprehensive review. *International Journal of Architecture and Planning*, *3*(2), 12–48. https://doi.org/10.51483/IJARP.3.2.2023.12-48

Richerson, J. (2022). Facilities management. In C. DeVereaux (Ed.), *Managing the arts and culture* (pp. 125–149). Routledge.

Roberts, M., Allen, S., & Coley, D. (2020). Life cycle assessment in the building design process – A systematic literature review. *Building and Environment*, *185*, 1–12. https://doi.org/10.1016/j.buildenv.2020.107274

SaberiKamarposhti, M., Kamyab, H., Krishnan, S., Yusuf, M., Rezania, S., Chelliapan, S., & Khorami, M. (2024). A comprehensive review of AI-enhanced smart grid integration for hydrogen energy: Advances, challenges, and future prospects. *International Journal of Hydrogen Energy*, *67*, 1009–1025. https://doi.org/10.1016/j.ijhydene.2024.01.129

Samarasekara, H. M. S. N., Purushothaman, M. B., & Rotimi, F. E. (2024). Interrelations of the factors influencing the whole-life cost estimation of buildings: A systematic literature review. *Buildings*, *14*(3), 1–47.

Scoggins, M., Booth, D. B., Fletcher, T., Fork, M., Gonzalez, A., Hale, R. L., Hawley, R., J., Roy, A., H., Bilger, E., E., Bond, N., Burns, M., J., Hopkins, K., G., Macneale, K., H., Martí, E., McKay, S., K., Neale, W., K., Paul, M., J., Rios-Touma, B., Russell, K., L., … & Wenger, S. (2022). Community-powered urban stream restoration: A vision for sustainable and resilient urban ecosystems. *Freshwater Science*, *41*(3), 404–419.

Serra, J., Iñarra, S., Torres, A., & Llopis, J. (2021). Analysis of facade solutions as an alternative to demolition for architectures with visual impact in historical urban scenes. *Journal of Cultural Heritage*, *52*, 84–92. https://doi.org/10.1016/j.culher.2021.09.005

Svensson, H. (2020). Universal design-universal access: Sweden as leaders in the built environment and transport. In C. Curtis (Ed.), *Handbook of sustainable transport* (pp. 90–99). Edward Elgar Publishing.

Thanu, H. P. (2022). *Building performance score model for assessing the sustainable performance in life cycle of building* [Doctoral dissertation, National Institute of Technology Karnataka, Surathkal].

Tu, H. M. (2020). The attractiveness of adaptive heritage reuse: A theoretical framework. *Sustainability*, *12*(6), 1–15.

Vijayan, D. S., Devarajan, P., Sivasuriyan, A., Stefańska, A., Koda, E., Jakimiuk, A., Vaverková, M., D., Winkler, J., Duarte, C., C., & Corticos, N. D. (2023). A state of review on instigating resources and technological sustainable approaches in green construction. *Sustainability*, *15*(8), 1–24.

Wang, J., Lu, X., Adetola, V., & Louie, E. (2024). Modeling variable refrigerant flow (VRF) systems in building applications: A comprehensive review. *Energy and Buildings*, *311*, 1–20. https://doi.org/10.1016/j.enbuild.2024.114128

Weeks, D. J., & Leite, F. (2021). Facility defect and cost reduction by incorporating maintainability knowledge transfer using maintenance management system data. *Journal of Performance of Constructed Facilities*, *35*(2), 1–12. https://doi.org/10.1061/(ASCE)CF.1943-5509.0001569

Wernersson, L., Román, S., Fuso Nerini, F., Mutyaba, R., Stratton-Short, S., & Adshead, D. (2024). Mainstreaming systematic climate action in energy infrastructure to support the sustainable development goals. *Climate Action*, *3*(1), 1–28.

Xu, Z., Li, J., Li, J., Du, J., Li, T., Zeng, W., … Meng, F. (2023). Bionic structures for optimizing the design of stealth materials. *Physical Chemistry Chemical Physics*, *25*(8), 5913–5925.

Yadav, A., & Bhatnagar, A. (2024). Sustainable digital manufacturing processes in Industry 4.0. In S. Shah, H. Nautiyal, G. Gugliani, A. Kumar, T. Namboodri, Y. K. Singla (Eds.), *Sustainability in smart manufacturing* (pp. 19–36). CRC Press.

Zhang, S., Ocłoń, P., Klemeš, J. J., Michorczyk, P., Pielichowska, K., & Pielichowski, K. (2022). Renewable energy systems for building heating, cooling and electricity production with thermal energy storage. *Renewable and Sustainable Energy Reviews*, *165*, 1–22. https://doi.org/10.1016/j.rser.2022.112560

Zhou, J. X., Shen, G. Q., Yoon, S. H., & Jin, X. (2021). Customization of on-site assembly services by integrating the Internet of Things and BIM technologies in modular integrated construction. *Automation in Construction*, *126*, 1–12. https://doi.org/10.1016/j.autcon.2021.103663

Chapter Eleven

General Summary of Stealth Construction

Abstract

The exploration of construction practices and stages intersects with the concept of stealth construction, where innovation drives progress and adaptation within the construction landscape. As we navigate the vast terrain of construction methodologies, from lean construction to smart construction, sustainable construction to risk management and other sustainable practices, the industry's pursuit of excellence unfolds through many disciplines. Stealth construction emerges as a paradigm challenging conventional notions of visibility, championing resilience and sustainability. It seamlessly integrates with these practices, enhancing their efficacy and augmenting their impact across the pre-construction, construction, and post-construction phases. From propelling nature-based solutions to addressing challenges associated with climate change and infrastructure resilience, stealth construction embodies a transformative force within the construction industry, reshaping the built environment while navigating obstacles towards sustainable development goals.

Keywords: Construction practices; stealth construction; sustainability; resilience; collaboration; sustainable development goals (SDGs)

Introduction

In construction, where every project unfolds through many stages and practices, pursuing excellence knows no bounds. From the initial conception to the final touches, the journey encompasses many methodologies and disciplines to achieve efficiency, safety, and longevity. Throughout this book, we have delved into the intricacies of stealth construction – a paradigm that challenges the conventional

Stealth Construction: Integrating Practices for Resilience and Sustainability, 259–277

Published under exclusive licence by Emerald Publishing Limited
doi:10.1108/978-1-83608-182-120251011

notions of visibility and champions the virtues of resilience and sustainability. As we draw towards the end of our exploration, it is fitting to reflect on the vast landscape of construction practices and stages that intersect with the rationale of stealth construction.

Navigating the Construction Landscape

The construction industry is like a fabric woven with diverse threads of methodologies, each contributing its unique hue to the final masterpiece. From lean construction, smart construction, sustainable construction, value management, risk management, whole life cycle, supply chain management, procurement, health and safety, benchmarking, tendering, and partnering, the array of practices reflects the industry's continuous quest for improvement and adaptation. These methodologies, spanning the pre-construction, construction, and post-construction phases, serve as the foundation upon which stealth construction builds its edifice. Throughout our journey, we have witnessed how stealth construction seamlessly integrates with these practices, enhancing their efficacy and augmenting their impact. As we bid farewell to this exploration, let us revisit the rich dynamism of construction practices and stages, each a chapter in the saga of innovation and progress driving the novel stealth practice in the construction industry.

Innovations and Evolutions

In an ever-evolving construction perspective, innovation is the driving force propelling the industry forward. From groundbreaking technologies to revolutionary methodologies, the quest for progress knows no bounds. Innovation takes centre stage within the scope of stealth construction as practitioners constantly seek new ways to conceal, streamline, and fortify their projects. Evolutions in materials science have led to the development of advanced composites and eco-friendly alternatives, enhancing durability and sustainability. Meanwhile, digital tools and Building Information Modelling (BIM) software have revolutionised the design and planning phases, enabling precise coordination and seamless execution. As we look to the future, the horizon is ripe with possibilities where innovation and evolution intertwine to shape the next generation of construction practices.

Sustainable Practices in Stealth Construction

In pursuing resilience and sustainability, stealth construction emerges as a beacon of progress, championing practices that minimise environmental impact and maximise resource efficiency. From selecting materials to managing waste, every aspect of the construction process is meticulously scrutinised through a sustainability lens. Green building materials, such as recycled steel and bamboo composite, offer durable and eco-friendly alternatives to traditional counterparts. Meanwhile, energy-efficient designs and renewable energy systems reduce operational costs and carbon footprints, ensuring a greener future for future generations. Through strategic planning and conscientious decision-making, stealth construction not only conceals its presence but also leaves a legacy of environmental stewardship.

The Human Element in Stealth Construction

In understanding human elements in stealth construction, the human element is the conductor and performer, guiding the integration of innovation and discretion. Stealth construction is a testament to the ingenuity and expertise of designers, architects, and construction professionals who orchestrate every aspect of the project with precision and finesse. Their vision and creativity shape the built environment, transforming blueprints into tangible structures seamlessly blending into their surroundings. Moreover, the human touch imbues each project with a sense of craftsmanship and attention to detail, elevating it from mere construction to a work of art.

Role of Designers, Architects, and Construction Professionals

When discussing stealth construction, the roles of designers, architects, and construction professionals take on a heightened significance as they navigate the delicate balance between concealment and functionality. Designers become the architects of invisibility, conceptualising innovative strategies to integrate buildings into their surroundings while maintaining aesthetic appeal seamlessly. Their creative vision extends beyond traditional notions of design as they explore innovative materials, spatial configurations, and landscaping techniques to achieve the desired level of camouflage. As the custodians of form and function, architects collaborate closely with designers to translate these concepts into detailed plans prioritising concealment and sustainability. Through strategic site planning, facade design, and material selection, architects ensure that stealth construction projects blend harmoniously with their environment, minimise environmental impact, and maximise resource efficiency.

On the construction front, project managers and construction professionals play a pivotal role in executing the vision with precision and discretion. Project managers oversee every aspect of the construction process, from procurement and scheduling to quality control and safety compliance, ensuring that stealth construction projects remain on track and within budget. Skilled artisans and tradespeople bring their expertise to the construction site, employing specialised techniques and technologies to execute tasks with meticulous attention to detail. From installing sound-absorbing materials to implementing advanced security features, construction professionals work tirelessly to conceal the inner workings of buildings while maintaining structural integrity and operational efficiency. Their commitment to excellence and dedication to their craft are essential components of stealth construction, ensuring that projects meet and exceed expectations regarding resilience, sustainability, and functionality.

In essence, the success of stealth construction projects hinges upon the seamless collaboration and participation among designers, architects, and construction professionals. They form a cohesive team united by a shared vision of innovation and discretion, where creativity meets practicality function. Through their collective expertise and ingenuity, they push the boundaries of what is possible in construction, creating spaces that disappear into their surroundings and leave a lasting impact on the built environment. In stealth construction, designers, architects, and construction professionals are not just about building structures – they are about shaping the future of urban landscapes and redefining the relationship between architecture and its environment.

Collaboration and Interdisciplinary Approaches

Stealth construction, collaboration, and interdisciplinary approaches are encouraged and essential for success. As discussed above, designers, architects, and other construction professionals must collaborate seamlessly, leveraging their expertise to balance concealment and functionality. Through open communication and mutual respect, teams bridge the gap between design and execution, ensuring that every aspect of the project aligns with the overarching objectives of resilience, sustainability, and discretion, even when technologies are integrated into the practices. Interdisciplinary collaboration fosters innovation and cross-pollination of ideas as diverse perspectives come together to tackle complex challenges and explore new opportunities. When isolated divisions are dismantled and a culture of innovation and teamwork is, stealth construction endeavours can unlock their complete capabilities, achieving outstanding outcomes that surpass anticipations and make a lasting impression on the constructed landscape.

Furthermore, interdisciplinary collaboration extends beyond the core project team to include stakeholders from diverse backgrounds, such as urban planners, environmental scientists, and community members. These stakeholders bring valuable insights and perspectives, enriching the dialogue and ensuring that projects are technically sound and socially and environmentally responsible. Urban planners provide valuable input on land use and zoning regulations, helping to optimise site selection and maximise the project's impact on the surrounding community. Environmental scientists offer expertise in sustainability and ecological design, guiding decisions on materials, energy systems, and landscaping strategies. Meanwhile, community members (users) provide valuable feedback and input throughout the design and construction process, ensuring that projects reflect the needs and values of the people they serve. Through collaboration and interdisciplinary approaches, stealth construction projects can transcend traditional boundaries, creating spaces that are not only visually striking but also culturally and socially relevant.

Theories and Models Supporting Stealth Construction Practice

These theories and models support the practice of stealth in the construction industry, as shown in Fig. 6.

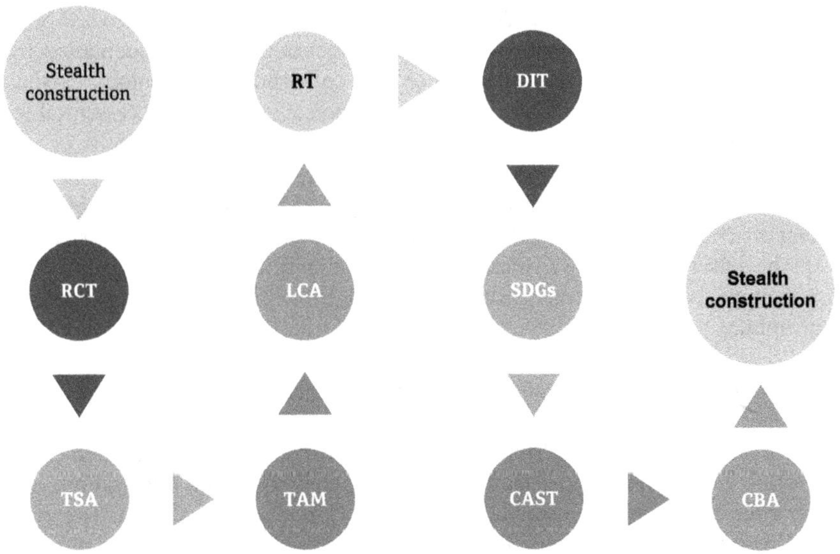

Fig. 6. Stealth Construction Theories and Models.

Relational cohesion theory (RCT), a cornerstone in organisational dynamics, underscores the importance of cultivating trust and fostering meaningful relationships with stakeholders throughout the implementation process. This approach seeks to enhance project outcomes and stakeholder satisfaction by prioritising communication and collaboration.

The systems approach (TSA), drawing insights from systems thinking, network theory, and social behaviour change, offers a comprehensive framework for bolstering resilience in infrastructure projects. By integrating best practices from diverse disciplines, this approach equips infrastructure systems with the capacity to withstand shocks, adapt to dynamic environments, and sustain functionality amidst disruptions.

Resilience theory (RT), rooted in the understanding of system dynamics, accentuates the capacity of infrastructure systems to absorb shocks, navigate evolving conditions, and uphold functionality amid adversities. By providing a conceptual framework, RT informs the design and management of infrastructure systems, guiding efforts to fortify their resilience and enhance their ability to serve communities reliably.

Life Cycle Assessment (LCA) emerges as a pivotal tool for evaluating the environmental footprint of infrastructure projects across their entire life cycle. By scrutinising environmental impacts from inception to disposal, LCA enables informed decision-making and facilitates comparisons between traditional and stealth construction strategies, thereby informing sustainable practices and mitigating ecological footprints.

The Technology Acceptance Model (TAM) offers valuable insights into adopting and utilising new technologies within the construction sector. By elucidating the factors influencing the acceptance of innovative construction technologies associated with stealth strategies, TAM drives technological advancements essential for fortifying infrastructure resilience.

Diffusion of Innovations Theory (DIT) sheds light on the spread and adoption of novel construction practices within the construction industry. This theory elucidates the drivers and barriers shaping their dissemination by examining the adoption dynamics of stealth construction strategies, informing strategies to expedite their uptake.

Sustainable Development Goals (SDGs) furnish a comprehensive blueprint for advancing global sustainability, including infrastructure resilience. By aligning research efforts with specific SDGs pertinent to infrastructure development, such as Goal 7 (Affordable and Clean Energy), Goal 9 (Industry, Innovation, and Infrastructure), Goal 11 (Sustainable Cities and Communities), Goal 12 (Responsible Consumption and Production), and Goal 13 (Climate Action) pertained to stealth construction, it underscores the pivotal role of resilient infrastructure in achieving overarching sustainable development objectives.

Complex Adaptive Systems Theory (CAST) provides a lens through which infrastructure systems are viewed as dynamic, self-organising entities responsive to internal and external stimuli. By embracing this perspective, studies elucidate how stealth construction strategies influence the resilience and adaptability of infrastructure systems, offering insights for enhancing their long-term sustainability.

Cost-benefit analysis (CBA) is a cornerstone in evaluating infrastructure projects' economic viability and resilience-enhancing potential. By comparing costs and benefits associated with stealth strategies, CBA informs decision-makers about the long-term economic feasibility and resilience benefits of different construction approaches, thereby guiding resource allocation and investment decisions.

Global Impact of Stealth Construction

Contributions to SDGs (Bali Swain & Yang-Wallentin, 2020; Henderson & Loreau, 2023)

Stealth construction, characterised by its emphasis on efficiency, innovation, and resource optimisation, significantly contributes to several SDGs, fostering a more sustainable future. To begin with, aligned with Goal 7 (Affordable and Clean Energy), stealth construction techniques prioritise integrating renewable energy sources such as solar panels, wind turbines, and geothermal systems into buildings and infrastructure, thereby reducing reliance on fossil fuels and promoting access to affordable, clean energy. Moreover, stealth construction is pivotal in advancing Goal 9 (Industry, Innovation, and Infrastructure) by driving technological advancements and fostering innovation in construction practices. Through the adoption of cutting-edge materials, techniques, and digital tools, stealth construction enhances the efficiency and resilience of infrastructure, laying the foundation for sustainable development and economic growth.

In Goal 11 (Sustainable Cities and Communities), stealth construction contributes to creating environmentally friendly urban spaces that prioritise the well-being of inhabitants. By implementing green building practices, optimising land use, and promoting mixed land-use development, stealth construction helps build resilient and inclusive cities conducive to sustainable living and community well-being. Furthermore, stealth construction aligns with Goal 12 (Responsible Consumption and Production) by promoting resource efficiency and minimising waste throughout the construction process. By prioritising the use of sustainable materials, reducing construction-related emissions, and adopting circular economy principles, stealth construction minimises environmental impact and fosters responsible consumption and production patterns.

Ultimately, stealth construction addresses Goal 13 (Climate Action) by mitigating greenhouse gas emissions and enhancing climate resilience. Through the implementation of energy-efficient designs, green roofs, and rainwater harvesting systems, stealth construction reduces the carbon footprint of buildings and infrastructure while enhancing their ability to withstand the impacts of climate change, thus contributing to global efforts to combat climate change and build a more sustainable future.

Contributions to Agenda 2063 (Aniche, 2023; Nwozor et al., 2021)

Stealth construction significantly contributes to the realisation of the goals outlined in Agenda 2063, the strategic framework for the socio-economic transformation of Africa over the next five decades. Integrating sustainable construction practices and advanced technologies, stealth construction is crucial in addressing key priorities such as infrastructure development, environmental sustainability, and economic growth. Through proactive measures such as green building design, energy-efficient construction techniques, and waste reduction strategies, stealth construction projects actively contribute to creating resilient and environmentally responsible infrastructure that meets the needs of present and future generations.

Furthermore, stealth construction fosters innovation and capacity building, aligning with Agenda 2063's emphasis on promoting science, technology, and innovation as development drivers. Stealth construction projects enhance infrastructure development efficiency, productivity, and quality by embracing cutting-edge technologies such as BIM, drones, and smart building systems. Sustainable construction practices promote job creation, skills development, and local economic empowerment, aligning with Agenda 2063's vision of building a prosperous and inclusive Africa. Through these contributions, stealth construction actively supports the realisation of Agenda 2063's aspirations for a united, thriving, and peaceful Africa, driven by its citizens and representing a dynamic force in the global arena.

Related Case Studies from Around the World

Moving towards stealth construction leaves a lasting mark on the global landscape through exemplary case studies that showcase its innovative approaches and tangible impacts. One notable case study is the Bosco Verticale (Vertical Forest) in Milan, Italy, designed by architect Stefano Boeri (Liu, 2023). This iconic project features residential towers adorned with lush vegetation, acting as natural air purifiers and carbon sinks. By integrating nature into urban environments, the Bosco Verticale actively contributes to climate action (SDG 13) by reducing carbon emissions and improving air quality. This case study demonstrates how stealth construction can effectively blend sustainability with aesthetic appeal, setting new standards for green urban development worldwide.

Another compelling example of stealth construction is the EDGE East Side Tower in Berlin, Germany, designed by BIG's renowned architecture firm (Bjarke Ingels Group) (EDGE company, 2023). This carbon-neutral tower showcases cutting-edge sustainable technologies and design principles, from its energy-efficient facade to its implementation of renewable energy sources like rooftop solar panels. The EDGE East Side Tower contributes to SDG 12: Responsible Consumption and Production by reducing its environmental footprint. Through its innovative approach to sustainable construction, this case study highlights the transformative potential of stealth construction in advancing global sustainability objectives and promoting resource efficiency.

Furthermore, the One Central Park project in Sydney, Australia, represents a pioneering endeavour in integrating nature into urban landscapes (Narwal, 2022). Designed by architect Jean Nouvel in collaboration with botanist Patrick Blanc, this project features lush vertical gardens and innovative sun shading systems, creating a sustainable oasis in the city's heart. One Central Park actively contributes to SDG 11: Sustainable Cities and Communities by enhancing biodiversity and improving air quality. This case study demonstrates how stealth construction projects can foster vibrant and resilient urban environments while promoting social cohesion and environmental sustainability. Through these diverse case studies, stealth construction emerges as a transformative force in shaping the built environment and advancing global sustainability goals.

Contribution to the Built Environment

Stealth construction fundamentally transforms the built environment, leaving an enduring mark through innovative approaches and sustainable practices. It enhances the built environment by seamlessly integrating buildings into their surroundings and prioritising resilience and sustainability, creating spaces that blend harmoniously with nature and promote environmental stewardship and community well-being. Through their strategic concealment and discreet execution, these projects contribute to the evolution of the built environment, setting new standards for architectural design and urban development.

One significant contribution of stealth construction to the built environment is its promotion of sustainable and eco-friendly building practices. These projects actively reduce environmental impact while maximising resource efficiency by incorporating green building materials, energy-efficient systems, and renewable energy sources. For instance, using recycled materials and sustainable construction methods minimises waste generation and reduces the demand for virgin resources, thus promoting responsible consumption and production (SDG 12). Through the commitment to sustainability, stealth construction projects inspire a shift towards greener building practices and create a more resilient and sustainable built environment.

Moreover, stealth construction projects actively shape the urban landscape, fostering vibrant and inclusive communities. By integrating nature into urban environments and promoting green spaces, these projects enhance the quality of life for residents and contribute to creating healthier and more liveable cities. For example, the Bosco Verticale (Vertical Forest) in Milan, Italy, serves as a green oasis in the city's heart, providing residents access to nature and improving air quality. Through their innovative design and community-oriented approach, stealth construction projects actively contribute to creating sustainable and resilient urban environments that enrich the lives of residents and future generations.

Propelling Nature-based Solutions

Stealth construction projects serve as catalysts for propelling nature-based solutions, harnessing the power of the natural environment to enhance sustainability and resilience. By integrating nature into the built environment, these projects actively contribute to advancing nature-based solutions, which utilise natural processes and ecosystems to address environmental challenges. Through their innovative design strategies and sustainable construction practices, stealth construction projects demonstrate the transformative potential of nature-based solutions in promoting ecological sustainability and mitigating the impacts of climate change.

Furthermore, stealth construction propels nature-based solutions by integrating green spaces and vegetation into urban environments. Projects such as the Bosco Verticale (Vertical Forest) in Milan, Italy, and the One Central Park in Sydney, Australia, showcase the effective use of vertical gardens and lush vegetation to enhance biodiversity, improve air quality, and mitigate urban heat island effects. By creating green oases within densely populated urban areas, these projects provide habitat for wildlife and offer residents access to nature, promoting health and well-being. Through their strategic incorporation of green infrastructure, stealth construction projects demonstrate how nature-based solutions can create sustainable and resilient cities.

Moreover, stealth construction projects actively leverage natural processes and ecosystems to enhance resilience and mitigate environmental risks. When adopting nature based approaches such as green roofs, rain gardens, and permeable pavements, they help projects manage stormwater runoff, reduce flood risk, and improve water quality. For example, the EDGE East Side Tower in Berlin, Germany, utilises green roofs and rainwater harvesting systems to capture and store rainwater, reducing pressure on municipal water supply systems and minimising the risk of urban flooding. Through their innovative use of nature-based solutions, stealth construction projects demonstrate how sustainable urban design can enhance resilience to climate change and promote communities' long-term health and vitality.

The Drive Towards Industry 5.0

Stealth construction projects are at the forefront of driving the construction industry towards Industry 5.0, a paradigm shift that emphasises the integration of advanced technologies with human-centred approaches. Embracing innovative design methodologies, cutting-edge construction techniques, and sustainable practices, these projects exemplify the evolution towards Industry 5.0, which prioritises collaboration, customisation, and sustainability. Through their seamless integration of technology and craftsmanship, stealth construction projects lead the way in reshaping the construction industry for the digital age.

One key aspect of Industry 5.0 that stealth construction projects embody is the integration of advanced technologies such as BIM, augmented reality (AR), and prefabrication techniques. When these technologies are leveraged, designers, architects, and construction professionals can optimise project planning, streamline construction processes, and enhance collaboration among project stakeholders. For example, BIM technology allows for the creation of digital models that simulate the entire construction process, enabling stakeholders to visualise and analyse project components in real-time. Similarly, the adoption of AR technology enables on-site workers to overlay digital information onto physical spaces, improving communication and reducing errors during construction. Through their proactive embrace of advanced technologies, stealth construction projects drive the construction industry towards greater efficiency, productivity, and innovation.

Moreover, stealth construction projects actively promote human-centred approaches that prioritise the well-being and safety of workers and communities. When a culture of collaboration, inclusivity, and empowerment is adopted, projects demonstrate the transformative potential of Industry 5.0 in promoting a more people-centric construction industry. For instance, through implementing lean construction principles and participatory design processes, stealth construction projects empower workers to contribute their expertise and insights to project planning and execution. Similarly, by prioritising health and safety measures and promoting a culture of accountability and transparency, these projects ensure the well-being of workers and communities throughout the construction process. Through their commitment to human-centred approaches, stealth construction projects drive the construction industry towards greater sustainability, resilience, and social responsibility.

Stealth and Climate Change

Stealth construction emerges as a potent tool in the fight against climate change, actively mitigating its impacts through innovative design strategies and sustainable building practices. By prioritising resilience and sustainability, these projects are pivotal in reducing carbon emissions, enhancing biodiversity, and fostering climate resilience in the built environment. Through a proactive approach to addressing climate change, stealth construction projects demonstrate the transformative potential of the construction industry in combating one of the most pressing global challenges of our time.

Another significant way stealth construction contributes to addressing climate change is by emphasising energy efficiency and renewable energy. When passive design principles, energy-efficient systems, and renewable energy sources such as solar panels and wind turbines are incorporated into construction projects, these projects actively reduce reliance on fossil fuels and minimise carbon emissions. For instance, integrating green roofs and energy-efficient facade systems helps regulate indoor temperatures, reducing the need for heating and cooling and lowering energy consumption. Similarly, adopting renewable energy technologies such as rooftop solar panels enables stealth construction projects to generate clean energy on-site, reducing their carbon footprint. Through their commitment to energy efficiency and renewable energy, stealth construction projects are crucial in reducing greenhouse gas emissions and mitigating climate change.

Moreover, stealth construction projects actively contribute to climate resilience by incorporating adaptive design strategies and nature-based solutions. Integrating green infrastructure, such as rain gardens, permeable pavements, and bioswales, these projects help manage stormwater runoff, reduce flood risk, and improve water quality. Additionally, incorporating green spaces and vegetation into urban environments helps mitigate urban heat island effects, enhance air quality, and provide wildlife habitat. For example, projects like the Bosco Verticale (Vertical Forest) in Milan, Italy, and One Central Park in Sydney, Australia, mentioned above, showcase nature-based solutions' transformative potential in enhancing climate resilience and promoting sustainable urban development. Through a proactive approach to climate resilience, stealth construction projects demonstrate the importance of building adaptive and resilient communities in the face of a changing climate.

Infrastructure Resilience and Sustainability

Stealth construction is crucial in enhancing infrastructure resilience and sustainability, ensuring that built environments can withstand and adapt to various challenges, including climate change, natural disasters, and urbanisation pressures. Through innovative design strategies, advanced construction techniques, and sustainable practices, these projects actively contribute to creating resilient and sustainable infrastructure that can withstand the test of time. Through prioritising resilience and sustainability, stealth construction projects help future-proof infrastructure and ensure the long-term viability of built environments.

One key aspect of infrastructure resilience and sustainability that stealth construction projects embody is their integration of green infrastructure and nature-based solutions. When green roofs, permeable pavements, and rain gardens are incorporated into construction, they help projects manage stormwater runoff, reduce flood risk, and improve water quality. Additionally, integrating green spaces and vegetation into urban environments helps mitigate urban heat island effects and improve wildlife habitat.

Moreover, stealth construction projects leverage advanced technologies and innovative design strategies to enhance infrastructure resilience and sustainability. These projects reduce energy consumption, minimise environmental impact, and improve operational efficiency by incorporating features such as energy-efficient systems, renewable energy sources, and smart building technologies. For instance, implementing BIM technology allows for creating digital models that simulate the entire construction process, enabling stakeholders to optimise project planning and streamline construction processes (Sun et al., 2021). Similarly, adopting renewable energy technologies such as solar panels and wind turbines helps reduce reliance on fossil fuels and minimise carbon emissions. Through their proactive embrace of advanced technologies and sustainable practices, stealth construction projects drive innovation and set new standards for resilient and sustainable infrastructure.

Carbon Emission and Energy Reduction

Stealth construction initiatives are pivotal in curbing carbon emissions and achieving substantial reductions in energy consumption, contributing significantly to sustainability objectives. By integrating innovative design principles and energy-efficient technologies, these projects actively reduce carbon footprints while promoting resource efficiency and environmental stewardship. By prioritising energy reduction and carbon emission mitigation, stealth construction initiatives exemplify sustainable development in the built environment.

As a strategy, stealth construction minimises carbon emissions through green building materials and sustainable construction practices. These initiatives significantly reduce construction-related emissions by opting for eco-friendly materials with low embodied carbon and incorporating sustainable construction techniques such as prefabrication and modular construction. Furthermore, implementing energy-efficient systems, such as high-performance insulation, LED lighting, and advanced HVAC (Heating, Ventilation, and Air Conditioning) systems, further enhances energy efficiency and contributes to carbon emission reductions (Simpeh et al., 2022). Through their proactive adoption of sustainable construction practices, stealth construction projects effectively mitigate carbon emissions associated with building construction and operation.

Moreover, stealth construction projects promote renewable energy integration to substantially reduce energy consumption and carbon emissions. By harnessing renewable energy sources such as solar, wind, and geothermal, these initiatives minimise reliance on fossil fuels and promote a transition towards clean and sustainable energy systems. For instance, installing rooftop solar panels and wind turbines allows stealth construction projects to generate onsite renewable energy, reducing grid dependency and carbon emissions associated with electricity consumption. Additionally, incorporating energy storage systems and smart grid technologies enables efficient management and utilisation of renewable energy resources, further optimising energy performance and contributing to carbon emission reduction goals. Through their commitment to renewable energy integration, stealth construction projects are vital in advancing sustainable development and combatting climate change.

Improving Traditional Infrastructure Practices

Stealth construction initiatives spearhead the advancement of traditional infrastructure practices by integrating innovative methodologies and sustainable principles, thereby enhancing resilience, efficiency, and longevity. Adopting cutting-edge technologies, such as BIM and prefabrication techniques, these projects streamline construction processes and optimise resource utilisation, improving project outcomes and reducing environmental impact (Zhou et al., 2021). By challenging conventional norms and embracing progressive approaches, stealth construction initiatives redefine traditional infrastructure practices for a more sustainable and resilient future.

One significant aspect in which stealth construction initiatives improve traditional infrastructure practices is implementing lean construction principles and methodologies. These initiatives enhance project delivery timelines and minimise construction-related costs by prioritising efficiency, productivity, and waste reduction. Through optimising material flows, reducing non-value-added activities, and promoting collaborative workflows, stealth construction projects achieve greater operational efficiency and project success. Additionally, modular construction techniques facilitate off-site fabrication and assembly, accelerating project timelines and reducing construction waste. Through their proactive embrace of lean construction practices, stealth construction initiatives drive innovation and efficiency in traditional infrastructure projects.

Moreover, stealth construction projects promote sustainability and resilience in traditional infrastructure by integrating nature-based solutions and green infrastructure components. These initiatives mitigate stormwater runoff, reduce flood risk, and improve water quality by incorporating green roofs, permeable pavements, and rain gardens. Furthermore, integrating green spaces and vegetation into urban environments enhances biodiversity, mitigates urban heat island effects, and promotes social well-being.

Challenges Associated with Stealth Construction

The impacts of stealth construction have been discussed above; however, for every novel practice, there are bound to be challenges. In integration, understanding, and execution, obstacles can always mitigate proper adoption in any sector. Some of the potential challenges are identified below from related practices.

Technological Integration (Shirowzhan et al., 2020)

Implementing stealth construction requires seamless integration of advanced technologies into traditional construction practices. This integration process may pose challenges such as compatibility issues between different software systems, lack of expertise among construction professionals in using new technologies, and the need for substantial investments in upgrading existing infrastructure to accommodate technological advancements.

Regulatory Compliance (Cezarino et al., 2022)

Stealth construction projects often involve unconventional design strategies and building methods, which may challenge navigating regulatory frameworks and obtaining necessary permits and approvals. Compliance with building codes, zoning regulations, and environmental requirements becomes crucial but may encounter resistance due to the novel nature of stealth construction approaches, requiring careful coordination and communication with regulatory authorities.

Cost and Budget Constraints (Hasani et al., 2021)

Stealth construction can entail higher upfront costs due to investments in innovative technologies, sustainable materials, and specialised expertise. Budget constraints may challenge balancing the desire for sustainability and resilience with the need to adhere to financial limitations. Additionally, uncertainties related to project scope, schedule overruns, and unforeseen expenses may further strain budgets, necessitating careful financial planning and risk management throughout the project lifecycle.

Supply Chain Disruptions (Yevu et al., 2021)

Stealth construction projects rely on a complex network of suppliers and contractors to procure materials and execute construction activities. Disruptions in the supply chain, whether due to natural disasters, geopolitical tensions, or global pandemics, can lead to delays, cost overruns, and logistical challenges. Securing a resilient and reliable supply chain becomes imperative to mitigate risks and ensure the timely completion of stealth construction projects.

Public Perception and Community Engagement (Zhang et al., 2020)

Implementing stealth construction may face resistance or scepticism from the public, local communities, and stakeholders who may be unfamiliar with or wary of novel construction approaches. Building trust, fostering transparent communication, and actively engaging with stakeholders throughout the project lifecycle become essential to addressing concerns, gaining support, and fostering a sense of ownership and collaboration.

Skills and Workforce Development (Awan & Sroufe, 2022)

Adopting stealth construction practices requires a skilled and knowledgeable workforce capable of implementing innovative design concepts and construction techniques. However, the construction industry may face challenges in attracting and retaining talent with the requisite expertise in sustainable construction, digital technologies, and interdisciplinary collaboration. Investing in workforce development, training programs, and knowledge transfer becomes critical to building a competent workforce capable of driving stealth construction initiatives forward.

Stealth Construction Across Construction Phases

In Pre-Construction

In the pre-construction phase, implementing lean, smart, and sustainable construction practices is pivotal for realising the objectives of stealth construction. Lean construction principles focus on minimising waste and maximising value, streamlining pre-construction processes by optimising resource allocation, and enhancing stakeholder collaboration. Smart construction technologies, such as BIM and digital twin systems, provide comprehensive insights into project planning and design, facilitating better coordination and decision-making. Sustainable construction practices, including passive design strategies and eco-friendly material selections, ensure that environmental considerations are integrated from the outset, minimising carbon footprint and promoting long-term resilience.

Furthermore, value management and risk management strategies are indispensable in the pre-construction phase of stealth construction projects. Value management techniques, such as value engineering workshops and value analysis studies, identify opportunities for cost savings and performance improvements without compromising quality. Risk management practices, including risk identification and mitigation planning, proactively address potential threats to project success, such as design flaws or supply chain disruptions. By systematically managing value and risk considerations, project teams can optimise project outcomes and minimise uncertainties, ensuring successful project delivery.

In addition, whole LCA and supply chain management are crucial components in achieving stealth construction objectives during the pre-construction phase. Whole LCA tools evaluate the environmental impacts of design decisions, ensuring that sustainable solutions are prioritised and implemented. Supply chain management practices ensure that materials and equipment sourced for the project meet stringent environmental and social criteria, enhancing project sustainability and minimising environmental impact. Project teams can optimise resource utilisation and minimise carbon emissions throughout the project lifecycle by integrating whole LCA and supply chain management into pre-construction planning.

Procurement, health and safety, benchmarking, tendering, and partnering practices also play significant roles in achieving stealth construction goals in the pre-construction phase. Strategic procurement practices ensure that materials and services are obtained efficiently and sustainably, aligning with project objectives. Health and safety measures safeguard workers and stakeholders, minimising accidents and disruptions during pre-construction. Benchmarking allows project teams to compare performance metrics against industry standards, identifying areas for improvement and enhancing project efficiency. Tendering processes facilitate fair and transparent selection of contractors and suppliers, ensuring project requirements are met. Partnering with stakeholders fosters collaboration and shared accountability, driving project success from the outset of the pre-construction phase.

During Construction

Integrating lean construction practices promotes efficiency and minimises disruptions during the construction phase. By eliminating waste and optimising workflows, construction teams can streamline operations, reducing unnecessary downtime and delays. Smart construction technologies play a pivotal role by enabling real-time monitoring and data-driven decision-making, allowing for agile adjustments to project schedules and resource allocation as needed. Furthermore, sustainable construction principles ensure that environmental considerations are upheld throughout construction. From eco-friendly material selection to energy-efficient building practices, sustainable construction minimises the project's ecological footprint while enhancing its long-term viability. Value management techniques further support this by ensuring that resources are allocated judiciously to activities that deliver tangible benefits, fostering a balance between environmental responsibility and project objectives.

Risk management strategies are essential for maintaining project momentum and mitigating potential disruptions during construction. By proactively identifying and addressing risks such as supply chain disruptions or unexpected site conditions, construction teams can minimise the likelihood of costly setbacks and maintain project timelines. Whole life cycle approaches consider the full spectrum of potential risks and opportunities, enabling informed decision-making that accounts for the project's lifespan. Also, effective supply chain management and procurement practices are critical for ensuring seamless construction operations. By establishing reliable supply chain networks and negotiating favourable procurement agreements, construction projects can minimise delays caused by material shortages or logistical challenges. Health and safety measures remain paramount throughout the construction phase, with proactive measures in place to protect workers and stakeholders from accidents and hazards.

In addition, benchmarking initiatives enable construction teams to continuously monitor performance against industry standards and best practices. Projects can enhance efficiency and quality throughout the construction process when areas for improvement and implementing targeted interventions are identified early. Tendering processes facilitate transparent and competitive bidding, ensuring that construction contracts are awarded to qualified vendors who offer the best value for the project. Also, partnering initiatives foster stakeholder collaboration, promoting shared goals and effective communication throughout construction. By leveraging all parties' collective expertise and resources, construction projects can overcome challenges more effectively and capitalise on delivering superior outcomes. Integrating these practices into construction operations allows projects to achieve stealth construction by minimising disruptions, optimising resource utilisation, and delivering sustainable, high-quality built environments.

In Post-Construction

In the post-construction phase, integrating lean construction practices facilitates the efficient handover of the completed project to stakeholders. By streamlining documentation processes and ensuring clear communication, lean construction minimises administrative delays and expedites the transition from construction to operation. Smart construction technologies play a crucial role by providing real-time monitoring and predictive maintenance capabilities, enabling proactive identification and resolution of issues that may arise after project completion.

Sustainable construction principles remain relevant in the post-construction phase by ensuring that the building continues to operate in an environmentally responsible manner. Through ongoing energy monitoring, waste management, and maintenance practices, sustainable construction promotes the long-term durability and resilience of the built environment. Value management methodologies further support this by optimising operational processes and resource utilisation, maximising the return on investment while minimising environmental impacts. Also, risk management strategies are essential for maintaining the functionality and safety of the building throughout its lifecycle. Construction teams can mitigate risks such as structural degradation or equipment failures by conducting regular inspections and implementing preventive maintenance measures. Whole life cycle approaches provide a holistic framework for decision-making in the post-construction phase, considering maintenance costs, energy efficiency, and adaptability to changing user needs.

Effective supply chain management and procurement practices remain imperative in the post-construction phase to ensure the timely availability of spare parts and materials for maintenance and repairs. Construction projects can minimise downtime and maximise operational efficiency by establishing reliable supplier relationships and implementing efficient inventory management systems. Health and safety considerations remain paramount, with ongoing training and protocols in place to protect occupants and maintain compliance with regulatory requirements. Additionally, benchmarking initiatives enable construction teams to benchmark the performance of the building against industry standards and best practices in the post-construction phase. Identifying areas for improvement and implementing targeted interventions and projects can optimise operational processes and enhance user satisfaction. Tendering processes facilitate the selection of qualified service providers for ongoing maintenance and upgrades, ensuring that the building effectively meets its occupants' needs. Partnering initiatives foster stakeholder collaboration, facilitating knowledge sharing and innovation in the post-construction phase. Furthermore, when the collective expertise and resources of all parties involved are leveraged, construction projects can address challenges more effectively and optimise the performance of the built environment over time. Ultimately, by integrating these practices into post-construction operations, projects can achieve stealth construction by maximising operational efficiency, ensuring sustainability, and maintaining the long-term functionality and safety of the building.

Looking Ahead

As the construction industry evolves, we must anticipate trends shaping how we build and develop infrastructure. One can expect many innovations and advancements that will revolutionise construction practices, driven by technological breakthroughs, sustainability imperatives, and changing market demands. Anticipating these trends is crucial for industry professionals to stay ahead of the curve and adapt their strategies to meet the challenges and opportunities.

One prominent trend expected to shape the future of construction further is the widespread adoption of advanced technologies such as BIM, artificial intelligence (AI), and robotics. These technologies have the potential to streamline project delivery, improve productivity, and enhance collaboration among project stakeholders even more than what is being actualised now. Through data-driven insights and automation, construction firms can optimise processes, reduce costs, and accelerate project timelines, ultimately delivering better outcomes for clients and communities.

Furthermore, sustainability will continue to be a driving force in shaping the future of construction practices. With increasing concerns about environmental degradation and climate change, there is a growing emphasis on green building practices, energy efficiency, and carbon neutrality. Anticipating this trend, construction firms invest in sustainable design solutions, renewable energy integration, and circular economy principles to minimise environmental impact and create resilient and future-proof buildings and infrastructure. By embracing sustainable construction practices, the industry can pave the way for a more sustainable and equitable future for future generations.

In addition, industry professionals are urged to take proactive steps in anticipating future trends in construction and embracing innovations that will shape the industry landscape. To remain competitive in a rapidly evolving industry, professionals must stay informed about emerging technologies, sustainability initiatives, and market trends. When industry forums are engaged, attending conferences, and participating in continuing education programs, professionals can stay ahead of the curve and position themselves as leaders in the field.

Conclusion

The discourse surrounding integrating innovative practices such as lean construction, smart construction, sustainable construction, and other practices into the fabric of the construction industry underscores a collective commitment to progress and adaptation. From anticipating future trends to implementing proactive measures during the pre-construction, construction, and post-construction phases, industry professionals are poised to navigate the evolving construction landscape with resilience and foresight. Through collaborative efforts and a dedication to continuous improvement, the industry is primed to embrace the challenges and opportunities that lie ahead, paving the way for a more sustainable, efficient, and technologically advanced future. As the call to action resonates among professionals to integrate emerging technologies, foster sustainability, and optimise project outcomes, the construction industry is moving towards innovation-driven growth and transformative change.

References

Aniche, E. T. (2023). African continental free trade area and African Union Agenda 2063: The roads to Addis Ababa and Kigali. *Journal of Contemporary African Studies*, *41*(4), 377–392.

Awan, U., & Sroufe, R. (2022). Sustainability in the circular economy: Insights and dynamics of designing circular business models. *Applied Sciences*, *12*(3), 1–30. https://doi.org/10.3390/app12031521

Bali Swain, R., & Yang-Wallentin, F. (2020). Achieving sustainable development goals: Predicaments and strategies. *International Journal of Sustainable Development & World Ecology*, *27*(2), 96–106.

Cezarino, L. O., Liboni, L. B., Hunter, T., Pacheco, L. M., & Martins, F. P. (2022). Corporate social responsibility in emerging markets: Opportunities and challenges for sustainability integration. *Journal of Cleaner Production*, *362*, 1–15. https://doi.org/10.1016/j.jclepro.2022.132224

EDGE company. (2023). *EDGE East Side Berlin*. https://edge.tech/developments/edge-east-side-berlin

Hasani, A., Mokhtari, H., & Fattahi, M. (2021). A multi-objective optimization approach for green and resilient supply chain network design: A real-life case study. *Journal of Cleaner Production*, *278*, 1–26. https://doi.org/10.1016/j.jclepro.2020.123199

Henderson, K., & Loreau, M. (2023). A model of Sustainable Development Goals: Challenges and opportunities in promoting human well-being and environmental sustainability. *Ecological Modelling*, *475*, 1–9. https://doi.org/10.1016/j.ecolmodel.2022.110164

Liu, Y. (2023). Analysis of the vertical forest of Milan in terms of high-rise architecture and biodiversity. *Highlights in Art and Design*, *3*(2), 47–52.

Narwal, H. T. (2022). *Engaging in more-than-conservation in multispecies cities: A study of One Central Park, Sydney* [Doctoral dissertation, Macquarie University].

Nwozor, A., Okidu, O., & Adedire, S. (2021). Agenda 2063 and the feasibility of sustainable development in Africa: Any silver bullet? *Journal of Black Studies*, *52*(7), 688–715.

Shirowzhan, S., Sepasgozar, S. M., Edwards, D. J., Li, H., & Wang, C. (2020). BIM compatibility and its differentiation with interoperability challenges as an innovation factor. *Automation in Construction*, *112*, 1–18. https://doi.org/10.1016/j.autcon.2020.103086

Simpeh, E. K., Pillay, J. P. G., Ndihokubwayo, R., & Nalumu, D. J. (2022). Improving energy efficiency of HVAC systems in buildings: A review of best practices. *International Journal of Building Pathology and Adaptation*, *40*(2), 165–182.

Sun, H., Fan, M., & Sharma, A. (2021). Design and implementation of construction prediction and management platform based on building information modelling and three-dimensional simulation technology in industry 4.0. *IET Collaborative Intelligent Manufacturing*, *3*(3), 224–232.

Yevu, S. K., Ann, T. W., & Darko, A. (2021). Digitalization of construction supply chain and procurement in the built environment: Emerging technologies and opportunities for sustainable processes. *Journal of Cleaner Production*, *322*, 1–14. https://doi.org/10.1016/j.jclepro.2021.129093

Zhang, Y., Xiao, X., Cao, R., Zheng, C., Guo, Y., Gong, W., & Wei, Z. (2020). How important is community participation to eco-environmental conservation in protected areas? From the perspective of predicting locals' pro-environmental behaviours. *Science of the Total Environment*, *739*, 1–10. https://doi.org/10.1016/j.scitotenv.2020.139889

Zhou, J. X., Shen, G. Q., Yoon, S. H., & Jin, X. (2021). Customization of on-site assembly services by integrating the Internet of Things and BIM technologies in modular integrated construction. *Automation in Construction*, *126*, 1–12. https://doi.org/10.1016/j.autcon.2021.103663

Index

www.ingramcontent.com/pod-product-compliance
Lightning Source LLC
Chambersburg PA
CBHW050635190326
41458CB00008B/2285
9781836081838